THE COMMISSION OF THE EUROPEAN COMMUNITIES

Testing and Evaluation of Solidified High Level Radioactive Waste

This report was prepared for the European Atomic Energy Community's Cost-sharing Research Programme on "Radioactive Waste Management and Disposal".

Action No. 5: "Testing and Evaluation of the Properties of Various Potential Materials for Immobilizing Highly Radioactive Waste".

Contract Nos:
WAS 321-83-53 UK(H)/WAS 268-81-55F
WAS 323-83-53 D(B)/WAS 322-83-53 D(B)
WAS 324-83-55 B/WAS 327-83-55 NL(N)

Contributing Authors

J.A.C. Marples A.R. Hall A. Hough K.A. Boult	AERE Harwell (UK)
Ch. Engelmann P. Trocellier	CEN Saclay (France)
N. Jacquet-Francillon E. Vernaz A. Magnier A. Terki	CEA Marcoule (France)
R. Conradt H. Roggendorf	Fraunhofer-Institut für Silicatforschung Würzburg (FRG)
G. Malow P. Offermann R.F. Haaker R. Muller P. Schubert	HMI Berlin (FRG)
H.W. Zandbergen F.C. Mijlhoff D.H.W. Ijdo	State University Leiden (The Netherlands)
R. de Batist P. Van Iseghem W. Timmermans	SCK/CEN Mol (Belgium)

Library of Congress CIP data is available

Radioactive Waste Management Series

TESTING AND EVALUATION OF SOLIDIFIED HIGH LEVEL RADIOACTIVE WASTE

edited by

A. R. HALL

Atomic Energy Research Establishment
Harwell, UK

published by
Graham & Trotman
for the Commission of the European Communities

Published in 1987 by

Graham & Trotman Ltd
Sterling House
66 Wilton Road
London SW1V 1DE
UK

Graham & Trotman
Kluwer Academic Publishers Group
101 Philip Drive
Assinippi Park
Norwell, MA 02061
USA

For the Commission of the European Communities
Directorate-General Telecommunications, Information
Industries and Innovation, Luxembourg

EUR 10852 EN

© ECSC, EEC, EAEC, Brussels and Luxembourg, 1987

ISBN 0-86010-893-7 (volume)
ISBN 0-86010-929-1 (series)

British Library Cataloguing in Publication Data

Testing and evaluation of solidified high level radioactive waste.
1. Radioactive waste disposal
I. Hall, A. R. II. commission of the European Communities
621.48'38   TD898

ISBN 0-86010-893-7

TD
898
.T47
1987

Legal Notice

Neither the Commission of the European Communities, its contractors nor any person acting on their behalf, make any warranty or representation, express or implied, with respect to the accuracy, completeness or usefulness of the information contained in this document, or that the use of any information, apparatus, methods or process disclosed in this document may not infringe privately owned rights; or assume liability with respect to the use of, or for damages resulting from the use of any information, apparatus, method or process disclosed in this document.

All rights reserved. No part of this publication may be reproduced, stored in a retrieval system, or transmitted in any form or by any means, electronic, mechanical, photocopying, recording or otherwise, without the prior permission of the publishers.

Printed and bound in Great Britain.

CONTENTS

		Page No.
1.	**INTRODUCTION**	1
	1.1 Scope of the Programme	1
	1.2 Abstracts	2
2.	**MATERIALS**	9
3.	**LEACHING MECHANISM THEORY**	12
	3.1 Determination of the Corrosion Mechanisms (FhG/ISC)	12
	3.1.1 Performance of Corrosion Tests	12
	3.1.1.1 Materials and Sample Preparation	12
	3.1.1.2 Corrosion Conditions	12
	3.1.1.3 Analytical Techniques	14
	3.1.2 Results	14
	3.1.2.1 Common Observations	14
	3.1.2.2 Static Corrosion Tests at Constant SA/V	15
	3.1.2.3 Periodically Interrupted Tests	26
	3.1.2.4 Static Corrosion Tests at Varied SA/V	34
	3.1.3 Summary	39
	3.2 Basic Mechanisms of Glass Leaching (AERE)	41
	3.2.1 General Theory of Diffusion-controlled Reaction	41
	3.2.2 Ion Exchange Theory	42
	3.2.3 Solutions to the Diffusion/Reaction Equations	43
	3.2.4 Comparison of the Theoretical Curves with Experimental Results	44
	3.2.5 Matrix Dissolution	46
	3.2.6 Temperature Dependence	47
	3.2.7 Effect of SA/V	48
	3.2.8 Saturation Effects	48
	References	63
4	**STUDIES OF SURFACE MORPHOLOGY AND CHEMICAL COMPOSITION OF SUB-SURFACE REGIONS**	68
	4.1 Hydrothermal Leaching of Simulated HLW Glasses (HMI)	68
	4.1.1 Preparation of Samples	68
	4.1.2 Experimental, and Selection of Leachants	68

		Page No.
4.1.3	Glass C-31-3-EC-SPF-Na	72
4.1.4	Leaching Model in NaCl Brine	96
4.1.5	Glass SON 68 18 17 L1 C2 A2 Z1 (Type R7T7)	103
4.1.6	A Comparison of C31-3 and SON 68 Glasses	107
4.1.7	Investigation of High Level Waste Glasses	108
4.2	Contribution of Nuclear Analytical Techniques to the Study of Aqueous Corrosion of Glasses (Saclay)	115
4.2.1	Introduction	115
4.2.2	Basic Principles and Analytical Capabilities of the Experimental Methods	116
4.2.3	Experimental Devices	120
4.2.4	Comparative Study of the Hydrated Layer of some Nuclear Glasses	124
4.2.5	Contribution to the Study of the Mechanism of Hydrated Layer Formation	128
4.2.6	Conclusions	130
4.3	The Use of Nuclear Techniques in the Study of Thermal Diffusion of Actinides and the Characterisation of Nuclear Glass Surfaces (Saclay)	153
4.3.1	Introduction	153
4.3.2	Study of Thermal Diffusion of Actinides in Glass SON 68	153
4.3.3	Characterisation of the Surface Region of Leached Glasses by Nuclear Techniques	156
4.3.4	Conclusions	157
	References	167

5 **PARAMETRIC STUDY OF CORROSION STABILITY** 173

5.1	Characterisation of Simulated Highly Active Waste Forms (SCK/CEN)	173
5.1.1	Study of the Corrosion Mechanisms in Static Distilled Water at 90°C	173
5.1.2	Influence of Temperature (40-200°C) on Corrosion in Static Distilled Water	188

	Page No.
5.1.3 Corrosion of Glass SAN 60 25 19 L_3 C_2 in Near-saturation Conditions	193
5.1.4 Corrosion at Constant Flow Rate	196
5.1.5 Influence of the pH of the Leachate	197
5.1.6 Corrosion in Clay-related Media	198
5.1.7 Characterisation of Glasses SM 58 LW 11 and SAN 60 25 19 L_3 C_2	205
5.2 Leaching of Americium from Glasses UK209 and C31-3EC (HMI)	223
5.3 Experimental Study of the Influence of Various Parameters on the Aqueous Leach Rates of Radioactive Glasses (CEA)	227
5.3.1 Introduction	227
5.3.2 Effect of Time	229
5.3.3 Effect of Temperature at Atmospheric Pressure	235
5.3.4 Effect of Pressure at Ambient Temperature	245
5.3.5 Effect of pH at Ambient Temperature	247
5.3.6 Effect of Type of Water	252
5.3.7 Effect of Crystallisation	255
5.3.8 Effect of Type of Leach Test and Frequency of Leachant Renewal	259
5.3.9 Conclusions	260
References	274
6 INFLUENCE OF REPOSITORY CONDITIONS ON THE LEACHING OF HIGH LEVEL WASTE FORMS	276
6.1 Leaching of Radioisotope-spiked Glasses under Simulated Repository Conditions (AERE)	276
6.1.1 Introduction	276
6.1.2 Flowing Leachant Experiments	277
6.1.3 Leaching in a Repository Simulation	283
6.1.4 The Effect of Backfill	287
6.2 Interaction between Repository Clay and Simulated HLW Forms (SCK/CEN)	296
6.2.1 In-situ Experiments	296
6.2.1.1 Surface Clay Quarry at Terhagen	296

			Page No.
	6.2.1.2	Underground Laboratory in Clay at 220 m Depth	299
	6.2.2 Laboratory Experiments		301
	6.2.2.1	Elemental Leaching Behaviour in Clay Media	301
	6.2.2.2	Influence of an External γ-radiation Field on Corrosion	303
	References		308
7	**RADIATION STABILITY (AERE)**		309
	7.1 Introduction		309
	7.2 Leach Rates		310
	7.3 Densities		310
	References		319
8	**MECHANICAL STABILITY**		320
	8.1 Fracture Studies (HMI)		320
	8.2 Canister/Glass Interaction (AERE)		326
	References		330
9	**CERAMICS (LEIDEN UNIVERSITY)**		331
	9.1 Hollandites		332
	9.2 Perovskite and Perovskite-like Phases		335
	9.3 The Phase Relations of $RuO2$		339
	References		350

1. INTRODUCTION

This document is the final report on the second joint five-year programme (1980-1985) under the Indirect Action Programme, Action No. 5. 'Testing and Evaluation of the Properties of Various Potential Materials for immobilizing Highly Radioactive Waste'. It is compiled from reports submitted by the following seven participating Laboratories:

Atomic Energy Research Establishment, Harwell (UK)
Centre d'Etudes Nucleaires de Saclay, (France)
Establissment CEA de la Vallee du Rhone, Marcoule (France)
Fraunhofer-Institut fur Silicatforschung, Wurzburg (FRG)
Hahn-Meitner-Institut fur Kernforschung, Berlin (FRG)
Rijksuniversiteit Leiden (The Netherlands)
Studiecentrum voor Kernenergie, Mol (Belgium)

Annual progress reports for the years 1981, 1982 and 1983 have been published under the same title as EUR reports 8424EN, 9268EN and 10038EN respectively. In some instances more detailed information is to be found in these reports than is given in this final report.

The work carried out within the framework of this second programme is a continuation and extension of that performed in the first indirect action programme for the period 1975-1979 (EUR 7138EN).

1.1 Scope of the Programme

The overall objective of the research undertaken is to test various European solidified high-level waste forms for thermal, mechanical and radiation stability and for their behaviour under attack by various aqueous media so as to predict their activity retention capabilities after disposal. This latter is probably the single most important aspect of the programme. Experiments have been performed under waste repository relevant conditions and under simplified laboratory conditions chosen to investigate the

mechanisms of attack. The research includes studies of fully
active waste forms as well as inactive simulates and simulates
spiked with specific active isotopes.

Although by far the greater emphasis has been placed on research
into vitrified wastes, a significant amount of work has been done
on ceramic waste forms, in particular the determination of phase
relationships and crystal structures of various perovskites and
hollandites.

1.2 Abstracts

A.E.R.E. Harwell (Sections 3.2, 6.1, 7 and 8.2)
A general theory of diffusion-controlled reaction is outlined.
Solutions to the equations are given and compared with alkali
leach rates and depth profiles. Good agreement between theory and
experiment is found; the effect of pH on alkali leaching is
satisfactorily explained.

Experiments at very high SA/V ratio have demonstrated that
saturation of the leachate with a particular species does not
inhibit further leaching of that species unless the precipitate
formed seals the glass surface.

Glasses containing simulated high level waste have been tested
under conditions likely to occur after underground disposal. The
leach rates obtained are much lower than those found at high water
flow rates. The leaching mechanism has been studied and the
release rates of six radiologically significant isotopes have been
measured. Of these, ^{99}Tc, ^{137}Cs, and to a lesser extent ^{90}Sr are
more readily released than the actinides, ^{237}Np, Pu and ^{241}Am: the
latter two in particular give very low leach rates. Results
suggest that cement (or concrete) may prove a better backfill than
bentonite.

Investigations of the radiation stability of ^{238}Pu doped glasses
have shown increases in the leach rate of up to x4 for some

glasses but no change in others. The densities changed slowly with dose, some increasing, some decreasing to saturation at a total change below 1%.

Less cracking occurred when a large block of glass was cast in a carbon fibre lined container than in an unlined one.

CEN Saclay (Sections 4.2 and 4.3)

Some original techniques of surface analysis based on direct observation of resonant nuclear reactions and elastic backscattering of ^{4}He-ions have been developed. These are particularly useful in the non-destructive determination of the concentration profiles induced by aqueous corrosion in the near surface region of glasses (explorable depth about one micrometer).

The analytical capabilities of the nuclear microprobe to obtain information concerning the distribution of the elements in the hydrated layer and the underlying pristine glass were also studied.

These methods were used to follow the behaviour of different glasses submitted to aqueous corrosion tests conducted at 90°C from 7 days up to 12 months. The pH and the composition of the leachates were both determined.

The results obtained demonstrate the good corrosion resistance of the French reference glass SON 68 18 17 L1C2A2Z1 compared with some other nuclear glass compositions.

The measurement of thorium and uranium diffusion coefficients in the glass SON 68 18 17 at 300°C was undertaken, using thin targets of natural oxides deposited on glass samples, by ^{4}He Rutherford backscattering.

The results obtained after thermal tests from 8 to 21 months at

300°C are in the range 1.65 to 14.3 x 10^{-20} $cm^2.s^{-1}$.
The resonant nuclear reactions of hydrogen, lithium, sodium and
aluminium were used to compare the surface compositions of two
glass samples (SON 68 18 17) leached respectively during 28 days
at 90°C (static test) and 100°C (continuous flowing test).

The characterization of a glass sample containing 0.85 wt % of
$^{237}NpO_2$ and leached during 15 days at room temperature, using
Rutherford backscattering spectrometry is described. A
significant enrichment of neptunium is shown.

C.E.A. Marcoule (Section 5.2)

The objective of this study, undertaken within the framework of
the CEC contract WAS 123-80-55F (1980-83), was to reveal the
sensitivity of leach rates of fission product glasses to the
separate effects of the following experimental parameters in the
water/glass system:

>Time
>Temperature
>Pressure
>Type of water
>Crystallisation
>Leach test method

The glasses used are borosilicates whose fission product oxide
contents vary from 12.7 to 28 wt %. Their compositions are given
in Table 2.1.

The tests have been made mainly on α, β and γ-active glasses
prepared and leached in the VULCAIN cell at Marcoule.

The effect of certain of the parameters has also been tested on
inactive glasses of the same composition by measurement of weight
loss or elemental analysis.

No surface analyses have been made.

The principal results obtained are as follows:

(a) For the least soluble species (Ce, Ru, Am), the leach test method has the most pronounced effect, particularly the frequency of leachate renewal. The less frequent the renewal the lower is the leach rate.

(b) When the water is renewed frequently, the type of water does not affect the leaching significantly.

(c) Hydrostatic pressure up to 200 bars appears to be without significant effect at ambient temperature.

(d) The pH tests (from 1 to 14) show the characteristic peculiar to F.P. glasses, which have relatively small leach rates in basic media (pH 7 to 14) and very much greater in acids of pH < 4. (The opposite is generally true for soda-lime-silica glasses).

(e) The phenomenon of non-congruent corrosion persists even after 500 days testing at ambient temperature for both americium and cerium.

(f) The effect of temperature is difficult to quantify, secondary complex physico-chemical mechanisms probably appear as the temperature is increased. From 50 to nearly 150°C the phenomenon manifests itself as a greatly reduced leaching for all the species determined, the apparent mean activation energy falling by a factor of approximately 4.

(g) The effect of crystallisation is not systematic and depends greatly on the composition of the glasses used.

(h) Finally, where direct comparison was possible (pressure particularly) no difference could be detected between the behaviour of active and inactive glass.

Fraunhofer-Institut für Silicatforschung (Section 3.1)

The topic of this contribution is the determination of the corrosion mechanism of the simulated HLW glass SM 58 LW 11 in a Q-brine typical of a rock-salt repository. Leaching behaviour was studied over a wide range of temperature and time; leach rates and concentration depth profiles were measured.

In addition, tests were conducted in which the reaction layer was removed and the sample either returned to the original leachate or placed into new leachant. By this means the relative influences of the growing reaction layer and the changing leachate composition could be separated.

It is concluded that the principal mechanisms are glass network dissolution, reaction zone and ion exchange zone formation. The reaction zone is no protection against further attack. The presence of some metals (Fe, Cu, Pb or Al) was found to enhance the corrosion rate.

Hahn-Meitner Institut, Berlin (Sections 4.1, 5.2, 8.1)

A detailed investigation of the surface morphology of two glasses (C 31 and SON 68) leached at 200°C in three different leachants (NaCl-, Q- and Z-brine) has been made, using SEM, EPMA and X-ray diffraction techniques. The crystalline phases which are formed at the surface of the leached glass have been identified. They include Analcime ($Na(AlSi_2O_6) \cdot H_2O$), various sheet silicates, zinc and calcium silicates and Ca-As apatite. A leaching model for glass in NaCl brine is elaborated.

Investigation of the leaching of americium from glasses UK 209 and C 31 in NaCl-brine, Q-brine and DIW have shown that there is a much greater retention of Am in the surface layer when Q-brine is used.

An extensive theoretical and experimental study of glass fracture is reported. Crack propagation studies of four waste glasses and one commercial glass were made using an elastic spherical indenter and a double torsional method. Nuclear waste glasses are less susceptible to slow crack propagation than is the commercial soda-lime silicate glass tested.

Small samples of fully active glasses made in VULCAIN were leached and their surface morphologies examined. Both as-received and annealed samples were tested in various leachants.

Leiden University (Section 9)

The aim of this research is the preparation of model compounds of phases with hollandite, perovskite and perovskite-related structures. The thermal stability, phase relationships, crystallographic properties and resistance to attack by water were studied.

More specifically, the compounds studied include:

(a)   Ba and Sr hollandites. It is concluded that hollandites have excellent stability in water, but excess Ba can lead to the formation of other, very soluble, compounds containing Sr and Cs.

(b)   Ca, Ba, Sr perovskites, Ca, Sr and Nd perovskite-related phases. The majority of these are very insoluble. However, some very soluble compounds were found, for example, Ca_3UO_6 and $SrMUO_6$ (M = Mg, Mn, Ca).

(c)   The systems $BaO-RuO_2-Fe_2O_3$
$BaO-RuO_2-Nd_2O_3$
$BaO-RuO_2-Al_2O_3$
$SrO-RuO_2-Nd_2O_3$
$SrO-RuO_2-Al_2O_3$
$SrO-TiO_2-Al_2O_3$
$SrO-TiO_2-Gd_2O_3$

Of approximately 50 compounds reported, some 17 were found to be readily attacked by water.

SCK/CEN, Mol (Sections 5.1 and 6.2)

Static leach tests have been made on six glass compositions at 90°C for times up to 2 years in DIW, clay-related media and leachants buffered at various pH. Surface area:volume ratios of $10\ m^{-1}$ and $100\ m^{-1}$ were used. The influences of redox conditions and of temperature were investigated in the case of the clay media. The main conclusions are:

*In the reference condition (DIW, 90°C) either diffusion processes or matrix dissolution control the corrosion of the main glass components Si, Na, B, depending on the SA/V ratio (100 or $10\ m^{-1}$, respectively). At $100\ m^{-1}$, the leachate is quickly saturated for elements such as Mg, Al, Ca, U, Si, but also (apparently) for B, Na and Li. Evidence has been obtained that these saturation phenomena do not necessarily correspond with a steady state situation.

*Mass losses are increased by the addition of clay particles to the leachate; the larger the clay concentration the greater is the increase.

*The temperature dependence of the corrosion in the interval 40-200°C in either DIW or clay media does not present an Arrhenius type relationship.

*The various waste forms (five waste glasses and one glass-ceramic) perform rather similarly in all experimental conditions considered. Some long term data (such as in the MCCI test), however, present a waste form dependent divergence.

Tests in the surface clay quarry at Terhagen are completed. Results for ten different waste forms are given.

The underground laboratory at Mol and the experimental
arrangements being made are described. This laboratory is
expected to be operational by mid-1985. The tests planned
include:

(a)   In-situ testing of the interaction between HLW forms and
clay in the underground facility at a depth of 200 m.

(b)   Lab-scale testing of the interaction between HLW forms and
clay using waste forms tracered with various isotopes.

(c)   Testing of the influence of an external radiation field on
the interaction between HLW forms and clay.

2. MATERIALS

Originally, five glass compositions were selected for
investigation. As the scope of the programme widened and as
improved glass formulations were developed, so additional
compositions were included by some contributors.

Compositions which have been investigated by more than one
participating laboratory are given in Tables 2.1 and 2.2.
Additional compositions are in some cases given in individual
chapters.

TABLE 2.1   Compositions of Test Materials

	UK 209	UK 189	C 31/3	B 1/3	VG 98/3	78/7	SON 58	SON 64	SON 64/G3	SON 68	SAN 60	SM 58 LW 11 (SCK)	SM 58 LW 11 (FI)
Fission product oxides	9.75	9.50	15.20	14.55	15.54	28.10	22.70	13.14	13.00	11.25	1.21	6.06	2.82
Waste oxides	◊	◊	*	+	+		‡	o	▽	**	△		•
Fe_2O_3	2.73	2.68	1.47	1.51	0.70	0.50	0.66	5.07	5.90	2.91		1.17	1.71
Cr_2O_3	0.56	0.55	0.42	0.43	0.24	0.20	0.20	0.43	0.50	0.51			0.28
NiO	0.36	0.36	0.23	0.24	0.21	0.20	0.10	0.24	0.20	0.74		0.11	0.43
MgO	6.34	6.23			0.40								
Al_2O_3	5.11	5.03					0.10	1.69	0.20	4.91			0.94
U_3O_8	0.06	0.058	0.46	0.47	1.21	2.40	3.60	0.45	0.90		0.02		0.12
Na_2O				2.2								3.67	3.37

Also : ◊ 0.1 % SO_4, 0.4 % ZnO, 0.23 % P_2O_5, * 2.2 % Na_2O, + 0.60 % P_2O_5, ▽ 0.4 % P_2O_5, 5.9 % Gd_2O_3, • 0.55 % SO_3, 0.77 % $F_?$
△ 0.15 % F^-, 0.62 % SO_4^{--}, 0.33 % Fe_2O_3, ** 0.52 % UO_2, 0.33 % ThO_2, ‡ 0.6 % P_2O_5, o 0.36 % F_2O_3, 0.43 % ThO_2,

Glass formers	UK 209	UK 189	C 31/3	B 1/3	VG 98/3	78/7	SON 58	SON 64	SON 64/G3	SON 68	SAN 60	SM 58 LW 11 (SCK)	SM 58 LW 11 (FI)
			φ	x		***				****			
SiO_2	50.88	41.51	34.75	28.00	41.84		43.60	7.16	44.20	45.48	43.41	56.87	56.87
B_2O_3	11.12	21.87	4.14	6.40	10.48		19.00	18.42	17.30	14.02	17.00	12.28	12.28
Na_2O	8.30	7.68	1.12	1.60	22.25		9.40	12.53	11.50	9.86	10.67	4.63	4.63
Li_2O	3.99	3.69	0.98	2.40						1.98	5.00	3.74	3.74
Al_2O_3			10.28	12.80	1.20	4.9				4.91	18.09	1.16	1.16
CaO			3.84	4.00	2.32					4.04	3.50	3.83	3.83
BaO			14.48	14.80									
TiO_2			2.80	4.56	3.52							4.45	4.45
MgO			1.44	1.2								2.05	2.05

Also : φ 0.8 % ZrO_2, 4.88 % ZnO, 0.48 % As_2O_3 *** 48.50 % P_2O_5, 15.2 % Fe_2O_3
x 0.8 % ZrO_2, 3.60 % ZnO, 0.40 % As_2O_3.   **** 0.28 % P_2O_5, 1 % ZrO_2, 2.5 % ZnO.

TABLE 2.2   Compositions of fission product oxides contained in the test materials

	UK 209	UK 189	C 31/3	B 1/3	VC 98/3	78/7	SON 58	SON 64	SON 64/G3	SON 68	SAN 60	SM58 LW11 (SCK)	SM58 LW11 (FI)
ZrO_2	1.43	1.39	3.32	2.45	2.35	3.73	3.14	1.96	1.97	1.65	0.144	1.14	0.81
BaO	0.38	0.38	0.33	0.70	0.64	1.65	1.00	0.62	0.63	0.61	0.052	0.65	0.10
Rb_2O	0.11	0.10	0.16	0.16	0.15	0.34	0.23	0.15	0.14	0.13			
SrO	0.32	0.32	0.59	0.59	0.50	0.84	0.66	0.41	0.14	0.34	0.056		0.05
Y_2O_3	0.17	0.17	0.34	0.34	0.32	0.47	0.37	0.23	0.23	0.20			
SeO_2						0.05		0.03	0.03				
MoO_3	1.77	1.74	2.36	2.36	2.17	3.48	0.30	1.89	2.00	1.70	0.161	1.19	0.28
MnO_2			0.34	0.34	0.40					0.74		0.18	0.67
CoO_3										0.12		0.11	
PdO	0.44	0.42	0.11	0.08	0.62	1.42	0.97	0.34	0.34				
Ag_2O			0.002	0.002		0.06	0.04	0.03	0.02	0.03			
CdO			0.004	0.004		0.09	0.06	0.04	0.04	0.03			
SnO_2			0.016	0.016		0.05	0.04	0.03	0.04	0.02			
Sb_2O_3			0.004	0.004		0.01	0.01	0.01	0.01	0.004			
TeO_2			0.26	0.26	0.24	0.57	0.50	0.31	0.31	0.23			
Cs_2O	0.77	0.75	1.70	1.70	1.35	2.44	1.80	1.13	1.13	1.30	0.215		0.16
La_2O_3	0.44	0.43	0.66	0.66	0.61	1.27	0.95	0.59	0.59	0.90	0.105		
CeO_2	0.99	0.97	1.31	1.31	1.56	2.49	2.07	1.29	1.30	0.93	0.118	0.65	0.18
Pr_2O_3	0.43	0.42	0.60	0.60	0.57	1.20	0.89	0.56	0.56	0.44	0.050		
Nd_2O_3	1.82	1.76	2.40	2.40	2.15	4.15	2.96	1.84	1.86	1.59	0.169	2.14	0.57
Tc_2O_3							0.66	0.41	0.41				
RuO_2	0.68	0.66	0.92	0.53**	1.06	2.16	1.86	0.63	0.64		0.074		
Rh_2O_3			0.24	0.15***	0.18	0.38	0.30	0.10	0.10		0.066		
Pm_2O_3					0.05	0.91	0.07	0.04	0.04				
Sm_2O_3					0.47		0.60	0.37	0.38				
Eu_2O_3					0.09	0.18	0.13	0.08	0.09				
Gd_2O_3					0.06	0.13	0.08	0.05					
Total	9.75	9.50	15.20	14.55	15.54	28.10	22.70	13.14	13.00	11.25	1.21	6.06	2.82

* PdO substituted by NiO   —   ** RuO_2 substituted by TiO_2 —   *** Rh_2O_3 substituted by Cr_2O_3.

3. LEACHING MECHANISM THEORY

3.1 Determination of Corrosion Mechanisms
(Fraunhofer Institut fur Silicatforschung)

The topic of this contribution is the determination of the corrosion mechanism of the simulated (ie inactive) HLW glass SM 58 LW 11. The experimental conditions are defined with respect to a hypothetical water access to a rock salt repository according to a recommendation of the Physikalisch-Technische Bundesanstalt Braunschweig.

3.1.1 Performance of Corrosion Tests

3.1.1.1 Materials and sample preparation

The HLW glass SM 58 LW 11 was produced by the Hahn-Meitner-Institut, Berlin, and received in three different batches (I to III). The theoretical composition and the compositions of the different batches determined by microprobe analysis are given in Table 3.1. Batch II is in best agreement with the theoretical composition. The deviations of batches I and III mainly refer to the SiO_2 and Al_2O_3 contents; batch I has an additional K_2O impurity level of 0.9 wt %. Samples were cut by a low-speed saw (600 grit, those of batch II and III were additionally polished to optical quality. All samples were etched in a $HF-HNO_3$ solution and pretreated with H_2O at 60 °C (0.5 to 1 h).

3.1.1.2 Corrosion Conditions

According to a recommendation of the Physikalisch-Technische Bundesanstalt Braunschweig, the so-called Q-solution [1] was used as leachant in most cases. Its composition is shown in Table 3.2. Additional amounts of NaCl according to [2] were added. Some experiments were carried out in NaOH or HCl solution, both 0.01 molar. The leachant volume was 90 cm^3. Unless stated differently, the sample surface area was slightly larger than 3 cm^2 leading to a sample surface area to leachant volume ratio s of 0.034 cm^{-1}.

TABLE 3.1 Composition of glass SM 58 LW 11 in wt%

component	theoretical	by microprobe analysis		
		batch I	batch II	batch III
SiO_2	56.8	52.9 ± 1.7	56.2 ± 1.7	54.8 ± 0.8
B_2O_3	12.3	*	*	*
Al_2O_3	2.1	4.4 ± 0.2	2.8 ± 0.2	2.9 ± 0.1
Li_2O	3.7	*	*	*
Na_2O	8.0	8.1 ± 0.4	7.8 ± 0.4	7.6 ± 0.1
K_2O	0.0	0.9 ± 0.1	b.d.	b.d.
Cs_2O	0.16	*	*	*
MgO	2.1	*	2.0 ± 0.2	2.0 ± 0.1
CaO	3.8	4.1 ± 0.4	4.1 ± 0.4	3.6 ± 0.1
SrO	0.05	*	*	*
BaO	0.10	*	*	*
CeO_2	0.19	*	*	*
Nd_2O_3	0.57	*	*	*
U_3O_8	0.12	*	*	*
TiO_2	4.5	4.5 ± 0.3	4.7 ± 0.3	4.0 ± 0.2
ZrO_2	0.81	0.8 ± 0.1	0.7 ± 0.1	0.7 ± 0.1
Cr_2O_3	0.28	*	*	*
MoO_3	0.28	*	0.3 ± 0.1	0.3 ± 0.1
MnO_2	0.67	*	0.6 ± 0.2	0.6 ± 0.1
Fe_2O_3	1.71	2.0 ± 0.1	1.7 ± 0.1	1.6 ± 0.1
NiO	0.43	*	0.4 ± 0.1	0.4 ± 0.1
ZnO	0.02	*	*	*
SO_3	0.55	*	0.3 ± 0.3	*
F_2	0.77	*	*	*
Sum	100.0			

* : not determined
b.d.: below detection limit

The temperatures ranged from 80 to 200°C. In most cases, a pressure of 13 MPa was applied.

TABLE 3.2   Composition of the Q-solution at 55°C after [1] in wt %

NaCl	KCl	$MgCl_2$	$MgSO_4$	H_2O
1.4	4.7	26.8	1.4	65.7

3.1.1.3   Analytical techniques

The effects of the corrosion process were determined by weighing, by chemical analysis of the leachates including solid residues, and by surface analysis of the glass samples. The applied chemical techniques were ICP (<u>i</u>nductive <u>c</u>oupled <u>p</u>lasma), AES (<u>a</u>tom <u>e</u>mission <u>s</u>pectroscopy for <u>c</u>hemical <u>a</u>nalysis) combined with Ar ion etching and the resonant nuclear reaction method $N^{15} + H \rightarrow C^{12} + \alpha + \gamma$ after Lanford et al [3], carried out at the Kernphysikalisches Institut der Universitat Frankfurt*

3.1.2 Results

3.1.2.1   Common observations

The main corrosion mechanisms of SM 58 LW 11 in Q solution are

the dissolution of the glass network.

the formation of a layer of reaction products (reaction layer) on top of the surface of the undissolved (residual) glass.

the alteration of a thin surface zone of the residual glass by ion exchange processes (ion exchange zone).

The material dissolved from the glass is partly present in the reaction layer, partly in the leachant, partly as a solid residue

---

*The generous help by Prof. Rauch and Mr. March is gratefully acknowledged.

of the leachant. As a clear coordination of elements to either of the latter two parts is not possible after cooling the solution to room temperature, they are combined to the term 'leachant including solid residues'. The situation after a corrosion experiment can be characterized by the following sketch:

leachant + residues reaction layer//ion exchange zone/ bulk glass

residual glass

The dotted stroke indicates that the reaction layer is porous and contains some leachant; the double stroke marks a sharp transition and the single stroke a gradual one. Due to the sharp transition between reaction layer and residual glass, the layer can very easily be removed. Thus, the amounts of dissolved glass can be determined directly by weighing (of the residual glass).

In connection with the above observations, some useful quantities for the presentation and discussion of the experimental results are defined: q in $mg.cm^{-2}$ is the mass loss per surface area determined directly by weighing the residual glass or indirectly by the boron or lithium release; its time derivative is the dissolution rate r in $mg.cm^{-2}.d^{-1}$; the network dissolution depth is d in μm; it can be calculated by $d = q/\rho$ with the density $\rho = 2.61$ $g.cm^{-3}$; s in cm^{-1} is the sample surface area to leachant volume ratio; c in $mg.cm^{-3}$ is the mass loss per leachant volume related to q by the equation $q.s = c$. The time t is given in units of 1 d.

3.1.2.2 Static corrosion tests at constant surface area to leachant volume ratio

Static corrosion tests, ie tests in a closed sample/leachant system without interim manipulations, were performed in Q-solution at 80, 120, 160 and 200°C and 13 MPa for test periods between 6 and 36 d. Some tests in NaOH or HCl solution (0.01 mol/l) were carried out for reasons of comparison. The mass loss data in Q- and NaOH solution (determined by weighing) are shown in Figure 3.1

in a double logarithmic presentation. The corresponding boron and lithium release data are in excellent agreement.

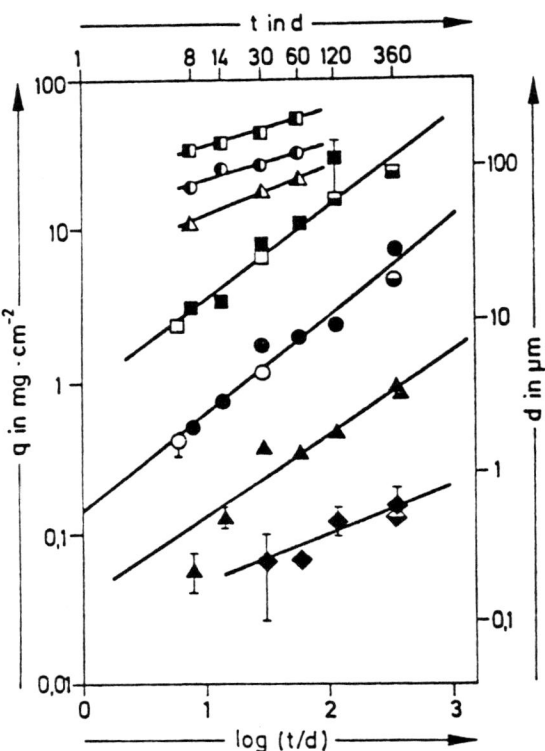

Fig. 3.1 Mass loss per surface area q and network dissolution depth d as a function of the corrosion time for samples of:

batch I in Q-solution: ◆ = 80°C, ▲ = 120°C
● = 160°C, ■ = 200°C

batch II in Q-solution: ○ = 160°C, □ = 200°C

batch III in Q-solution: ⬥ = 80°C, ⬤ = 120°C
⊖ = 160°C, ⊟ = 200°C

batch I in NaOH solution: ▲ = 120°C, ◐ = 160°C
(0.01 mol/1) ◨ = 200°C,

$s = 0.034$ cm^{-1}, $P = 13$ MPa

Time dependence of the network dissolution

According to Figure 3.1, the respective data sets at a given temperature are fairly well represented by straight lines. This corresponds to an empiric time law.

$$q \sim t^\beta \qquad (1)$$

The exponents β, ie the slopes of the straight lines, allow a first approximation of the time dependence of the network dissolution process. In Q-solution, they range from $\beta \approx 0.3$ at 80°C to $\beta \approx 0.6$ at 160 and 200°C; in NaOH solution $\beta < 0.3$ is valid for all temperatures, although very high initial rates occur. In HCl solution, slopes $\beta < 0.4$ are found for $T > 120°C$ and $\beta = 0.8$ at 80°C.

Temperature dependence of the network dissolution

By an evaluation of the data in Figure 3.1 according to the Arrhenius relation

$$q \sim \exp(E_A/RT) \qquad (2)$$

the activation energies E_A are determined as a function of time. The results are shown in Figure 3.2. After an initial period, a long term limit of the activation energy seems to be reached in Q-solution; $E_A = 69 \pm 3$ kJ/mol (valid only between 120 and 200°C; the 80°C values deviate systematically). The E_A values in NaOH solution (and as well in HCl solution for $T > 120°C$) are much smaller: $E_A = 21 \pm 3$ kJ/mol. The former value is typical of glass dissolution processes, the latter of diffusion processes in aqueous solutions*.

---

*A rate controlling diffusion step across a porous reaction layer (ie along the pores filled with leachant) is expected to yield the corresponding activation energy for diffusion in aqueous media. Tests to check the role of the layers formed in NaOH solution are underway. On the other hand the high E_A value in Q-solution is a first hint to the inability of the reaction layers between 120 and 200°C to protect the glass (see 3.1.2.3).

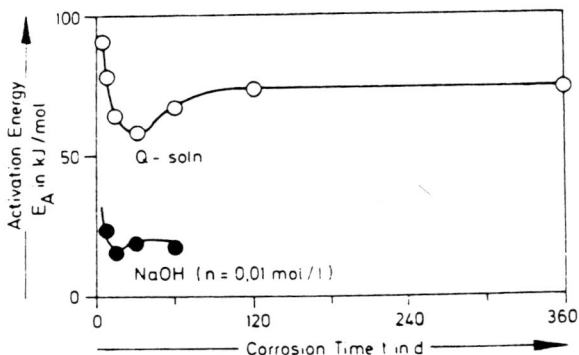

Fig. 3.2 Activation energy E_A for the corrosion processes in Q-solution (O) for $T > 120°C$ and in NaOH solution (●), as a function of time.

Pressure dependence of the network dissolution

Some additional tests were performed in Q-solution in order to check the influence of the pressure on the corrosion process. The results of three comparative tests show only a minor pressure effect, if any. The following mass losses per surface area in $mg.cm^{-2}$ were determined:

at 160°C, s = 0.034 cm^{-1}, 30 d,   0.6 MPa: 1.4 ± 0.3;
                                        13 MPa: 1.8 ± 0.3;

at 200°C, s = 0.034 cm^{-1}, 30 d,   1.6 MPa: 6.8 ± 1.2;
                                        13 MPa: 8.5 ± 1.1;

at 200°C, s = 0.064 cm^{-1}, 8 d,   1.6 MPa: 1.5 ± 0.15;
                                        52 MPa: 1.8 ± 0.15*.

Effect of crystallization

During the production of HLW glass on a technical scale, crystallization phenomena may occur. Therefore, the effect of crystallization on the corrosion behaviour of SM 58 LW 11 was investigated. Samples were heat treated between 600 and 900°C and then examined by microscope and microprobe. The formation of at least three different kinds of crystallites were observed: spherulites containing > 94.5 wt% of SiO_2, causing circumferential cracks in the glass matrix, a bright phase with a Si to total O ratio of 1:3 (typical of chain silicates) and a SiO_2 content of approximately 52 wt%, and a further phase still unspecified. Besides these, $Li_2B_4O_7$ was detected by X-ray diffraction. The crystallization behaviour of SM 58 LW 11 is summarized in Figure 3.3. Corrosion tests with samples heat treated for 7 d at 725°C show a strong local corrosion around the spherulites (see Figure 3.4) and higher weight losses than uncrystallized samples. Due to the residues of Q-solution in the cracks and fissures, the effect could not be quantified.

Influence of metallic container materials

It is well known [4] that traces of metal ions can largely influence the corrosion of glass. More recently, Buckwalter and Pederson [5] reported that the corrosion of a HLW glass in water at 90°C is only slightly enhanced in the presence of container

---

*These tests were performed in high pressure autoclaves with a leachant volume of approx. 9 cm^3.

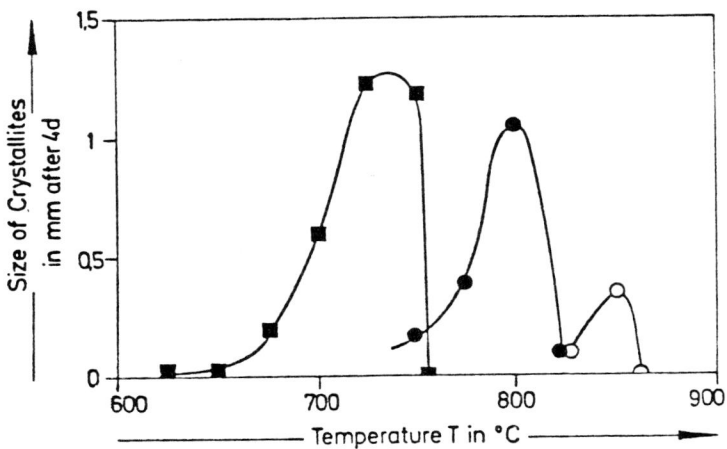

Fig. 3.3 Size (diameter) of the spherulites (■), the bright phase (●) and a further phase still unspecified (O) after 4 d of heat treatment at different temperatures

metals like Cu, Ti, or Zn but drastically reduced by Pb or Aℓ. Similar tests in Q-solution at 200°C lead to different results: the presence of steel, Cu, Pb or Aℓ (in the order of increasing effect) greatly enhance the corrosion. The mass losses per surface area (means of results by weighing, by boron and by lithium release) in mg.cm^{-2} after 3 d are:

4.3 ± 0.3	without metal
13.7 ± 0.3	in the presence of steel
17.2 ± 0.7	in the presence of Cu
19.4 ± 0.4	in the presence of Pb
27 ± 7	in the presence of Aℓ

Fig. 3.4 Partial view of the surface of a sample tempered for 7 d at 725°C and corroded for 30 d at 200°C in Q-solution

The results demonstrate the importance of the new type of so-called integrated corrosion tests, where container and backfill materials are included in the system.

Characterization of the ion exchange zone

The concentration profiles of the elements Si, Aℓ, Ti, B, Nd, Ca, and Na in the ion exchange zone were determined by ESCA in combination with Ar ion etching. The proton concentration profiles were determined by the nuclear reaction technique with N^{15} (see 3.1.1.3). The investigated samples were some of those from Figure 3.1. Figure 3.5 a-b shows the concentration profiles of Na and Ca as a function of temperature and time. These elements are depleted in the ion exchange zone. The depletion depth depends strongly on temperature but only slightly on time. A similar behaviour is found for B and Nd, which are also depleted, and for Si, Aℓ and Ti, which are enriched.

The proton concentration profiles support the above findings. In Figure 3.6 the proton concentration is given in mol-% of the cation in the glass; for simplicity, the glass is treated as pure SiO_2 (leading to a maximum error of less than 10%). From

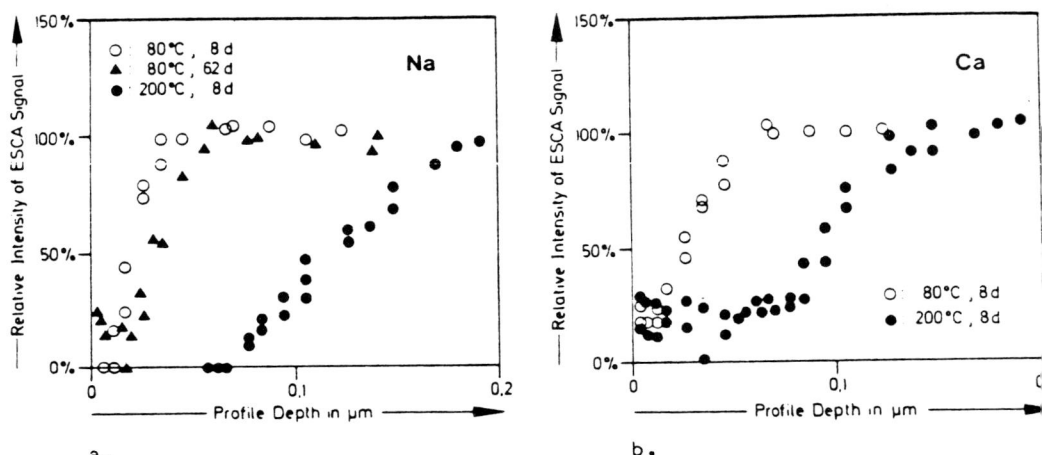

Fig. 3.5a-b Concentration profiles of Na (a) and Ca (b) in the ion exchange zone of samples (batch I) corrode in Q-solution for 8 d at 80°C (0), at 200°C (●), and for 62 d at 80°C (▲).

Figure 3.6 an activation energy for the proton incorporation can be calculated leading to 37 kJ/mol between 120 and 200°C; again, the 80°C values deviate systematically.

Figure 3.7 shows the amounts of incorporated protons (minus the proton content of the uncorroded glass) in the ion exchange zone as a function of time. After an initial period, a constant long term value seems to be reached.

Compared to the dissolution depths shown in Figure 3.1, the depths of the corresponding ion exchange zones are negligibly small. The ion exchange process does not affect the mass balances

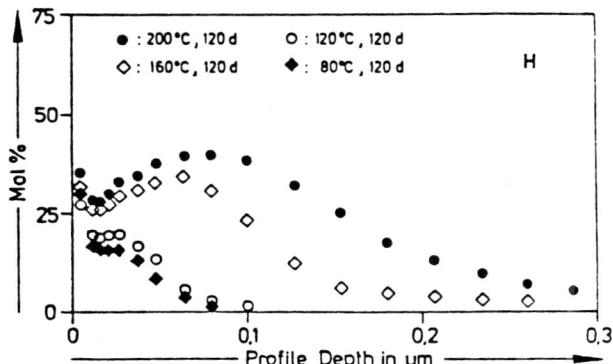

Fig. 3.6 Concentration profiles of H in the ion exchange zone of samples (batch II) corroded in Q-solution for 120 d at 80°C (◆), 120°C (○), 160°C (◇), and 200°C (●).

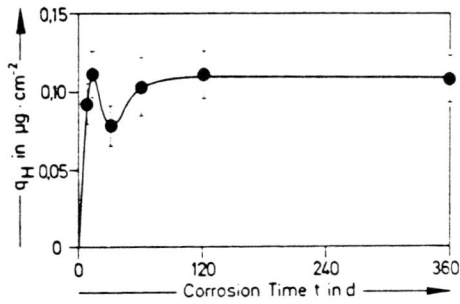

Fig. 3.7 Amounts of incorporated protons q_H in the ion exchange zone of samples (batch I) corroded in Q-solution at 160°C, as a function of time.

significantly, which explains the good agreement between boron and lithium release data with direct mass loss determinations.

Interim conclusions for the corrosion mechanisms

Although the results from the static long term tests provide a broad data base useful for further and more detailed investigations, the possible conclusions with respect to the corrosion mechanisms and to the long term behaviour are rather limited. The issue is demonstrated in Figure 3.8: the data obtained during 1 year do not provide a sufficient data base to distinguish between a long term behaviour proportional to t, or to $\sqrt{t}$, respectively; 2 or 3 years tests are not expected to change this situation.

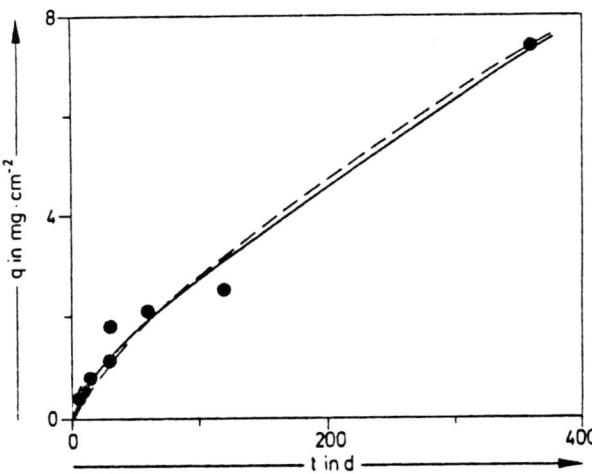

Fig. 3.8 Mass loss per surface area q of samples corroded in Q-solution at 160°C, as a function of time;
● : experimental data;
—— : time law with long term behaviour ~ t;
---- : time law with long term behaviour ~ $\sqrt{t}$

Therefore, the static long term experiments need to be supported
by other tests providing criteria to determine the rate
controlling mechanisms and the correct type of time law for long
times.

A careful examination of the results in Figures 3.1 and 3.2
reveals an incompatibility: permanently temperature-dependent
exponents β in eq. (1) lead to unreasonable conclusions for the
activation energies E_A in eq. (2); these would become infinitely
small or infinitely large. On the other hand, the assumption of a
finite long term value for E_A requires temperature-independent
exponents β, ie the same type of law for all temperatures where
the long term value of E_A is valid. This type of time law has to
be determined. For this purpose, several processes have to be
taken into consideration.

1. The attack of the Q-solution could by itself provide a still
unknown mechanism. For this case, no estimation of the long term
behaviour is possible at the moment.

2. The reaction layer could act as a rate controlling barrier.
The long term law would be proportional to $\sqrt{t}$ in this case [6].
The same dependence would result for the case of a readsorption of
corrosion products on the surface of the residual glass.

3. The ion exchange zone could act as a rate controlling
barrier. Due to the strong network dissolution, however, this is
only expected for the very beginning of the process. The long
term dependence would be proportional to t. The same conclusion
holds for the case when a temporally linear hydration process at
the boundary between ion exchange zone and bulk glass competes
with a diffusion process across the ion exchange zone [7].

4. The accumulation of reaction products in the leachant could
influence the driving force of the network dissolution process.
For the long run, assuming thermodynamic equilibrium between the
leachant and the crystalline solids in the leachant, a linear time

law is expected [8].

The following experiments were designed to provide criteria to dismiss or confirm some of the ideas summarized above.

3.1.2.3 Periodically interrupted tests

The whole test program consisted of 3 tests at 120 and 3 tests at 200°C (all at 13 MPa), with 5 samples per test. The tests were interrupted in intervals of approx. 40 days for interim manipulations and examinations. The following types of manipulations during such an interruption were performed.

type A reaction layer is wiped off directly; residual glass sample and layer are dipped back into the old leachate;

type B reaction layer is wiped off, dried at 110°C and weighed; residual glass sample and layer are dipped back into the old leachate;

type C sample covered with reaction layer is exposed to a fresh leachant; old leachate is analysed.

type D reaction layer is wiped off, dried at 110°C, weighed, and analysed; residual glass sample is exposed to a fresh leachant; old leachate is analysed;

type E sample covered with reaction layer remains in the old leachate.

Tests of type C were recently reported by another laboratory [9].

Experiences with earlier tests had shown that temperature is the major source for experimental errors. The precision achievable with the experimental set up used here was no better than ± 1 to 1.5% of the absolute temperature leading to a precision of the mass loss data of about ± 20%. In order to establish a higher precision, the actual temperature of each autoclave was recorded throughout the test. Thus the mass loss data can be corrected for temperature by the equation

$$\Delta q/q = (E_A \cdot \Sigma_j \Delta T_j \Delta t_j)/(RT_n^2 \cdot t) \qquad (3)$$

where $\Delta q/q$ is the relative error of mass loss due to a temperature deviation ΔT from the nominal value T_n. The indices j mark the respective time intervals Δt_j. For the activation energy E_A, a mean value of 76 kJ/mol is chosen. Table 3.3 shows the schedule for these tests and reports the real mean temperatures (based on 3 control measurements per week and autoclave). In many cases, only minor deviations from the nominal temperature occur; in some cases, however, the correction according to eq. (3) is substantial.

TABLE 3.3   Schedule of the periodically interrupted tests and real mean temperatures in °C during the intervals;
T_1 to T_3: nominal temperature 120°C;
T_4 to T_6: nominal temperature 200°C

Period	I	II	III	IV	V	VI	VII
Δt	47	42	39	43	42	42	42
t	47	89	128	171	213	255	297
T_1	*	120.8	122.5	121.1	120.3	121.3	121.6
T_2	*	120.0	122.0	120.6	120.3	121.3	120.6
T_3	*	116.6	117.1	117.7	117.6	117.8	117.1
T_4	197.7	192.8	203.1	203.7	203.7	204.0	201.9
T_5	202.4	201.9	194.3	193.8	193.7	194.2	195.6
T_6	200.6	200.2	199.5	200.2	200.3	200.5	201.5

* = not determined.

Role of the reaction layers and of an exchange of the leachant

Figures 3.9 and 3.10 summarize the main result of the periodically interrupted tests and compare them to the results from static tests (see 3.1.2.2). A first and most essential result is the good agreement between the type E values and the regression curve of the static tests. This means that the interruptions (cooling down to room temperature and heating up again after some days) do not affect the corrosion process in the quasistatic type E test. Consequently, the same is expected to hold for the other tests; thus they can be interpreted as if the manipulations had taken place in situ.

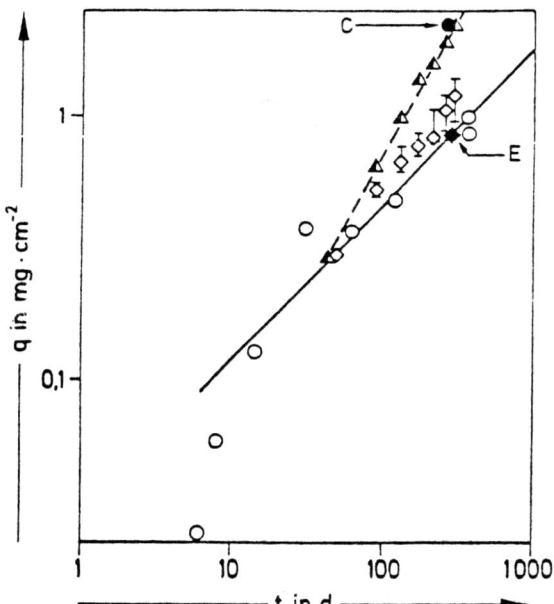

Fig. 3.9 Normalised weight loss of SM 58 corroded at 120°C as a function of time;
○ , ——— :static tests
◇ :stripped samples in old brines (means of types A & B)
● :covered samples in fresh brines (type C)
▲ :stripped samples in fresh brines (type D)
◆ :covered samples in old brines (type E)

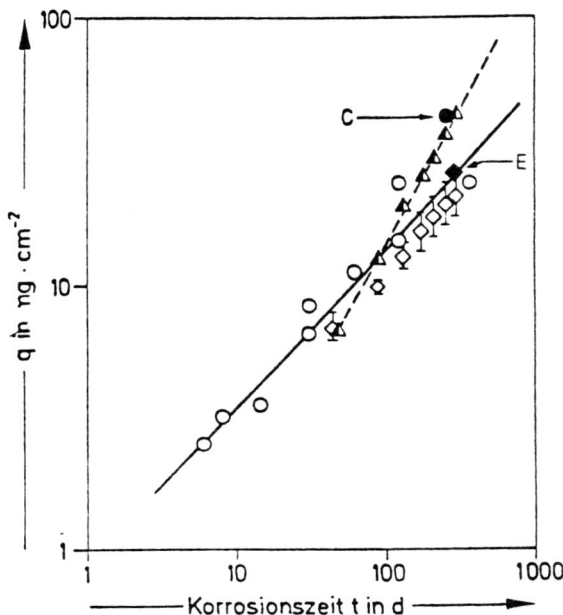

Fig. 3.10   Conditions as in figure 3.9, but for 200°C

A comparison of the mass loss data from type A and B tests leads to A:B = 0.95 ± 0.06 at 120°C and 0.99 ± 0.02 at 200°C. Therefore, these results are presented by the same symbol in Figures 3.9 and 3.10. A comparison between the static tests and the type A,B tests shows no significant differences, either. In particular, exactly the same time law for the corrosion process is valid. This means firstly that the reaction layers are not protective against the attack of an old leachant and secondly that the deviations from a direct proportionality to t are not due to these reaction layers. The idea of a diffusion barrier can be dismissed between 120 and 200°C.

As Figures 3.9 and 3.10 show the data from type C and D tests are very different from the rest; however, they agree among each other; (6.65 ± 0.2) and (129 ± 4) mg after 297 d at 120 and 200°C respectively. Lithium and boron analysis data (not available at all interruptions) show that this is also valid for the interim intervals:

C:D = 0.99 ± 0.10 (Li, 200°C, based on data from 5 intervals)
      1.07 ± 0.14 (B, 200°C, based on data from 3 intervals)
      1.13 ± 0.20 (Li, 120°C, based on data from 4 intervals
                                  1 outlier)

(The boron results at 120°C are somewhat doubtful.) The conclusion is that the reaction layers are not protective, either, against the attack of a fresh leachant.

A further conclusion can be drawn from a comparison of type A, B to type C, D data. While the type A, B data follow the non-linear time law of the static tests, a linear time law is valid for the type C, D data with constant corrosion rates of 0.0076 and 0.145 $mg.cm^{-2}.d^{-1}$ at 120 and 200°C, respectively. Thus, the deviations from linearity cannot be attributed to properties of the Q-solution itself; the previous hypothesis (pages 24-26) can be dismissed. They are, however, due to the accumulation of reaction products in the leachant (see point 4, page 25).

Role of the ion exchange zone

In a further test series, the influence of the ion exchange zone on the corrosion process was checked. Some samples were precorroded for 30 d in a 0.01 molar HCl solution at 200°C, 13 MPa and s = 0.034 cm^{-1}. From former microprobe measurements, this pretreatment is expected to generate ion exchange zones > 1 µm. An equivalent test in Q-solution generates zones ≤ 0.3 µm (see Figure 3.6). The precorroded samples were manipulated like in a type D test; the reaction layers were wiped off and the residual glass samples were exposed to a fresh Q-solution (30 d, 200°C, 13 MPa). As a reference, some uncorroded glass samples were also exposed to a fresh Q-solution. The mass losses per surface area in $mg.cm^{-2}$ are:

20.4 ± 0.7   after the HCl pretreatment
4.0 ± 0.3   after the further corrosion in Q-solution
4.0 ± 0.3   for the reference samples after the corrosion in
              Q-solution.

The conclusion is that a deep ion exchange zone is not able to
protect the samples any better than a small one. In other words,
the ion exchange zones are not protective. They do not act as a
diffusion barrier.

Consequences for the corrosion mechanism

Summarising the above results, the corrosion behaviour of the
glass SM 58 LW 11 in Q-solution between 120 and 200°C can be
characterized as follows:

1. The basic mechanism is the network dissolution proceeding
    linearly with time.

2. Deviations from linearity are due to the accumulation of
    reaction products in the leachant.

3. Neither the reaction layers nor the ion exchange zones act
    as diffusion barriers. They do not affect the temporal
    progress of the corrosion process.

The open questions are how the accumulated reaction products
influence the corrosion process and what consequences for the long
term corrosion behaviour follow. This problem is treated in
3.1.2.4. but first some further results from the periodically
interrupted tests are presented.

Regeneration of lost reaction layers

In the type D tests, the wiped off reaction layers from each
interval were dried, weighed, and analysed. The chemical analysis

revealed that the layers contained residues of the Q-solution.
After determining their Mg contents, their masses were corrected
for the component $MgCl_2.2H_2O$ which is, according to additional
screening experiments, the most likely Mg containing component in
a reaction layer dried at 110°C. After 47 and 171 d at 120 and
200°C, respectively, constant layer regeneration rates appeared
with respective values of 0.003 and 0.006 $mg.cm^{-2}.d^{-1}$,
corresponding to 40 and 4% of the dissolution rates,
respectively.

Mass balances and behaviour of individual elements

According to section 3.1.2.1 a mass balance requires a complete
chemical analysis of the leachants including residues, and of the
reaction layers. (The ion exchange zones need not be taken into
consideration.) The analysis was performed along three different
paths; the reaction layers were dissolved and analyzed; after a pH
measurement*, the leachants were acidified and brought to 250 ml,
then separated from solid residues and analyzed; the solid
residues were also dissolved and analyzed. The establishment of a
mass balance was only achieved for the elements Si, Ti, Zr, Nd,
and Sr. The Li and B data were discussed earlier (see page 30).
For the other elements, the amounts in the leachant were too
small, or the impurity level of the chemicals used for the Q-
solution was too high (eg for Cs), or some impurities were
introduced from the autoclave system (eg for the Fe, where the
amounts found by analysis were double the amounts calculated by
mass loss data). The mass balances for Si, Ti, Zr, Nd, and Sr are
presented in Table 3.4.

Si is mainly present in the leachant including residues; only a
small part of the total Si builds up the reaction layers.
Nevertheless, it is a major constituent of the layers. Ti and Zr
are mainly present in the reaction layers under static conditions

---

*The final pH values were 5.7 ± 0.3 in congruence with the initial
ones.

TABLE 3.4 Cumulative mass balances for glass components in type C, D, and E tests at different temperatures after 297 d; masses in mg;
I   determined by analysis of the leachant including solid residues
II   determined by analysis of the reaction layer
III calculated from the total mass loss

Oxide	T(°C)	type	I	II	III
SiO_2	120	C	4.1	0.03	3.6
		D	3.7	1.3	3.8
		E	1.3	0.06	1.4
	200	C	66	0.2	67
		D	66	3.3	71
		E	20	2.1	40
TiO_2	120	C	0.12	0.11	0.26
		D	0.04	0.23	0.28
		E	0.01	0.05	0.10
	200	C	5.1	0.5	4.9
		D	2.4	3.1	5.2
		E	0.7	2.7	2.9
ZrO_2	200	C	0.53	0.14	0.86
		D	0.24	0.44	0.91
		E	0.04	0.39	0.51
Nd_2O_3	200	C	0.54	0.00	0.70
		D	0.59	0.00	0.74
		E	0.30	0.00	0.42
SrO	200	C	0.06	0.00	0.06
		D	0.06	0.00	0.07
		E	0.03	0.00	0.04

(type E) and during the exposure of a stripped sample to a fresh brine (type D), whereas more Ti and Zr is released to a fresh brine, when the sample is still covered with its old reaction layer (type C). Ti and Zr are also major constituents of the layers. Nd, which can be regarded as a substitute for U in the glass, is only present in the leachants and cannot be detected in the reaction layers. The same is valid for Sr, which is an important element because of its high γ activity in a real waste glass.

3.1.2.4 Static corrosion tests at varied surface area to leachant volume

Role of the accumulation of reaction products in the leachant – theoretical aspects

The results of the preceding section prove that the accumulation of reaction products in the leachant essentially influences the corrosion progress. As an over-simplification, the problem can be treated like the dissolution of a crystalline phase, eg $NaCl$ in H_2O. The corresponding rate equation for the rate r_G of the dissolving glass (suffix G = glass, k = a rate constant) is

$$s_G \cdot r_G = k(c_S - c). \qquad (4)$$

Similar approaches for the dissolution of glass are found in the literature, eg [10]. As a direct consequence of eq. (4), r_G becomes zero for long times when c reaches the saturation value c_S. The meaning of c_S for a dissolving glass is not explained by itself; the metastable solubility of amorphous SiO_2 may be regarded as an upper limit. As Harvey [11] points out, silicate solutions tend to supersaturate and c_S is not unlikely to be reached during the dissolution progress. And indeed, a drop of r_G to zero can be observed in corrosion experiments, especially for high s ratios, see eg [12].

Harvey [11] emphasizes that besides the metastable amorphous SiO_2, there must exist at least one thermodynamically stable SiO_2 containing phase in the system. Therefore, we can conclude that a drop of the dissolution rate r_G to zero is an intermediate event, at most, but no long term behaviour [13]. Assuming a SiO_2 consuming phase P precipitating at $c_P < c_S$ and consuming SiO_2 at a rate r_P, the concentration change in the leachant will become

$$\begin{aligned} dc/dt \bigg|_{c > c_P} &= s_G r_G - (\partial/\partial t)(s_P \cdot \int r_P dt) \\ &= s_G r_G - [s_P r_P + (\partial s_P/\partial t) \cdot \int r_P dt] \end{aligned} \qquad (5)$$

The second term in eq. (5) becomes important when the situation $c > c_p$, eg $c = c_S$ occurs; it takes into account that contrary to s_G, s_p may undergo a considerable growth with time. Thus, c is balanced at the value c_p if sufficient time is given to the system, no matter how small r_G is. For long times, a constant dissolution rate $r_\infty \sim (c_S - c_p)$ is established. The model by Grambow [8] leads to a similar conclusion for the long term behaviour. Grambow presents the dissolution rate r_G by an initial rate r_o and the affinity A, A < 0, of the irreversible dissolution reaction. A is a complicated function of the actual composition of the leachant. The rate equation in a somewhat simplified form is

$$r_G = r_o [1-\exp(A/RT)]. \qquad (6)$$

For long times, a limiting value for A, $A_\infty \neq 0$, is reached leading to the constant long term rate

$$r_\infty = r_o [1-\exp(A_\infty/RT)] \qquad (7)$$

Role of the accumulation of reaction products in the leachant — experimental observations

In order to check the effects of accumulated reaction products in Q-solution, tests were carried out at 200°C and a varied s ratio between approx. 0.003 and 0.238; in one test series $s > 15$ cm^{-1}. Most of the tests were performed under equilibrium pressure, some 30 d values were obtained from 13 MPa tests. As shown in Figure 3.11, the amounts c of the dissolved glass in 7 and 30 d experiments approach a limiting value of approx. 0.3 mg.cm^{-3}* for increasing s.t. This limit is clearly exceeded when the process is allowed to continue (arrow in Figure 3.11). In Figure 3.12,

---

*This limit corresponds to 0.17 mg.cm^{-3} SiO$_2$ which is, indeed, a reasonable solubility value for amorphous SiO$_2$ in Q-solution (see discussion of literature in [14]). This coincidence must, however, not be over-interpreted.

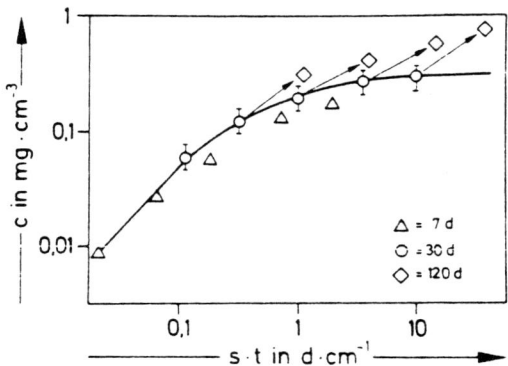

Fig. 3.11 Amount c of dissolved glass as a function of the parameter s.t for corrosion tests in Q-solution at 200°C; the arrows show the corrosion path for t > 30 d at s = const.

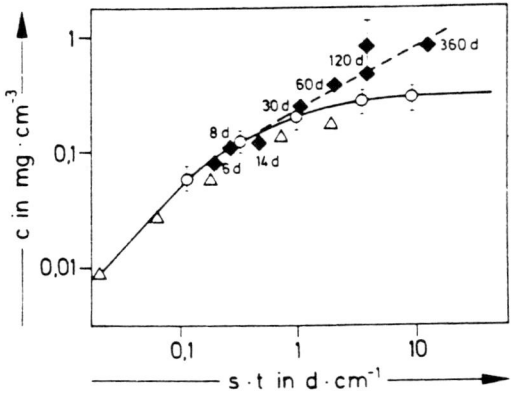

Fig. 3.12 Amount c of dissolved glass as a function of the parameter s.t for corrosion tests in Q-solution at 200°C;
△ , ○ , ———: tests at constant time (7 and 30 d)
◆ , ----: tests at constant s (0.034 cm^{-1})

the results from the 7 and 30 d experiments at varied s are compared to data from Figure 3.1 (constant $s = 0.034$ cm^{-1}, 13 MPa, different times). The approach to a limiting value observed in the first case does not occur in the second one.

In a further test series (see Figure 3.13), samples were corroded for 8, 30 and 120 d at 200°C and equilibrium pressure (one 30 d test at 13 MPa, too); the sample surface area was 3 cm^2, the leachant volume 90 cm^3. By an addition of 0.5 g of glass powder with a mean particle size of 5 μm, s was increased to $s >$ 15 cm^{-1}.

Under such conditions, a rapid saturation of the leachant was expected. For $t \leqslant 30$, a stop of the dissolution process would still be a possible interpretation. However, the increase between 30 and 120 d is indisputable. This increase is directly proportional to t, which is a first experimental evidence for a long term law linear in t. The long term rate can be estimated as $r_\infty = 0.0054$ mg.cm^{-2}.d^{-1}.

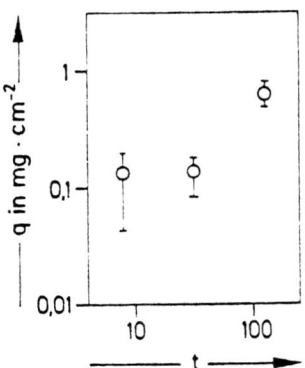

Fig. 3.13 Mass loss per surface area q after corrosion tests at 200°C in 90 cm^3 of Q-solution with additional 0.5 g of powdered glass, as a function of time; the sample surface area was 3 cm^2

The results for corrosion under saturation or near-saturation conditions discussed above are now compared with results at very low s (s = 0.003 cm^{-1}) ie for corrosion under conditions far from saturation. These latter results are presented in Figure 3.14.

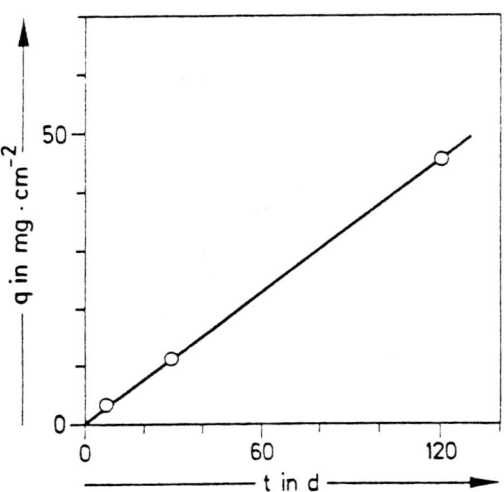

Fig. 3.14 Mass loss per surface area q after corrosion in Q-solution at 200°C and s = 0.003 cm^{-1}, as a function of time

The mass loss increases linearly with t at the initial rate r_o = 0.4 mg.cm^{-2}.d^{-1} which is the same value as that derived from the static 6 and 8 d tests at s = 0.034 cm^{-1} (see Figure 3.1). Using eq. (7) and the values for r_o and r_∞, the long term value for the affinity (see page 35) can be calculated as A_∞ = − 55 kJ/mol. For 120 and 160°C, the initial rates can be calculated from the data in Figure 3.1. Using the presumptive long term value for the activation energy from Figure 3.2 (E_A = 69 ± 3 kJ/mol), the long term rates at 120 and 160°C can also be estimated. Table 3.5 summarizes the respective results.

TABLE 3.5 Initial and estimated final corrosion rates r_o and r_∞ of SM 58 LW 11 in Q-solution

	120°C	160°C	200°C
r_o in mg cm^{-2}.d^{-1}	0.0070	0.069	0.40
r_∞ in mg cm^{-2}.d^{-1}	0.00015	0.0011	0.0054

3.1.3 Summary

The purpose of the reported work was to determine the corrosion behaviour of the inactive HLW glass SM 58 LW 11 in Q-solution at temperatures up to 200°C and elevated pressures up to 13 MPa. In particular, a parameteric study on the effects of time, temperature, pressure, crystallization and metallic impurities was performed. Further tests helped to identify the rate determining steps in the entire process and the most likely long term corrosion law. The results are summarized below.

1. The main corrosion mechanisms are:

- the dissolution of the glass network.

- the formation of a reaction layer on top of the surface of the residual glass.

- the formation of an ion exchange zone in the surface of the residual glass.

The reaction layer is only loosely attached to the residual glass.

2. Under static corrosion conditions, the network dissolution proceeds according to a non-linear function of time with exponents between 0 and 1, depending on the corrosion conditions. The approximate dissolution depths after 1 year are 0.6 μm at 80°C and 100 μm at 200°C. The corresponding ion exchange zones are as thin

as 50 and 250 nm, respectively, with no significant time dependence.

3.   The network dissolution process depends strongly on temperature. The activation energy in the initial stage is 79 kJ/mol; after 1 year a constant long term value of 69 kJ/mol seems to be reached (both valid between 120 and 200°C).

4.   A pressure dependence of the network dissolution cannot yet be proved or disproved. Between equilibrium pressure and 13 MPa, an increase in network dissolution of about 25% should be taken into account.

5.   Crystallization can have hazardous effects on the corrosion behaviour of the glass, when the crystallites formed between 650 and 750°C reach a size of 0.5 to 1 mm.

6.   The presence of steel or of metallic Cu, Pb, and Al (in the order of increasing effect) enhances the corrosion velocity by a factor of 3 to 6.

7.   The basic corrosion mechanism in Q-solution is a network dissolution directly proportional to t, it can be observed under conditions far from saturation, eg in the initial phase of corrosion, at small ratios of surface area to leachant volume, or when the leachant is changed. The initial rates are 0.0070, 0.069, and 0.40 $mg.cm^{-2}.d^{-1}$ at 120, 160, and 200°C, respectively.

8.   The reaction layers formed during corrosion are not protective against the further attack of the Q-solution, neither under static conditions nor under exchange of the leachant.

9.   The depths of the ion exchange zones do not affect the network dissolution velocity, ie the ion exchange zones are not protective, either.

10.   The deviations from the initially linear time law under

static conditions (see item 2) are due to the accumulation of corrosion products in the leachant.

11. For large ratios of surface area to leachant volume, an apparent upper limit for the amount of dissolvable glass is observed. This limit corresponds to a drop of the network dissolution rate to zero. For longer corrosion times, this limit is exceeded. A total stop of the dissolution process must not be taken into account.

12. The most likely long term law for the network dissolution is again a linear time law, however, at a reduced rate. The estimated long term rate at 200°C is 0.0054 $mg.cm^{-2}.d^{-1}$.

3.2 BASIC MECHANISMS OF GLASS LEACHING (AERE HARWELL)
(The fundamental processes pertaining to the corrosion of alkali borosilicate glasses by aqueous media)

Glass formulations proposed for the incorporation of highly-active waste are in most cases of the alkali borosilicate type. When such glasses contact aqueous media the immediate effect is the removal of alkalis from the near-surface region and this process appears to be a necessary precursor to any subsequent sub-surface corrosion processes. Alkali removal takes place by ion-exchange. Firstly the more general case of any diffusion-limited reaction will be considered.

3.2.1 General theory of diffusion-limited reaction

This process is considered in three stages.

(a) Inward diffusion of some reactive species A.

(b) Reaction of A with a particular element M in the glass structure to give a reaction product P and an altered glass structure G. ($A + M \rightarrow P + G$) A is absorbed and immobilised and M is released as the mobile P.

(c) Diffusion of the reaction product P outwards into the leachate.

If [A], [M], etc. are the concentrations of A, M, etc. with initial values $[A_o]$ in the leachant and [Mo] in the glass and k is a reaction constant, then the collision frequency between the mobile A and the immobile M is proportional to [A][M] and the rate of absorption of A will equal the rate of generation of P. In the absence of any diffusion transport,

$$\left(\frac{d[P]}{dt}\right)_x = -\left(\frac{d[A]}{dt}\right)_x = k\,[A][M] \tag{8}$$

The Ficklan diffusion equation is modified to include the reaction term, resulting in the equations

$$\frac{\partial[A]}{\partial t} = \frac{\partial J_A}{\partial x} - k[A][M] \tag{9}$$

$$\frac{\partial[P]}{\partial t} = \frac{\partial J_P}{\partial x} + k[A][M] \tag{10}$$

where J_A and J_P are the fluxes of A and P. In the simpliest case, the interdiffusion of uncharged species with constant diffusion coefficients D_A and D_P, the fluxes are

$$J_A = D_A \frac{\partial[A]}{\partial x} \quad \text{and} \quad J_P = D_P \frac{\partial[P]}{\partial x} \tag{11}$$

3.2.2 Ion exchange theory

Ion exchange theory is well known and the mathematical development will not be repeated here. Briefly, the driving force for the diffusion of an ionic species is governed by the electrochemical potential and an ionic concentration gradient results in an electrical potential gradient ($\delta\phi/\delta x$) in the diffusion layer. Where two ionic species are interdiffusing, both species contribute to $\delta\phi/\delta x$. Electrical equilibrium has the effect of accelerating the slower-moving species and decelerating the faster moving one, so that the fluxes come into balance. The diffusion coefficients can be replaced by a single 'effective diffusion coefficient' D, which is concentration dependent.

A frequently-quoted equation is:

$$D = \frac{D_{H+}D_{M+}([H+] + [M+])}{[H+]D_{M+} + [M+]D_{H+}} \quad (12)$$

If, as is probable, D_{H+} is much larger than D_{M+} then eq. (12) reduces to

$$D = D_M (1 + [H+]/[M+]) \quad (13)$$

3.2.3 Solutions to the diffusion-reaction equations

Equations 9 to 11 were solved numerically for [M], the concentration of unreacted species M remaining in the glass, using an explicit finite-difference method. Figure 3.15 shows the depth profile obtained plotted in dimensionless units, ie with abscissa of unit $x(k[M_o]/D_A)^{\frac{1}{2}}$ and ordinate $[M]/[M_o]$. When plotted in this way the shape of the curve is identical whatever values of $[A_o]$, $[M_o]$, k and D_A are taken. The position of the curve depends on $[A_o]$, $[M_o]$ and t. This position can be characterised by the distance to the point of maximum slope, which is related to the depth of attack. This distance is approximately

$$\ell = 1.3 \sqrt{([A_o]D_A t/[M_o])} - 0.45 \sqrt{(D_A/k[M_o])} \quad (14)$$

The value of the maximum slope varies slightly with the depth of penetration but is approximately:

$$1/3 \sqrt{([M_o]k/D)} \quad (cm^{-1}) \quad (15)$$

and occurs at about $[M]/[M_o] = 0.42$.

- 43 -

Incorporating eq. (13) into eq. (11) does not affect the results significantly and makes only a small difference to the calculated leach rate. D_{M+} is also affected by the structural changes which take place in both the reaction zone and the gel layer. The bulk diffusion coefficient for alkali ions in unleached UK189 glass, calculated from electrical resistivity measurements, is 10^{-19} $cm^2.s^{-1}$ at 25°C which is 6 or 7 orders of magnitude lower than the effective diffusion coefficient for alkali leaching. One effect of such a structural dependence is that released alkali ions are prevented from diffusing inwards into the unaltered glass and so are forced to diffuse outwards into the leachate. The model can readily be adapted to include a structure-dependent D, but there is no information available by which such a dependence could be defined.

The quantity Q leached after time t may be obtained by integrating the area above the curve (Figure 3.15). It was found to be within ± 5% of

$$Q = 2 \, [M_o] \left(\frac{[A_o] Dt}{\pi \, [M_o]} \right)^{\frac{1}{2}} \quad (16)$$

The similarity between this expression and the familiar solution to the Ficklan equation can readily be seen, and has led many researchers to deduce from experimentally-measured leach rates that leaching is controlled by a simple diffusion process. However, simple diffusion leads to a concentration depth profile of shape $erfc(x/2 \sqrt{Dt})$ which is totally different from those found experimentally.

3.2.4 Comparison of the theoretical curves with experimental results

Depth profiles

Depth profiles of the concentration have been measured for a number of elements and many examples have been published [15-20].

Two examples of such profiles are given in Figures 3.16 and 3.17. Figure 3.16 shows the sodium profile data obtained by Conradt et al (Figure 3.5(a)) together with computed profile curves. The total alkali in this glass $[M_o]$ was 0.0126 mole cm^{-3} and the leachant pH was ca. 5, so that $[A_o] = 10^{-8}$.mole.cm^{-3}.

It has been demonstrated by mass balance calculations [16] that in simple glasses the protons are consumed only by alkali ion exchange. Assuming this to hold true for the more complicated nuclear waste borosilicate glasses the profiles shown in Figure 3.16 were calculated using the constants shown in Table 3.6.

TABLE 3.6   Constants used in computing the curves in Figure 3.16

Test Conditions	D cm^2.s^{-1}	k (cm3.mole^{-1}.s^{-1})
80°C 62 days	1.3 x 10^{-12}	120
200°C 8 days	2.5 x 10^{-12}	1800

These calculations are only intended to illustrate how the model can be used to interpret such depth profiles, the actual values obtained are unlikely to be meaningful, particularly as the depth profiles are not in good agreement with the total weight loss data for the same samples.

Figure 3.17 shows one of the hydrogen depth profiles published by Engelmann [17]. The measured hydrogen concentration at the surface was ca. 20% higher than the sodium concentration in the unleached glass, but there are two possible sources of additional hydrogen; hydrogen is usually present in glass from the chemicals used in its manufacture, also some of the water which diffuses into the depleted layer may remain. The computer-generated curve in Figure 3.17 assumes that all the hydrogen arises from ion exchange.

Effect of pH on alkali leaching

Samples of glass UK189 were leached in buffer solutions of various pH between 4.2 and 6.9 at 20°C for up to 1 hour, and for 24 hours at pH 9.2.

The results are given in Figure 3.18 and the slopes of these lines are plotted as log (slope) vs pH in Figure 3.19. Equations 14 and 16 predict that alkali leaching would be proportional to $\sqrt{[H+]}$, so that the points in Figure 3.19 might be expected to fall on a straight line of slope $\frac{1}{2}$. The discrepant result at pH 9.2 may be due to network dissolution caused by the prolonged leaching time needed to obtain a result.

3.2.5 Matrix dissolution

In experiments with flowing leachant or small surface area/volume ratio (SA/V), it was found that boron, molybdenum, magnesium and strontium leached congruently with the alkalis over a wide range of pH and temperature. Some typical results are shown in Figures 3.20 - 3.24. The most probable explanation is as follows. Water is able to diffuse much more readily into glass from which the alkalis have been removed. In fact it has been observed by several experimenters that each alkali atom is replaced by one hydrogen atom and one water molecule. Thus the water concentration within the depleted surface layer is equal to the original alkali concentration, ie many orders of magnitude greater than the proton concentration in the leachate which controls alkali leaching, (see eq. 16). This water is available to react rapidly with other weakly-bonded elements in the glass structure, releasing them to diffuse into the leachate. In this way the leaching of these elements would keep pace with the alkalis even if their diffusion coefficients were 8 orders of magnitude smaller than the alkali inter-diffusion coefficient.

As the various elements are removed from the glass the resulting voids are occupied by water. The depelted surface layer is a hydrated matrix, mainly of silica, alumina, rare earth and iron oxides. Differential scanning calorimeter measurements showed that the water is only weakly bonded, being identical in this respect with hydrated silica gel; the layer has rightly been named the 'gel layer'.

It is evident from depth profiles and gel layer densities that Si and Aℓ are removed from within the gel layer. Si leaches congruently with alkalis at high pH when the rate of advance of the alkali ion-exchange zone is slow, but lags behind initially when pH is low and the ion-exchange zone is advancing too rapidly for the network dissolution process to keep up. Examples of this behaviour can be seen in Figure 3.20. Because of the large content of very attack-resistant elements (rare earths, Fe and to a lesser extent Aℓ), gel layers arising from typical HLW glasses retain their structual identity even when greatly depleted in Si and can contain up to 40% water. Such a highly-depleted gel layer is no barrier to diffusion; the resulting increase in D_{eff} for all reaction products causes a transition from $\sqrt{t}$ dependence to t dependence and the leach rates for alkalis, B, Mo, Si, Mg and many others become constant. Mathematically this transition could be described by a moving-boundary diffusion equation [21] but empirically the behaviour is sufficiently well described by the expression $Q = At + b\sqrt{t}$. Constant leach rates are achieved rapidly at high temperature, but may take years at ambient temperatures. Results quoted in the literature for Soxhlet tests at 100°C are invariably in the constant leach rate region.

3.2.6 Temperature dependence

The diffusion coefficients and reaction constants will have different activation energies for each element, nevertheless the leach rate of each element is affected by the matrix dissolution rate which in turn is linked to the combined leach rates of all elements. Hence any attempt to calculate activation energies for the leaching of individual elements may simply result in an activation energy for the whole leaching process. Lithium leaching in the initial ($\sqrt{t}$-dependent) stages gave an activation energy of 76 kJ.mole^{-1}. Figure 3.25 shows Arrhenius plots for some Si and Al results. Initially, both Si and Al have activation energies of 80 kJ.mole^{-1}, decreasing to 53 and 42 respectively. Activation energies for total weight loss from the same glass tested in various leachants were 50 - 65 kJ.mole^{-1}.

3.2.7 Effect of SA/V

The effect of varying the SA/V in the range 3 to 30 m^{-1} is shown in Figure 3.26 (pH 4.2 buffered leachant) and Figure 3.27 (DIW leachant). These demonstrate that the effect of the changing pH as reaction products accumulate is far greater than any effect due to incipient saturation. This indicated that experiments at much higher SA/V would be needed before any true saturation effects could be expected. These are described in the next section.

3.2.8 **Saturation Effects**

The objectives of these experiments are (a) to determine the extent to which reaction products are soluble in the leachate, (b) to investigate whether leach rates are affected significantly by a leachant containing high concentrations of reaction products, and (c) to discover whether leaching ceases when the leachant is saturated with reaction products. The experiments, which began in the period 1983-84, have been continued using the same techniques. Glass samples were in the form of crushed and sieved granules of a narrow size range. These were leached in Teflon containers with just sufficient deionised water to fill the interstices and cover the sample. In recent tests, small coupons of solid glass were included to measure the leach rates and study surfaces. Results were previously reported [22] for two granule sizes, corresponding to SA/V ratios of 7800 and 20000 respectively. Both SA/V ratios yielded similar leach rates and the lower SA/V tests have been discontinued because saturation would only be reached after impossibly long times.

Solids precipitated from the leachates were filtered off and weighed, precipitates and solutions were analysed separately by emission spectroscopy.

Results

Analysis of leachates from a selection of the additional tests made during the current year are given in Tables 3.7 to 3.10. These results, together with those previously obtained, are presented graphically in Figures 3.28 to 3.31. In these graphs data points represent the quantity leached per unit area, normalised to the glass composition by dividing by the appropriate fraction concentration in the original glass. The left-hand ordinate scale is thus a notional weight per unit area of glass equivalent. If the quantities for two or more elements coincide then these elements have leached congruently. The right-hand ordinate scale represents (approximately) the ratio of the concentration of each element in the leachate to its concentration in the original glass. Thus if the concentration in the leachate for a particular element were to reach the original glass concentration then the value on this scale would be unity.

Discussion

The leaching process is one of diffusion-limited reaction (see section 3.2). Many such reactions are occurring simultaneously within the near-surface region of the glass and each reaction is governed initially by a $D\sqrt{t}$ relationship. The diffusion coefficient (D) depends on both the diffusant and the matrix. In the more familiar case of a thick sample in a low SA/V experiment, an increase in $Q/\sqrt{t}$ (the parabolic rate) occurs as D increases due to matrix dissolution and the process moves from $\sqrt{t}$ to t dependence. In the present experiments, this effect is offset by the effect of reduced surface area as the reaction zone advances into the granules, so that for the most rapidly leached elements the parabolic rate appears substantially constant. In the most concentrated leachate so far encountered (150°C for 31 days on glass UK189) almost 50% of the boron and 62% of the molybdenum

TABLE 3.7   **Analysis of leachates from 45°C tests on UK209 glass**
Values of mg.m^{-2} of elements (mean of two tests)

Test time(days)	319		686		730	
SA/V(m^{-1})	19710		19070		17680	
pH	10.1		10.3		10.3	
	Solution	Solid	Solution	Solid	Solution	Solid
Si	12.5	375	17.9	426	16.0	572
B	188	-	279	-	288	-
Na	185	-	262	-	293	-
Li	71.2	-	94.7	-	96.0	-
Mg	0.33	105	0.52	128	0.65	120
Al	0.27	37	0.45	52.4	0.41	63.3
Sr	0.31	-	0.052	0.59	0.066	2.0
Mo	72.2	-	100	-	115	-
Total Solids by weight		1580		2623		2384
Total oxides by analysis	1145	998 (63%)	1648	1223 (46%)	1737	1543 (65%)

TABLE 3.8   **Analysis of leachates from 90°C, 7 day tests**
Values in mg.m^{-2} of elements (mean of two tests)

Glass	UK189		UK209	
SA/V(m^{-1})	23640		23840	
pH	9.8		10.3	
	Solution	Solid	Solution	Solid
Si	5.47	41.4	19.2	622
B	755	-	217	-
Na	308	-	272	-
Li	165	-	56.7	-
Mg	1.04	21.9	0.16	148
Al	0.65	5.55	0.31	86.6
Sr	0.15	0.38	0.034	0.33
Mo	166	-	80.9	-
Cr	ND	-	ND	-
Fe	ND	-	ND	-
Total solid by weight	-	173	-	2280
Total oxides by analysis	3466	136 (78.6%)	1291	1740 (76.3%)
Coupon wt loss		162		152
Coupon after scraping		1490		1350

TABLE 3.9 Analysis of leachates from 90°C, 120 day tests
Values in mg.m^{-2} of elements

Glass SA/V(m^{-1}) pH	UK189 20235 9.1		UK209 19570 9.7	
	Solution	Solid	Solution	Solid
Si	12.4	2586	31.2	2460
B	2420	–	690	–
Na	980	–	560	–
Li	389	–	120	–
Mg	0.79	754	0.22	429
Al	3.58	474	1.40	440
Sr	0.68	13	0.15	1.1
Mo	700	–	270	–
Cr	34.4	–	16.2	–
Fe	0.11	5.8	0.10	23
Total solid by weight		9853		8905
Total oxides by analysis	11100	7700 (78%)	3738	6839 (77%)
Coupon wt loss	15570		6716	
Coupon after scraping	50050		13000	

TABLE 3.10 Analysis of leachates from 150°C, 31 day tests
Values in mg.m^{-2} of elements (mean of two tests)

Glass SA/V(m^{-1}) pH	UK189 20420 8.2		UK209 20920 9.7	
	Solution	Solid	Solution	Solid
Si	20.0	6800	43.9	682
B	4930	–	548	–
Na	2270	–	453	–
Li	540	–	62.0	–
Mg	0.48	1090	0.298	90
Al	2.17	1810	0.604	178
Sr	0.127	104	0.034	0.11
Mo	1200	–	198	–
Cr	23.0	–	3.99	–
Fe	5.14	105	0.73	30
Total solid by weight		27300		2620
Total oxides by analysis	22010	20050 (73.4%)	2920	1990 (76%)
Coupon wt loss	Gain		1630	
Coupon after scraping	10500		8280	

available had dissolved into the leachate. This represents a reduction in effective surface area for these elements to 63% and 52% respectively of the original area.

In low SA/V leachates, where the concentration of dissolved elements is low, the 'soluble' elements (Na, Li, B, Mo) always leach congruently. In the present experiments, increasing departure from congruence was observed as the leachates became highly concentrated. Table 3.11 shows the highest concentrations found for each temperature and glass composition.

TABLE 3.11   Highest concentrations found in solution (molar)

Glass	UK189			UK209		
Temp (°C)	45	90	150	45	90	150
Time (days)	500	120	31	730	120	31
Si	4.6E-3	0.01	0.015	0.01	0.022	0.039
B	0.56	4.9	10	0.5	1.3	1.0
Na	0.13	0.9	2.0	0.22	0.5	0.4
Li	0.21	1.2	1.7	0.24	0.35	0.2
Mg	2.7E-3	6E-4	2.3E-4	4E-4	1.6E-4	2.5E-4
Al	9.3E-4	3.2E-3	2E-3	3E-4	1.1E-3	5.5E-4
Sr	5.7E-5	1.7E-4	1.8E-5	1.3E-5	3.4E-5	1E-5
Mo	0.01	0.16	0.24	0.02	0.056	0.04
Cr	–	0.014	6.6E-3	–	6.2E-3	3E-5
Fe	–	4E-5	1.7E-3	–	3.5E-5	2.7E-4

In Figure 3.32, the three solid lines leading from the apex represent the element concentrations expected from congruent dissolution of UK189, MW and UK209. These are all on the alkali-rich side of the mixed-alkali tetraborate composition. The circled data points on this diagram represent the compositions of various leachates. There is no recognisable correlation with either temperature or glass composition but the more highly-concentrated leachates tend increasingly towards the boron-rich side of the diagram.

The dotted curves in Figure 3.32 show the results of some room-temperature solubility experiments in the system 1:1 Na/Li, B,

H_2O. There is some common-ion effect between alkali hydroxide and tetraborate, also a much larger effect between boric acid and the mixed-alkali tetraborate.

Comparison with 90°C and 150°C leachate compositions show that in many cases tetraborates should crystallise from the leachates on cooling but none of the 45°C leachates had reached a sufficiently high concentration for this to happen.

Crystals grown from certain salient compositions (marked A, B, C on Figure 3.32 were analysed, with the following results:

A $7\ Na_2B_4O_7 \cdot 10H_2O + 1\ Li_2B_4O_7 + 5H_2O$
B $5\ Na_2B_4O_7 \cdot 10H_2O + 1\ Li_2B_4O_7 \cdot 5H_2O$
C H_3BO_3

Analysis C is surprising, in view of the reported (very soluble) compounds $Na_2B_{10}O_{16}$ and $Li_2B_{10}O_{16}$ and $Li_2B_4O_7$, but this has not been investigated.

As reported previously [22], precipitates of hydrous magnesium aluminium silicate form in the leachates. Similar precipitates are found by other workers when leaching glasses containing no magnesium, but using a magnesium-rich brine as leachant [23].

Table 3.12 shows the compositions of precipitates produced in a selection of tests: there is considerable variation in composition and no clear correlation between test conditions and composition can be seen.

The concentrations of Si, Mg, Aℓ and Sr found in solution are in all cases very low (Table 3.12) compared with the total quantities leached. The total quantities leached from glass UK209 show no unexpected trends (Figures 3.29 and 3.30). By contrast, the release of these elements from glass UK189 shows an unusual effect (Figures 3.28 and 3.31) at first, only minute amounts are found but then a rapid increase occurs after ca. 150 days at 45°C or 25 days at 90°C. The change in conditions responsible for triggering this increase is not understood.

TABLE 3.12 Mole % compositions of precipitates in leachates (excluding H_2O)

Temperature (°C)	45	90	90	90	150
Time (days)	319	7	30	120	31
Glass UK189					
SiO_2	50.0	59.5	53.3	69.9	75.4
MgO	45.3	36.4	44.7	23.5	13.9
Al_2O_3	4.7	4.1	2.5	6.6	10.4
Fe_2O_3	<DL	<DL	<DL	(0.04)	0.3
Temperature (°C)	45	90	90	90	150
Time (days)	319	7	30	120	31
Glass UK209					
SiO_2	71.6	74.1	73.1	77.2	77.0
MgO	24.5	20.5	17.5	15.5	11.6
Al_2O_3	3.9	5.4	8.6	7.1	10.5
Fe_2O_3	<DL	<DL	0.8	0.2	0.9

Figures 3.33 and 3.34 show weight loss leach rates for various tests of 30 days' duration. The differences between Soxhlet and low SA/V static tests are predictable from the pH differences. The differences between high and low SA/V static tests cannot be explained solely on the basis of soluble species, nor on the basis of decreasing effective surface areas. Although the precipitated silicates are not generally very strongly adherent to the sample surface, it appears that they do in some way modify the release of soluble species. The extraordinary result for glass UK209 where after 30 days there was les leaching at 150°C than at 90°C strongly suggests a protective surface coating. The surface coating on the 150°C samples was hard and compact. EPMA examination showed it to be of similar composition to the precipitate.

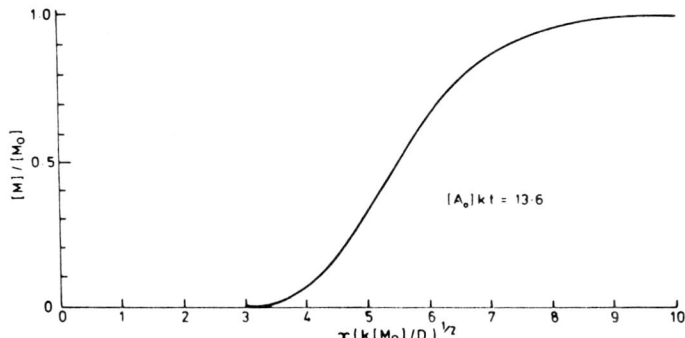

Fig. 3.15 Depth profile solution to equation 2

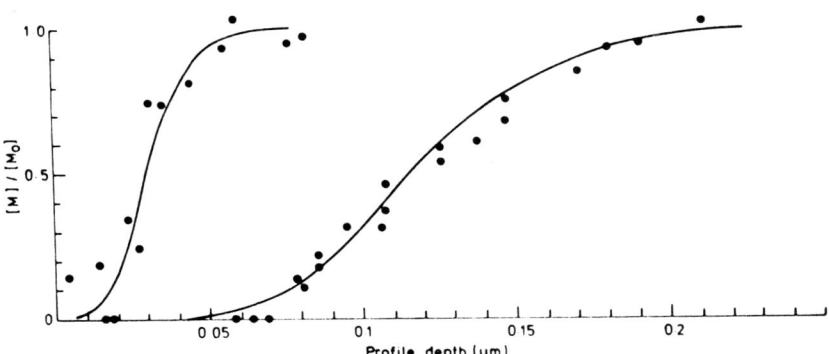

Fig. 3.16 Comparison with data from Fig. 3.5a

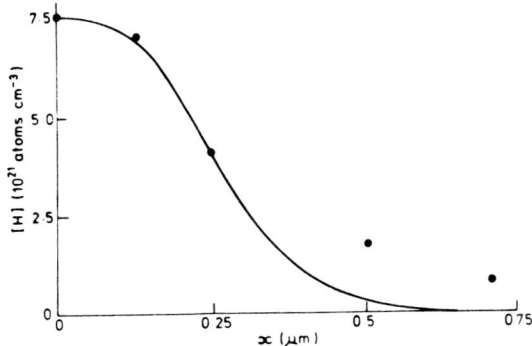

Fig. 3.17 Comparison with data from Reference 17

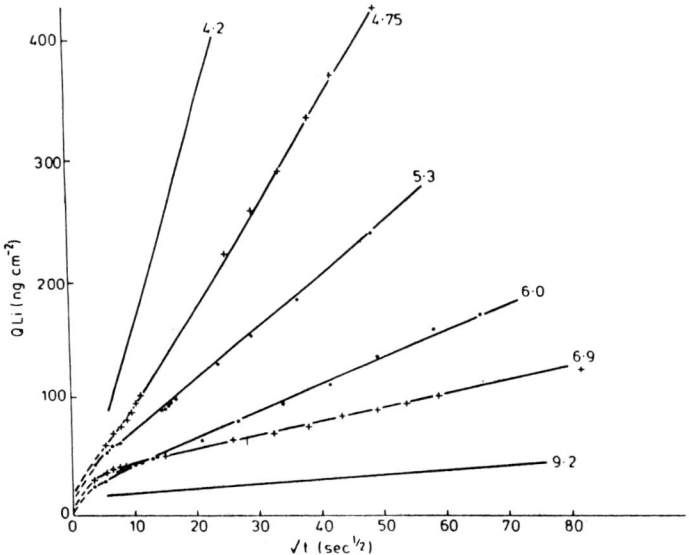

Fig. 3.18 Lithium leaching at various pH, Glass UK189 at 20°C

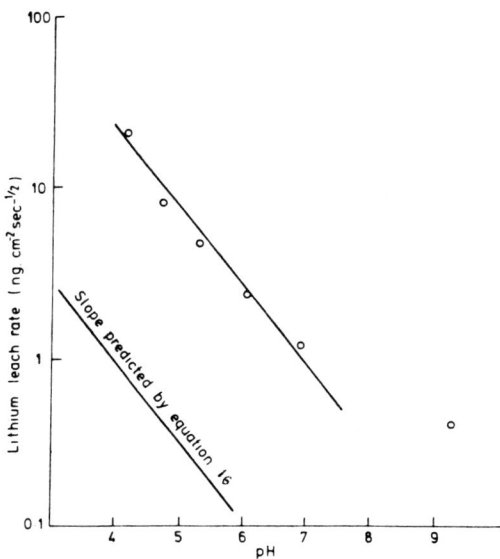

Fig. 3.19 Lithium leach rates at various pH, Glass UK189 at 20°C

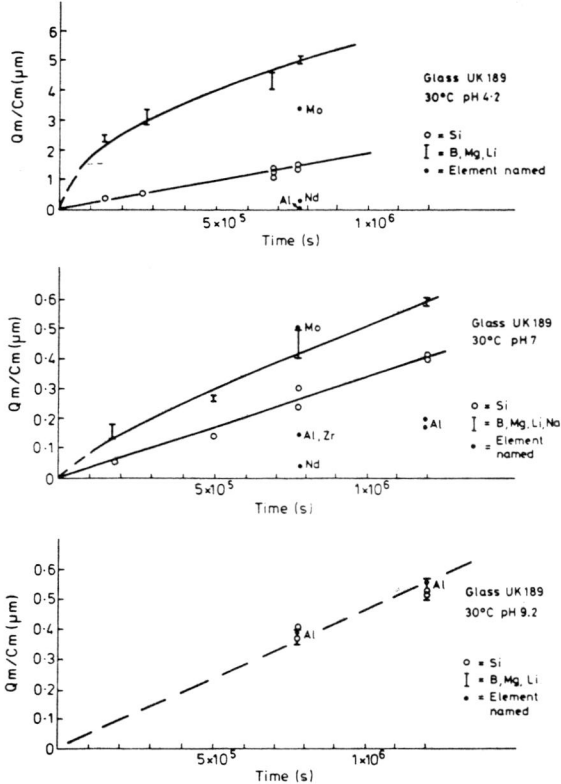

Fig. 3.20 Leaching of UK189 at 30°C, various pH

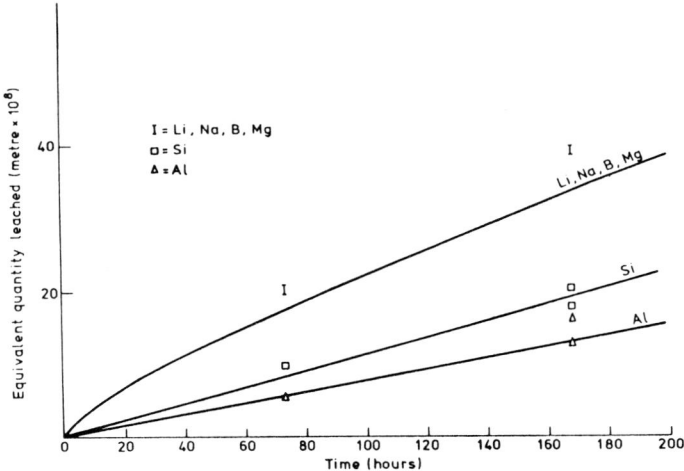

Fig. 3.21   Leaching of UK189 at 31°C

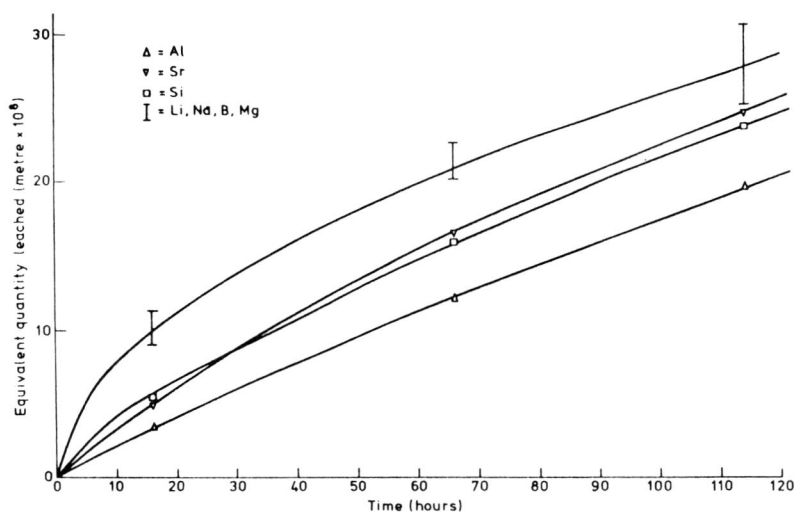

Fig. 3.22   Leaching of UK189 at 39°C

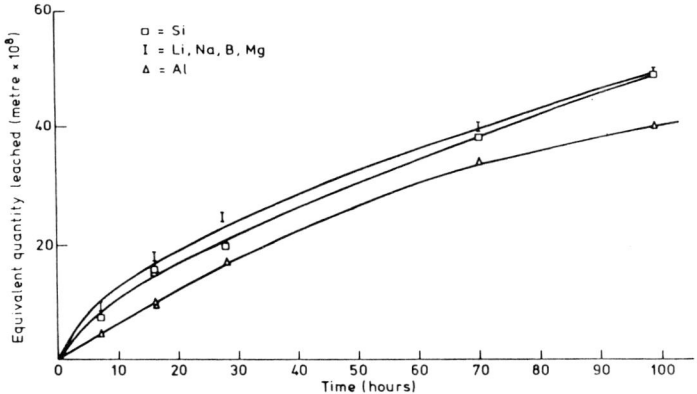

Fig. 3.23   Leaching of UK189 at 50°C

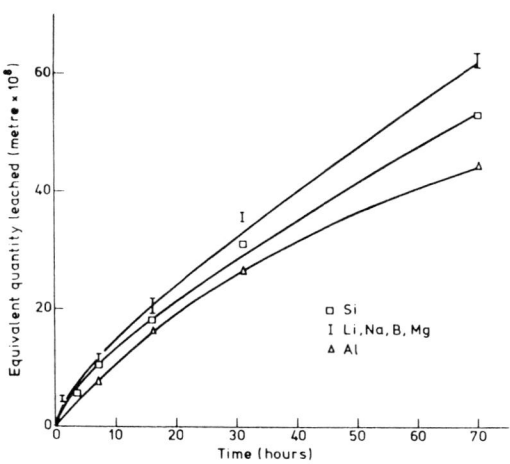

Fig. 3.24   Leaching of UK189 at 58°C

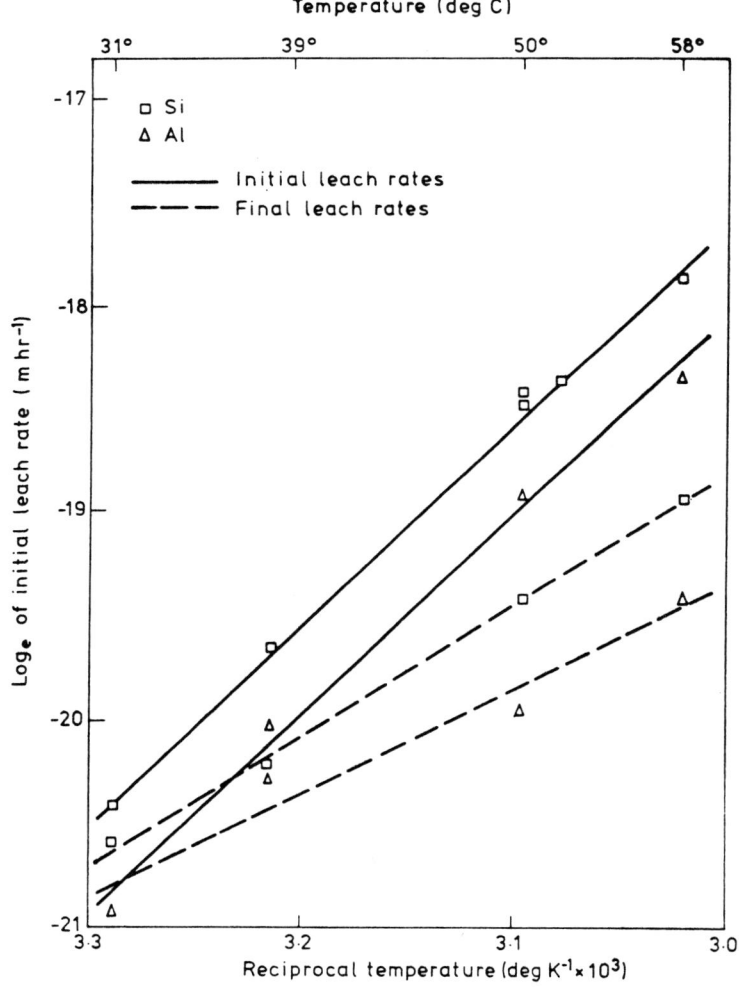

Fig. 3.25 Arrhenius plots of Si and Al leach rates

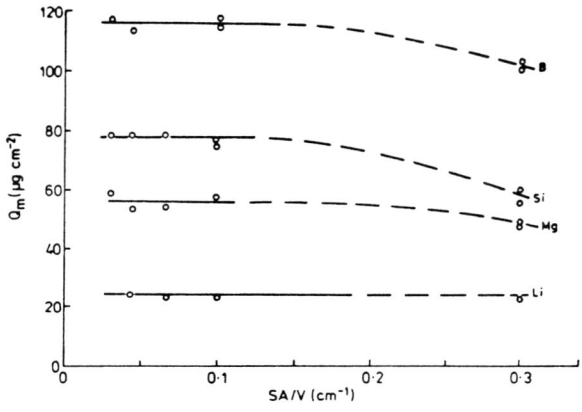

Fig. 3.26 Effects of SA/V, UK189 10 days at 30°C, pH 4.2

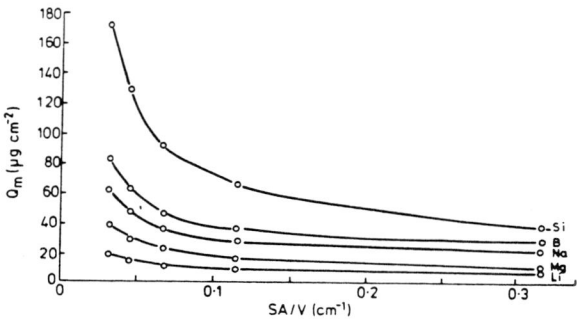

Fig. 3.27 Effects of SA/V, UK189, 73 days at 30°C in DIW

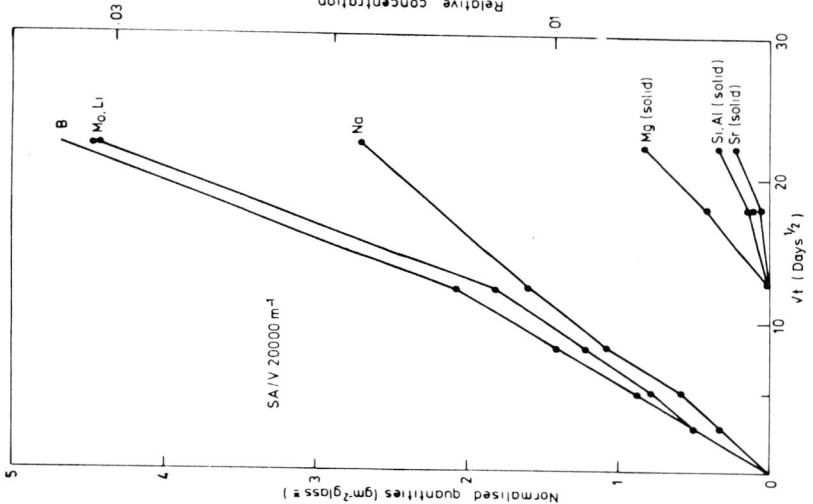

Fig. 3.29 Leaching of UK189 at 45°C, high SA/V

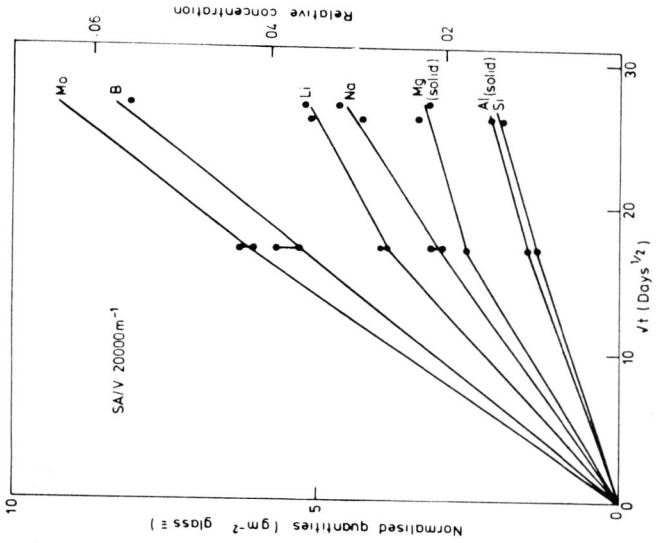

Fig. 3.28 Leaching of UK209 at 45°C, high SA/V

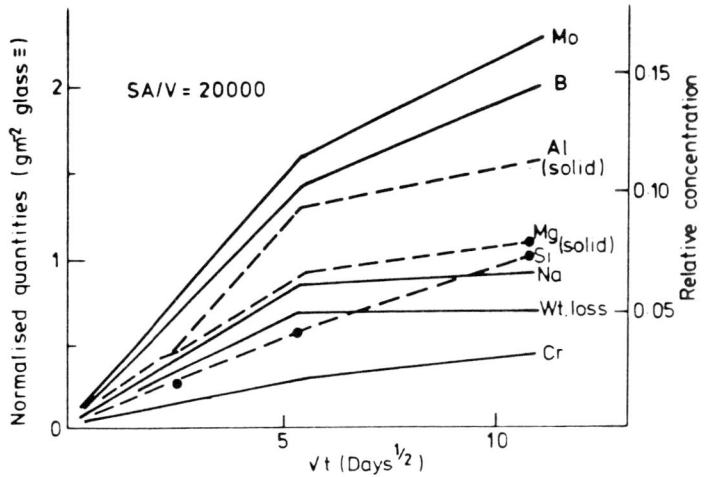

Fig. 3.30 Leaching of UK209 at 90°C, high SA/V

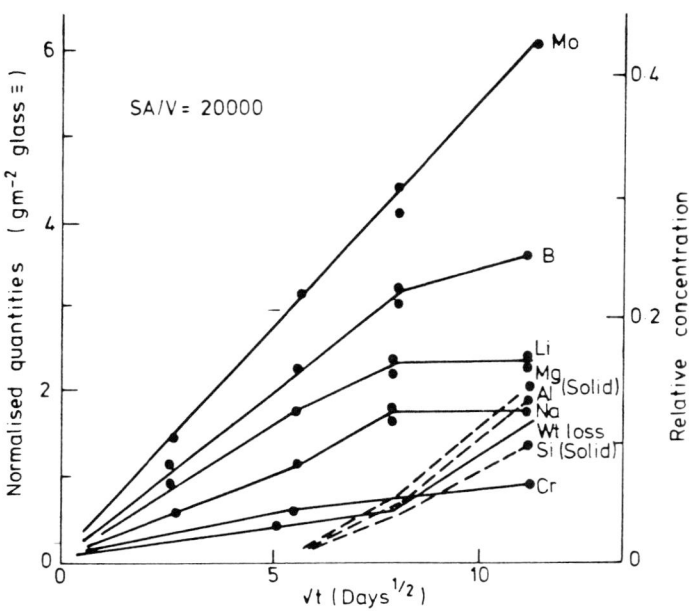

Fig. 3.31 Leaching of UK189 at 90°C, high SA/V

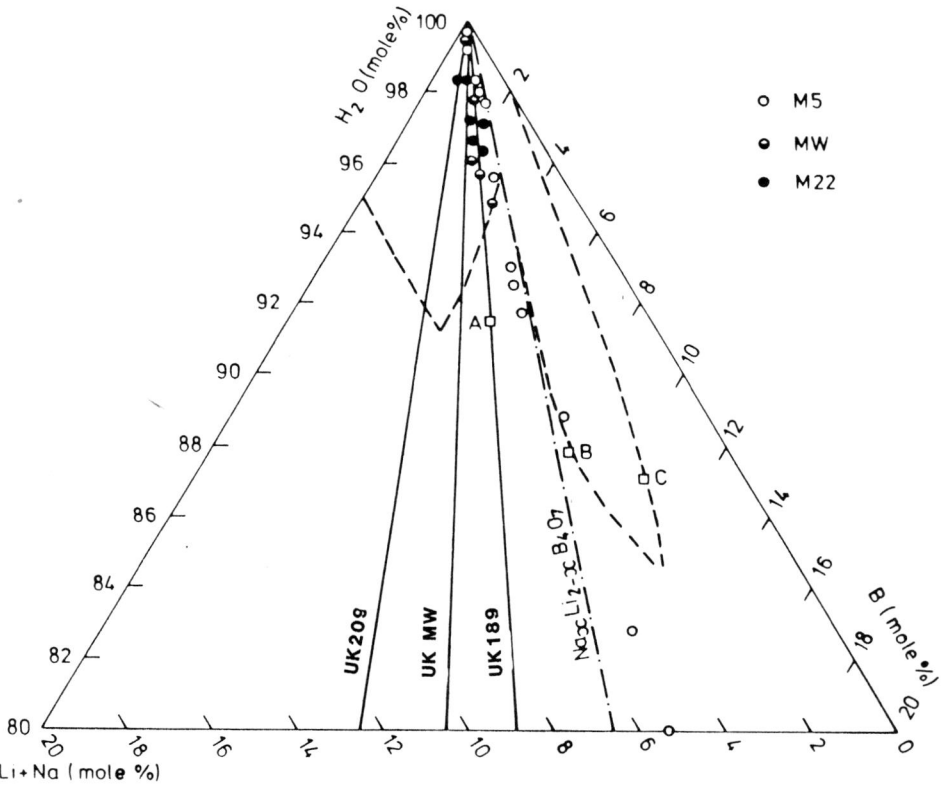

Fig. 3.32 Leachate compositions and solubilities

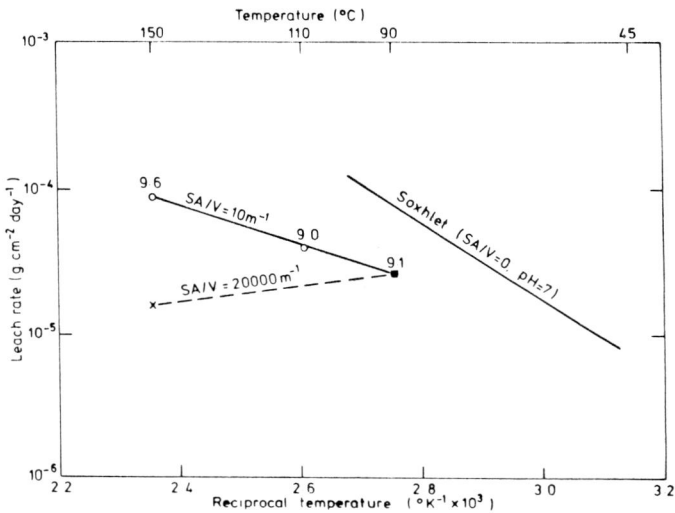

Fig. 3.33 Leach rates from various tests (UK209)

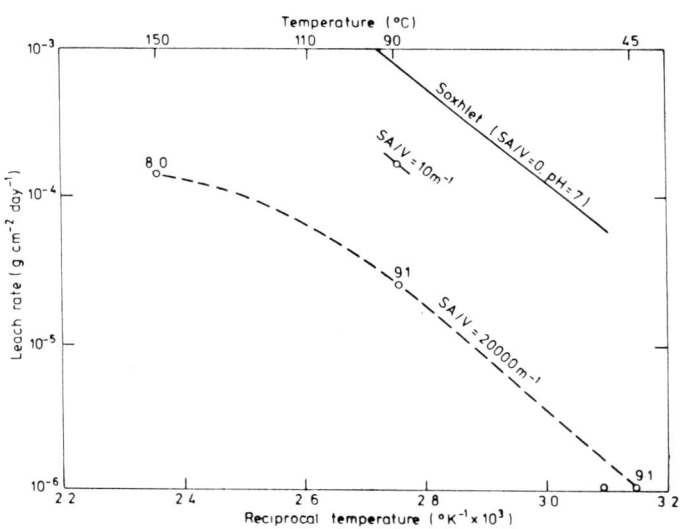

Fig. 3.34 Leach rates from various tests (UK189)

REFERENCES – CHAPTER 3

[1] J. D'Ans: Die Loslichkeitsgleichgewichete der Systeme der Saize ozeanischer Salzablagerungen. Verlagsgesellschaft fur Ackerbau mbH, Berlin 1933.

[2] R. Conradt, H. Engelke, and A. Kaiser: Investigation of the solution behaviour of NaCl in the quinary system $NaCl-KCl-MgCl_2-MgSO)_4-H_2O$ at different temperatures. Mat. Res. Soc. Symp. Proc. 11 (1982) p. 487-490.

[3] W.A. Lanford, K. Davis, P. Lamarche, T. Laursen, R. Groleau, and R.H. Doremus: Hydration of soda-lime-glass. J. Non-Cryst. Solids 33 (1979) p. 249-266.

[4] E. Wiegel: Uber die Beeinflussung der Heiβauslaugung von Silikatglasern durch Metallspuren. Glastechn. Ber. 34 (1961) p. 259-268.

[5] C.Q. Buckwalter and L.R. Pederson: Inhibition of nuclear waste glass leaching by chemisorption. J. Amer. Soc. 65 (1982) p. 431-436.

[6] W.L. Kuhn and R.D. Peters: Leach models for a commercial nuclear waste glass. Mat. Res. Soc. Symp. Proc. 15 (1983) p. 167-174.

[7] A.A. Belyustin and M.M. Shul'ts: Interdiffusion of cations and the accompanying processes in the surface layers of alkali silicate treated with aqueous solutions. Sov. J. Glass Phys. Chem. 9 (1983) p. 1-24.

[8] B. Grambow: A general rate equation for nuclear waste glass corrosion. Mat. Res. Soc. Symp. Proc. (1985) in press.

[9] L.A. Chick and L.R. Pederson: The relationship between reaction layer thickness and leach rate for nuclear waste glasses. Mat. Res. Soc. Symp. Proc. 26 (1984) p. 635-642.

[10] R.M. Wallace and G.G. Wicks: Leaching chemistry of defence borosilicate glass. Mat. Res. Soc. Symp. Proc. 15 (1983) p. 23-28.

[11] K.B. Harvey: A semiunified approach to leach testing of nuclear waste forms. Nucl. Chem. Waste Manage. 4 (1983) p. 201-205.

[12] R. de Batist, P. van Iseghem and W. Timmermans in: Testing and evaluation of solidified high-level waste forms. Joint annual report 1982 No. EUR 9268 of the Commission of the European Communities. Ch. Engelmann ed., Luxemburg 1983, p. 57-110.

[13] R. Conradt, H. Roggendorf and H. Scholze: A contribution to the modelling of the corrosion process for HLW glasses. Mat. Res. Soc. Symp. Proc. (1985) in press.

[14] R. Conradt, H. Roggendorf and H. Scholze: Chemical durability of a multicomponent glass in a simulated carnallite/rock salt environment. Mat. Res. Soc. Symp. Proc. 26 (1984) p. 9-15.

[15] L.R. Pederson et al. Nucl. Tech. 62 p. 151-158 (1983).

[16] B.C. Bunker et al. J. Non-Cryst. Solids 58 p. 295-322 (1983).

[17] C. Engelmann et al. Verres Refract. 35 p. 486-490 (1981).

[18] R. Conradt and H. Roggendorf in 'Testing and Evaluation of Solidified High Level Waste Forms'. Ed. C. Engelmann. EUR 9268EN (1984).

[19] G. Malow in Materials Research Society Symposium Proceedings Vol. 11. Ed. W. Lutze, P. 25 (1982).

[20] D.E. Clark. Ibid p. 71.

[21] H.S. Carslaw and J.C. Jaeger. The Conduction of Heat in Solids. 2nd Edition. Clarendon Press Oxford (1959) p.388.

[22] J.A.C. Marples et al in 'Testing and Evaluation of Solidified High Level Waste Forms'. Ed. G. Malow, EUR 10038EN (1985) p. 86-97.

[23] D.M. Strachan in Materials Research Society Symposium Proceedings, Vol. 11. Ed. W. Lutze, p. 181 (1982).

4. STUDIES OF SURFACE MORPHOLOGY AND CHEMICAL COMPOSITION OF SUB - SURFACE REGIONS

4.1 HYDROTHERMAL LEACHING OF SIMULATED HLW GLASSES (HMI)

The aim of the research work at the HMI is to produce data to calculate the activity release from waste products under the attack of salt solutions such as sodium chloride and carnallite 'brines'.

The programme comprises the investigation of both simulated inactive and highly radioactive samples.

In addition to the studies of surface attack of waste glasses reported in this Section, the HMI contract called for

- Leach tests of actinide-spiked samples (Section 5.2).

- Studies of the mechanical properties of waste products (Section 8.1).

4.1.1 Preparation of samples

The composition of the glasses is given in Table 4.1. The glasses were composed of a frit and a waste oxide mixture. Glass beads having size and shape comparable to the beads to be produced in the German prototype vitrification plant PAMELA [1], chips and powder were used for static hydrothermal leaching experiments in Teflon lined autoclaves.

4.1.2 Experimental and selection of leachants

The German final repository concept foresees the waste disposal in salt formations. The heat release from HLW glass blocks will be controlled by appropriate design of the emplacement geometry. Present model calculations comprize temperatures up to 200°C at the canister/salt interface. The hydrostatic pressure upon the blocks is assumed to be 150 bar maximum.

Table 4.1a Compositions of test materials (wt.%)

	C 31/3	SON 68	UK 209	B 1/3	VG 98/3	SON 64
Fission product oxides	15.20	11.25	9.75	14.55	15.54	13.14
Waste oxides						
Fe_2O_3	1.47	2.91	2.73	1.51	0.70	5.07
Cr_2O_3	0.42	0.51	0.56	0.43	0.24	0.43
NiO	0.23	0.41	0.36	0.24	0.21	0.24
MgO			6.34		0.40	
Al_2O_3			5.11			1.69
U_3O_8	0.46		0.06	0.47	1.21	0.45
Na_2O				2.2		

Also: * 2.2% Na_2O, ** 0.52% UO_2, 0.33% ThO_2, 0.1% SO_4, 0.5%ZnO, 0.23% P_2O_5, + 0.60% P_2O_5, o 0.36% F_2O_3, 0.43% ThO_2

	C 31/3	SON 68	UK 209	B 1/3	VG 98/3	SON 64
Glass formers		****		x		
SiO_2	34.75	45.48	50.88	28.00	41.84	47.16
B_2O_3	4.14	14.02	11.12	6.40	10.48	18.42
Na_2O	1.12	9.86	8.30	1.60	22.25	12.53
Li_2O	0.98	1.98	3.99	2.40		
Al_2O_3	10.28	4.91		12.80	1.20	
CaO	3.84	4.04		4.00	2.32	
BaO	14.48			14.80		
TiO_2	2.80			4.56	3.52	
MgO	1.44			1.2		

Also: 0.8% ZrO_2, 4.88% ZnO, 0,48% As_2O_3
**** 0.28% P_2O_5, 1% ZrO_2, 2.5% ZnO
x 0.8% ZrO_2, 3.60% ZnO, 0.40% As_2O_3

Table 4.1b Compositions of fission products in the test materials (wt.%)

	C 31/3	SON 68	UK 209	B 1/3	VG 98/3	SON 64
ZrO_2	3.32	1.65	1.43	2.45	2.35	1.96
BaO	0.33	0.61	0.38	0.70	0.64	0.62
Rb_2O	0.16	0.13	0.11	0.16	0.15	0.15
SrO	0.59	0.34	0.32	0.59	0.50	0.41
Y_2O_3	0.34	0.20	0.17	0.34	0.32	0.03
MoO_3	2.36	1.70	1.77	2.36	2.17	1.89
MnO_2	0.34	0.74		0.34	0.40	
CoO_3		0.12				
PdO	0.11	0.33*	0.44	0.08	0.62	0.34*
Ag_2O	0.002	0.03		0.002		0.03
CdO	0.004	0.03		0.004		0.04
SnO_2	0.016	0.02		0.016		0.03
Sb_2O_3	0.004	0.004		0.004		0.01
TeO_2	0.26	0.23		0.26	0.24	0.31
Cs_2O	1.70	1.30	0.77	1.70	1.35	1.13
La_2O_3	0.66	0.90	0.44	0.66	0.61	0.59
CeO_2	1.31	0.93	0.99	1.31	1.56	1.29
Pr_2O_3	0.60	0.44	0.43	0.60	0.57	0.56
Nd_2O_3	2.40	1.59	1.82	2.40	2.15	1.84
Tc_2O_3						0.41
RuO_2	0.92		0.68	0.53**	1.06	0.63
Rh_2O_3	0.24			0.15***	0.18	0.10
Pm_2O_3					0.05	0.04
Sm_2O_3					0.47	0.37
Eu_2O_3					0.09	0.08
Gd_2O_3					0.06	0.05
Total	15.20	11.25	9.75	14.55	15.54	13.14

* PdO substituted by NiO - **RuO_2 by TiO_2 - ***Rh_2O_3 substituted by Co_2O_3.

The composition of the leachants (brines) can be derived from observations and mining experience. Ten compositions representing almost all solutions ever observed were given by A.G. Herrmann [2] and discussed (1980) in the PTB* when a group of experts suggested their use for leaching experiments.

Some of the above mentioned conditions were simulated in leaching experiments. Glass samples were leached at 200°C at equilibrium pressure, i.e. ≈ 15 bar for saturated salt solution. The ratios of sample surface area to solution volume were between 1:75 cm^{-1} and 1:0.01 cm^{-1}.

The weight losses of the glass beads and chips were estimated and the solutions were used for quantitative analysis. The surface layers on the glass samples were mechanically removed and quantitatively analysed by ICP emission spectroscopy. In parallel experiments the layers were not removed but prepared for investigations by scanning electron microscope (SEM), microprobe analysis (EPMA) and X-ray diffraction (XRD).

Some leaching experiments were carried out with glass powder of grain size ≤ 60 μm in order to increase the area of corroded glass, i.e. the reaction progress and as a consequence get larger amounts of reaction products and bigger crystals. In every case the surface area of the powder was much larger than that of the chip. In most charges the glass powder weight to solution volume was 11.5 g/l, so that the sample surface area to solution volume ratio was approximately 3.6 cm^{-1}. Solution volumes ranged from 76 ml to approximately 360 ml. Teflon reaction vessels were cleaned by a procedure similar to the MCC method.

As a consequence of the results from screening tests [3], only a few solutions, NaCl-, Q- and Z-solutions, have been selected for a more detailed investigation of the leaching process. Their

---

*Physikalisch-Technische Bundesanstalt, Braunschweig, FRG.

compositions are given in Table 4.2.

TABLE 4.2   Compositions of brines used for leaching experiments

Solution	Moles per 1000 moles H_2O			
	NaCl	KCl	$MgCl_2$	$MgSO_4$
$NaCl-H_2O$ [a]	113.4	–	–	–
Q-brine [b]	6.8	17.4	77.3	3.2
Z brine	1.0	2.6	111.2	2.1

a: This solution is saturated at 55°C. An extra 60 g NaCl per litre is added so that the solution would be saturated at 200°C.
b: This solution is saturated at 55°C. An extra 71 g NaCl per litre is added so that the solution remained saturated at 200°C with respect to NaCl.

Leaching solutions were prepared from deionized water. In the glass powder experiments leaching solutions were NaCl brine saturated at 200°C with or without small amounts of $MgCl_2$ and also a quinary brine having KCl, $MgCl_2$, $MgSO_4$ and NaCl saturated with NaCl at 200°C. pH values were measured and corrected for highly concentrated salt solutions at the beginning and end of the experiments at room temperature using a Ross electrode. The reacted glasses were removed from the brines by vacuum filtration using filter paper. Except where noted otherwise this produced a clear solution that was in some cases tinted. Solutions and residual salt were diluted and saved for ICP analysis. The reacted glass powders and chips were thoroughly washed with deionized water then dried at approximately 70°C, and stored in a dessicator prior to surface analysis.

4.1.3 Glass C-31-3-EC-SPF-Na

Investigation of surface layers [4]

Surface layers are a common feature of leached borosilicate waste glasses. They are of particular importance to the resistance of waste glasses against water attack. Therefore, in addition to the overall analysis, SEM- and EPMA-investigations were performed to study in detail the morphology and structure of the surface layers. Again glass beads were used for hydrothermal leaching experiments under simulating final repository conditions. Brines of NaCl, KCl, $MgCl_2$ and $MgSO_4$ were used at 200°C. Figure 4.1(a) exhibits an SEM photomicrograph of a sectioned glass bead after 10 d leaching in NaCl-solution. The leached surface consists of two different layers. The inner layer is about 15 μm thick and looks like a typical gel layer with drying cracks, whereas the outer one appears more dense and is relatively thin with a thickness of about 1.5 μm.

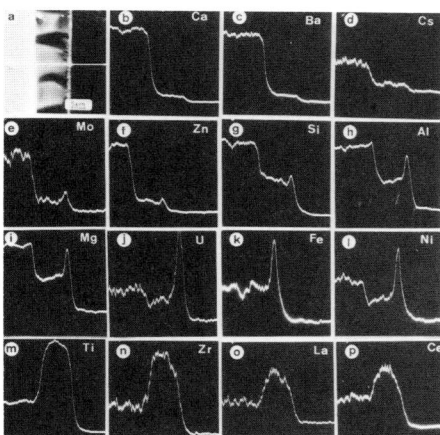

Figure 4.1
SEM photomicrograph and X-ray line scans of a glass bead leached 10 d at 200°C in saturated NaCl-solution.
(a) SE-micrograph with line scans position;
(b)-(p) line scans of elements as indicated [4].

Figures 4.1(b) to (p) show EPMA concentration profiles in the form of X-ray line scans obtained at the white line marked in Figure 4.1(a). As seen, elements such as Ca, Ba, Cs are nearly completely leached from the layers, whereas Mo and Zn are strongly depleted in the inner layer, but their concentrations increase again in the outer layer. Si, Al, Mg are also depleted in the inner layer but again have a higher concentration in the outer layer, whereas U, Ni, Fe are enriched in the outer layer when compared with the pristine glass. The elements Ti, Zr, La, Ce are enriched in both layers.

After 30 d leaching in NaCl the appearance of the surface changes (Figure 4.2(a)). The outer layer appeared to be thicker whereas the inner layer seems to become denser and thinner. X-ray line scans of the elements Si, Al, Mg, Zr, U indicated two concentration maxima, Figures 4.2(b-f), which implies a periodically alternating composition in the scale.

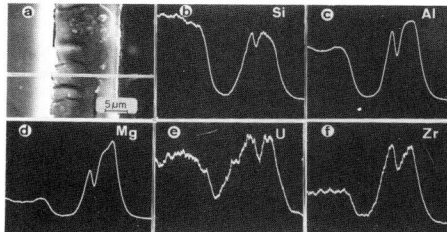

Fig. 4.2
SEM photomicrograph and X-ray line scans of a glass bead leached 30 d at 200°C in saturated NaCl-solution.
(a) Secondary electron micrograph with line scan position.
(b)-(f) line scans of elements as indicated [4].

The Q-leached surfaces showed an enrichment of Cl, S and K which are components of these leachants. Obviously, they contributed to deposit in the layers. These deposits probably precipitate at higher temperatures, ie ≈ 200°C but not during the cooling period of the autoclave. This was inferred from the observation that the phases, containing brine constituents, formed near the boundary of the pristine glass rather than on the outer surface of the scale. Figure 4.3 shows an SEM-photomicrograph of the cross-sectioned surface of a bead leached for one year in quinary brine (from the ASSE II salt mine). Figure 4.3(b) shows an energy-dispersive X-ray spectrum. The electron beam position is indicated by the black point in Figure 4.3(a). As seen the layer contains S and Ba, which is probably $BaSO_4$.

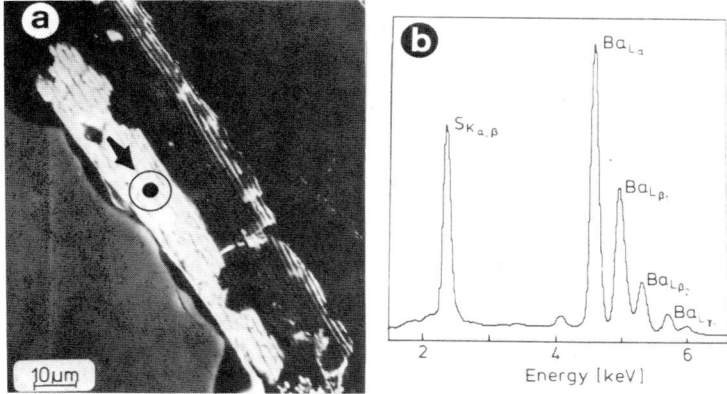

Fig. 4.3 (a) SEM photomicrograph of a cross-sectioned glass surface leached one year at 200°C in quinary brine. (b) EDX spectrum of the phase indicated by the black point in Figure (a) [4].

The results from semiquantitative wavelength dispersive EPMA of the surface layers are given in Table 4.3. For the purpose of comparison the last column contains the values of the surface layer formed during leaching in deionized water. These results are based on our earlier work [5]. As seen the elements Ca, Ba, Cs, Mo and Zn are strongly depleted in the layers leached in salt solutions, whereas Zn is strongly enriched in the outer scale of the water leached surface. Al and Mg are enriched in the Q- and Z-leached surface and also in samples leached 30 d in NaCl, whereas the samples leached in water are higher in Al and Mg after 3 d, Ti and Zr concentrations were high in layers formed from all solutions. La, Ce, Nd and Pr are enriched only in the layers leached in NaCl and are strongly depleted in Q- and Z-leached surfaces. Therefore most of the elements are enriched in the H_2O-leached surface and least of all in Q- and Z-leached surfaces. In other words, in Q- and Z-brines most elements of the glass were leached. This means that the tendency toward selective leaching decreases and congruent dissolution increases in the following order: H_2O, NaCl-, Q-, Z-solution. Nevertheless, as mentioned previously, the specific weight losses do not show differences beyond one order of magnitude. In all types of surface layers, independent of the salt brine, chlorine was present, and S and K in layers leached in Q- and Z-brine. When comparing the concentrations of Si and Mg in the layers it was found that the Q-leached layer contained only 1.6 wt.% Si and 13 wt.% Mg in contrast to ≈ 16 wt.% Si and 0.8 wt.% Mg in the unleached glass. This means that only 10% of the original silicon is in the layer, but the Mg-content was increased about a factor of 10. In terms of the glass structure then only every 10th SiO_4-tetrahedron, the main network former, is still present but certainly unable to preserve the glass structure. Therefore, when leaching the glass in Q-brine the glass network is nearly completely destroyed and the layer formed on the surface is assumed no longer to have structural similarities with the original glass structure. This presumption is further supported by the fact that the surface layers are partly crystalline and do not look homogeneous. There are enrichments of some elements, as seen in Figures 4.4(b-i) showing steps in X-ray line scans.

The surface layer of the one year leached sample adhered very strongly to the surface and could not be removed, whereas after short leaching times layers peeled off more readily. Longer leaching periods yielded denser surface layers.

Obviously, with increasing leaching time the mechanical stability of the surface layer increases but the layer does not become thicker after 30 days. The specific weight loss after one year was about 5 to 6 $mg.cm^{-2}$, the same as after 3-10 d leaching. The leaching rate slowed down close to zero.

Surface morphology and phase identification

NaCl-brine

In addition to the concentration profiles of some elements which were measured on cross sectioned beads, surface morphology of

TABLE 4.3 Relative composition of surface layers on glass, formed during leaching at 200°C in various solutions [4]

Element	NaCl-sol./ 10 d. inner	outer	NaCl-sol./ 30 d. inner	outer	Q-sol./ 10 d. inner	outer	Z-sol./ 10 d. inner	outer	H_2O_3/ 3 d. inner	outer
Ca	--	--	--	--	--	--	--	--	-	-
Ba	--	--	--	--	--	--	--	--	-	-
Cs	--	--	--	--	--	--	--	--	--	--
Mo	--	--	--	--	--	--	--	--	--	--
Zn	--	-	--	--	--	--	--	--	-	+ +
Si	--	-	--	--	o	-	--	--	--	-
Al	--	-	-	+	+	+ +	o	+ +	-	+ +
Mg	-	o	-	+ +	+ +	+ +	+ +	+ +	-	+ +
U	--	+ +	-	+	--	--	--	--	o	+
Fe	o	+ +	-	+ +	-	-	--	--	o	+ +
Ni	-	+ +	-	+ +	+ +	+ +	-	o	-	+ +
Ti	+ +	+ +	+ +	+ +	+ +	+ +	o	+ +	+ +	+ +
Zr	+ +	+ +	-	+ +	+ +	+ +	+	+ +	-	+ +
La	+ +	+ +	+ +	+ +	--	--	--	--	+	+
Ce	+ +	+ +	-	-	--	--	--	--	-	-
Nd	+ +	+ +	+ +	+ +	--	--	--	--	+ +	+ +
Pr			+ +	+ +	--	--	--	--	+	+
Cr					+ +	+ +	+ +	+ +	o	o
Cl	+	+ +	o	+ +	+ +	+ +	+ +	+ +		
S					+	+ +	+ +	+ +		
K					+	o	+ +	+		

- - strongly depleted: $c_l/c_g < 0.5$
- depleted: $0.5 < c_l/c_g < 1$
- o   unchanged: $c_l/c_g = 1$
- +   enriched: $1.5 > c_l/c_g > 1$
- + + strongly enriched: $c_l/c_g > 1.5$

c_l : concentration of element in the surface layer

c_g : concentration of element in the unleached glass.

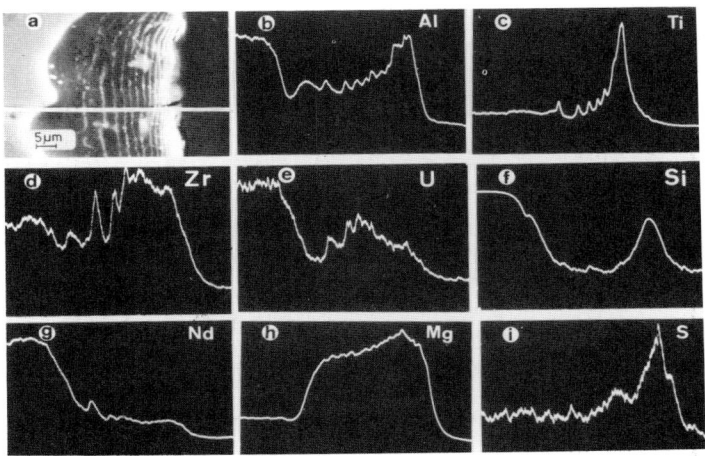

Fig. 4.4 SEM photomicrograph and X-ray line scans of a glass bead leached one year in quinary brine at 200°C. (a) SE micrograph with line scan position. (b)-(i) line scans of elements as indicated [4]

leached chips were investigated by SEM without further preparation [6]. Figure 4.5 shows the surface after 10 days and Figure 4.6 after 30 days. They are covered with a layer of crystals of various types. The cubes in Figures 4.6a and b were identified as analcime, $Na(AlSi_2O_6) \cdot H_2O$. Furthermore two other crystalline phases can be seen, the light clusters in the middle of Figure 4.5b and the light small needles in the foreground of Figure 4.5(c). Qualitative EDS analyses yielded the following elements with decreasing intensities [6]:

clusters	Si, Al, Mg, Fe, Ti, Mn
neeldes	U, Ti, Si, Na, Fe, Zn, Ca.

Figure 4.5(c) shows the cross section of the surface layer after 10 days. The bottom is pristine glass. Next is the amorphous 'gel' layer detached from the glass and cracked as a result of drying. The uppermost layer is mostly crystalline with large analcime crystals on top. The presence of crystals after 5 h leaching shows that solubility limits are reached shortly after the process has begun [6]. In the course of the leaching process,

the layer grows up and the amorphous gel does not grow around the crystals (as shown elsewhere [6]). This indicates that the amorphous layer is growing only inwards, replacing the glass phase. The crystal phases in Figures 4.5 and 4.6 are already visible after three days but they are small and continue to grow as leaching continues. X-ray patterns and semi-quantitative analyses of the crystalline species were obtained, but it was not possible to identify the crystalline phases except for analcime.

Fig. 4.5(a-d) SEM micrograph of C31-3-EC-SPF-Na glass surface leached 10 days at 200°C in saturated NaCl solution

Chips and spheres yielded insufficient amounts of surface layer material for powder X-ray diffraction technique. To get more corroded glass material in short times glass powder with a grain size of ≤ 60 μm in addition to chips was used. SEM characterizations were carried out on carbon coated chips and powders. When possible phases were also analysed by ICP. X-ray diffraction patterns were collected using a powder diffractometer.
Bulk mounts and preferentially oriented 'clay mounts' of the type normally used for X-ray identification of phyllosilicates (sheet silicates) were used. The procedures and criteria of Brown [7] were used in phyllosilicate analysis.

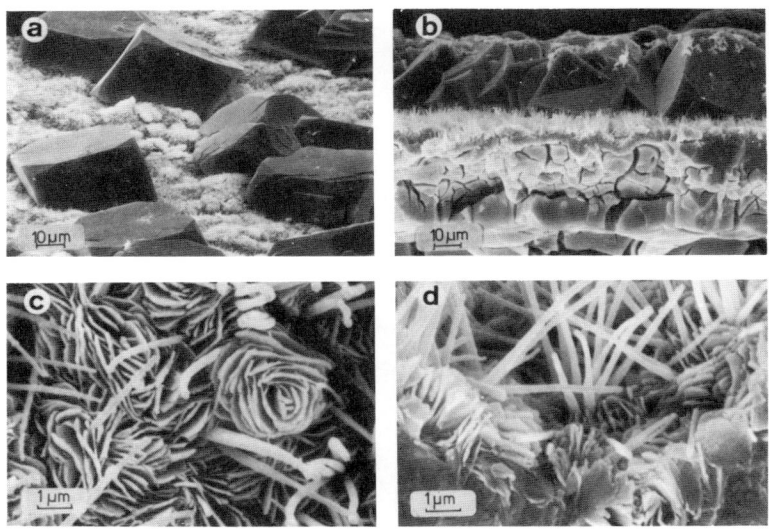

Fig. 4.6(a-d) SEM microcraph of C-31-3-EC-SPF-Na glass surface
leached 30 days at 200°C in saturated NaCl solution

A summary of experimental conditions, results and lists of
observed phases is given in Table 4.4. Included in Table 4.5 are
the solution compositions after leaching as measured by ICP
optical spectroscopy. Leaching experiments involving C31-3 glass
leached in NaCl and NaCl-0.002M $MgCl_2$ brines were similar in terms
of degree of reaction and phase formation; therefore, they are
discussed together.

It is apparent that the glass surface has a high affinity for Mg
since after 21 days almost no Mg is found in solution, Table 4.5.
A phyllosilicate, Figure 4.7(a), is present in surface layers
leached in both brines. X-ray data for this phase are summarized
in Table 4.6. Based on the criteria of Brown [7], it appears to
be a vermiculite since (a) the Mg saturated material is unaffected
by glycerol, and (b) heating to 500°C causes the basal (001) d-
spacing to collapse to 10Å. With ions other than Mg, swelling
behaviour is observed, Table 4.6, which suggests that the clay has

TABLE 4.4   An overview of leaching conditions and results for charges where C31-3 glass was leached with NaCl and NaCl-0.002M $MgCl_2$ brines. All experiments were completed at 200°C

	NaCl			NaCl brine with 0.002 M $MgCl_2$
	1	2	3	
S:V (cm^{-1})[a]	3.6	2.1	0.1	3.6
Time (d)	7-21	90	21	21
Reaction progress $(g \cdot l^{-1})$	10-11.5	11.5	0.3	9.4
Initial pH	8.9-9.1	9.1	9.1	9.3
Final pH	9.5	9.9	9.2	9.0
Final Si concentration (ppm)	25	27	11	
phases	analcime 14 Å clay zinc silicate[b] Ca U silicate[b,c] Ca silicate[b,c] Ca As apatite[b] $CaMoO_4$ $BaMoO_4$ $Ba(Cr,Mo)O_4$[b] Fe hydroxide (or oxide)			analcime 14 Å clay Ca-As-apatite $Ba(Cr,Mo)O_4$ Fe hydroxide (or oxide)

a: Minimum surface area.
b: Not observed in charge number NaCl 3.
c: Not observed in charge number NaCl 2.

TABLE 4.5 Ion concentrations of the leachate as measured by ICP and relative release factors for the reaction of C31-3 glass with NaCl and Q brines at 200°C, S:V = 3.6 cm^{-1}

Element	NaCl brine ppm	RR[a]	Q brine ppm	RR
Si	26	0.01	14	0.13
Zn	2.3	0.01	23	0.86
Ca	58	0.17	4	0.21
Mg	0.5	0.005	--	--
Ba	1020	0.59	2.5	0.028
Sr	24	0.49	2.6	1.00
Mo	39	0.19	13	1.2
Nd	--	--	10	0.76
B	161	1.00	8.6	1.00

Element	NaCl brine 0.002 M MgCl$_2$ ppm	RR	NaCl brine 0.01 M MgCl$_2$ ppm	RR
Si	b	--	48	0.12
Al	0.6	0.01	0.7	0.01
Mg	0.4	--	45	--
Zn	6.0	0.02	112	1.2[c]
Mo	24	0.15	33	0.83
Nd	1.4	0.01	1.4	0.01
B	127	1.00	32	1.00

a: RR_i is defined as the fraction of element "i" in solution devided by the fraction of boron which is in solution. Boron is not thought to enter crystalline or amorphous phases to an important extent, thus RR is useful in determining which elements are solubility limited.

b: The concentration of Si measured for this charge, 6.6 ppm, appears to be suspiciously low. It may have been affected by storage.

c: This value for Zn probably is too large. It should not have a RR greater than 1.0.

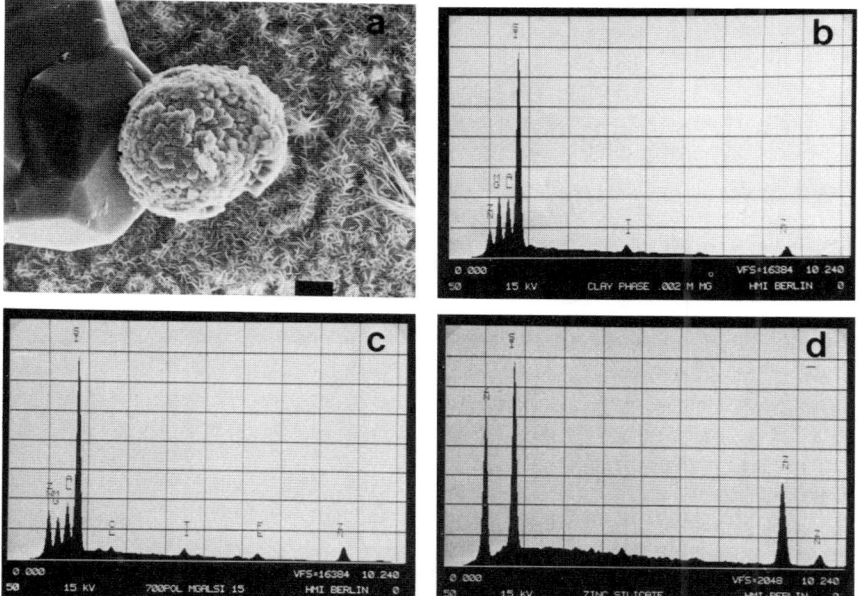

Fig. 4.7 (a) details of the surface of a C31-3 glass wafer which was leached in NaCl brine (200°C, S:V greater than 2.1 cm^{-1}, 90 d). The phase covering most of the micrograph is the phyllosilicate. The phyllosilicate has the same apearance when the glass is reacted with NaCl having 0.002 M MgCl$_2$. In the centre of the photograph is a nodule of zinc silicate, a large analcime crystal is on the left and a cluster of needle shaped Ca uranyl silicate crystals is on the right. (The black bar at the right bottom equal 10 μm).
(b) is an EDS spectrum of the phyllosilicate shown in (a).
(c) is an EDS spectrum of the phyllosilicate phase that formed when NaCl having 0.02 M MgCl$_2$ was the leachate. Comparison with 4.1 (b and c) suggests that the composition of the phyllosilicate may reflect the relative amounts of Mg and Zn in the system.
(d) is an EDS spectrum of the zinc silicate.

a layer charge that is intermediate between normal montmorillonite and normal vermiculite. Vermiculite has the structural formula $(Na,\frac{1}{2}Mg)_{1.4}(Si,Al)_8(Mg,Al)_6 O_{20}(OH)_4$, where Na^+ and Mg^{2+} in the first brackets are exchangeable. Montmorillonite phases have the same general stochiometry but have a lower ion exchange capacity

and a greater amount of swelling with water and organic liquids
[7]. The two phyllosilicates are similar and the distinction is
not always clear or even always worth making. In as leached
samples, the phyllosilicate phase has a 12Å d spacing which is
characteristic of sodium vermiculite or sodium montmorillonite.
In NaCl brine, the phyllosilicate contains significant amounts of
Mg, Al, Zn and Si, Figures 4.7(b) and (c). In NaCl having 0.002 M
$MgCl_2$, the phyllosilicate has approximately the same Al and Si
contents, more Mg and less Zn. It might be best not to regard the
compositions which are represented by the spectra given in Figures
7b and 7c as belonging to phases with fixed compositions. It is
possible that the phyllosilicate which first begins to precipitate
from both solutions is Mg rich, but as the reaction proceeds, Zn
begins to substitute for Mg. This interpretation would be in
agreement with the observation that Mg encounters a solubility
control earlier than does Zn [8].

Analcime, Figures 4.8(a) and (b), is the major phase which appears
in NaCl brine. Analcime crystals, $(NaAl)_{1-x}Si_{2+x}O_6 \cdot nH_2O$, are in
some cases larger than 100 micrometers and constitute the most
abundant phase on C31-3 surface layers. Semiquantitative electron
probe microanalyses on crystal faces of samples indicate this
phase to have a variable Si:Al ratio. Factors which control the
composition of analcime were not studied in detail, but a high
degree of reaction progress and the initial presence of $MgCl_2$
appear to contribute to high Si:Al ratios, Table 4.7. It also
appears that the Si:Al ratio of analcime is a reflection of the
Si:Al ratio of the parent glass. Low alumina glasses such as SON
68 and UK 209 react with NaCl brine to produce analcime with high
Si:Al ratios. This conclusion is in agreement with the studies of
Saha [9] on the hydrothermal decomposition of glasses in the
system $NaAlSiO_4-NaAlSi_3O_8$. The range of composition of naturally
occurring analcime is much more restricted than what is shown in
Table 4.7, typically 2.2:1 - 2.4:1 [10], suggesting that the
unusual synthetic compositions are readily formed from glasses but
thermodynamically unstable with respect to other phases.

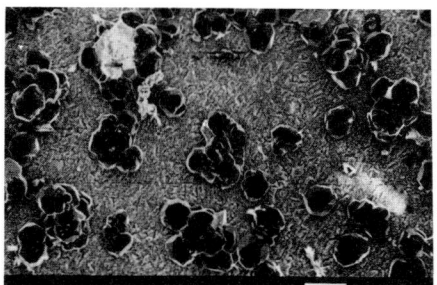 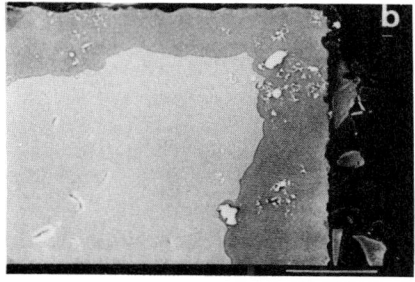

Fig. 4.8   C31-3 glass leached in NaCl. Scanning electron micrographs and an EDS spectrum of the sample shown in Figure 4.7(a).
(a) is of an 'as leached' surface and (b) is of a polished section. In (b) the bright area is fresh glass, the light grey area is 'reacted glass' the dark grey area is analcime and black areas are epoxy resin. (a) and (b) show that a significant part of the surface becomes covered by analcime. (The white bars equal 100 μm)

TABLE 4.6   Basal d-spacing for the phyllosilicate which results from reaction of C31-3 glass with NaCl and NaCl-0.002 M $MgCl_2$

	Phyllosilicate d-spacing[a]	
	NaCl brine	NaCl-brine with 0.002 M $MgCl_2$
As leached:		
air dried	12.0 Å	12.0 Å
H_2O[b]	15.0	NM[c]
ethylene glycol[b]	16.5	16.5
glycerol[b]	NM	15.5
heated 500°C 1 hr	10.0	NM
Saturated with exchangeable Mg:		
air dried	14.5	14.5
ethylene glycol	16.5	NM
glycerol	14.5	15.5

a: d-spacings are rounded to the nearest 0.5 Å.
b: The liquid with which the clay has been saturated.
c: NM = not measured.

TABLE 4.7   Estimated mean Si:Al ratios of analcime rims for selected charges having different degrees of reaction progress, different glasses or different solution compositions. The largest standard deviation in the Si:Al ratio is 0.1

charge	glass	degree of reaction (g/l)	analcime Si:Al
60 day, chip		0.9	2.0
90 day, chip		0.15	2.0
1 year, chip	glass C31-3	0.47	2.2
21 day, powder		9.8	2.2
90 day, powder		11.5	2.3
21 day, powder NaCl-0.002 MgCl$_2$		9.4	2.5
SON 68		2.6	3.3
UK 209		?a	3.4

a: Unknown but less than 11.5 g/l.

Surface layer morphologies suggest that analcime may form a protective rim. Figure 4.9 shows a micrograph of a polished glass grain and X-ray line scans for Al and Si after 7 d leaching in NaCl [8]. The bright area in the middle of the grain being unleached glass. The inner area is a gel layer and the outer dark area is analcime. The results of the 1 year experiments with glass beads seem to support the assumption of the protective analcime layer.

Figure 4.10 shows an optical micrograph of the 1 year leached bead. The dark spots statistically distributed on the surface are almost always covered with analcime crystals [8]. Figures 4.5(a) and 4.6(a) are SEM micrographs in which the dark analcime crystals can clearly be seen. The cross sectioned bead with surface layer on top is shown in Figure 4.11(a). Various regions can be distinguished. The bright zone is the pristine glass, followed by the 'reaction zone', which is covered by the surface layer. After removing the surface layer and the 'reaction zone' mechanically, the remaining surface is shown in Figure 4.11(b). The 'tree stumps' are probably surface spots, which were covered by analcime crystals formed upon leaching. As a consequence the glass might be protected at these spots.

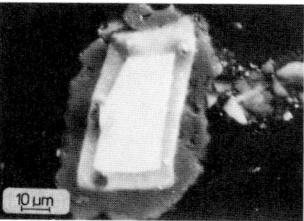 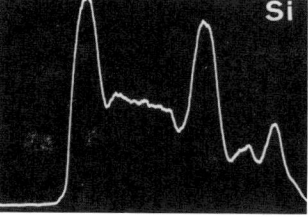 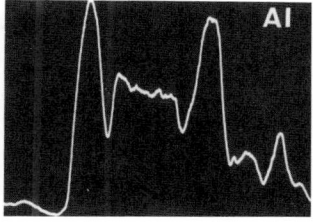

Fig. 4.9 SEM micrograph of a glass grain leached at 200°C in saturated NaCl brine and Si and Al X-ray line scans [8]

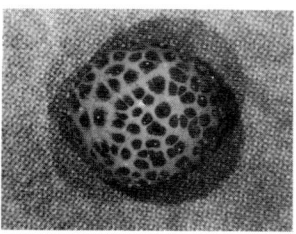 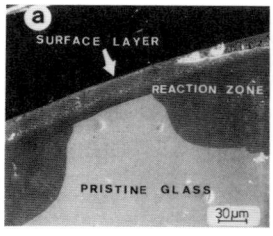 

Fig. 4.10 Glass bead leached 1 year in NaCl-brine at 200°C

Fig. 4.11 SEM micrographs of the bead shown in Fig. 10, a: cross section, b: surface after removing the surface layer and 'reaction zone [8]

Zinc silicate, Figures 4.7(a) and (d), has been detected in surface layers formed with saturated NaCl brine. Quantitative analysis using EPMA indicate a Zn:Si ratio of nearly 2:1 (67.3 wt.% ZnO, 28.6 wt.% SiO_2, total = 94.1%). This phase is initially present as spherical clusters of needles. After longer leaching times (90 d, S/V = 2.1 cm^{-1}), this phase has a hexagonal crystal morphology which is characteristic of willemite, Zn_2SiO_4. It is possible that the zinc silicate which initially precipitates is hemimorphite, $Zn_4Si_2O_7(OH)_2H_2O$ and that this phase is slowly replaced by willemite. The experimental conditions (200°C – equilibrium vapour pressure of saturated NaCl brine) are close to those reported by Roy and Mumpton [11] for the decomposition reaction: hemimorphite → 2 willemite + 2H$_2$O. The solubility of willemite is pH dependent, this may explain its absence at low pH.

A hexagonal Ca-As rich phase appears when NaCl brine is used as the leachate, Figure 4.12. EPMA on crystal faces indicate a 5:3 ratio of Ca to As and the presence of Cl. This phase is almost certainly an arsenate apatite, Ca$_5$(AsO$_4$)$_3$(OH,Cl). Other phases observed in NaCl brine include barium molybdate (BaMoO$_4$), calcium molybdate (CaMoO$_4$), barium chromate-molybdate solid solution (Ba(Cr,Mo)O$_4$) an iron hydroxide (or oxide), a calcium uranyl silicate and a calcium silicate. Barium molybdate occurs as yellow crystals measuring up to about 1 mm. A scanning electron micrograph and EDS spectra of the calcium uranyl silicate and calcium silicate phases are given in Figure 4.13. These two phases have different morphologies and signifcant differences in their U and Z contents (Figure 4.13(b and c).

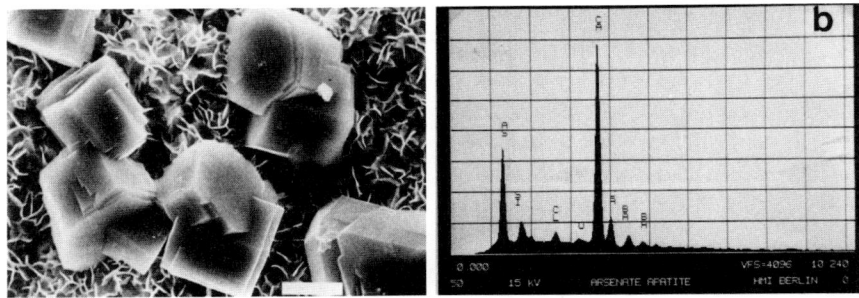

Fig. 4.12 A scanning electron micrograph of the arsenate apatite. (a) (the white bar equals 10 µm) and (b) its EDS spectrum

Quinary brine

For quinary brines and NaCl brine having 0.01 M MgCl$_2$, a summary of experimental conditions, results and lists of observed phases is given in Table 4.8. Comparison of Tables 4.4 and 4.8 indicates that with quinary brine (pH 5-7) C31-3 glass reacted more than one order of magnitude slower than with NaCl brine (pH 9 - 9.5). As a means of obtaining a higher degree of reaction, a quinary brine having 50% more water than usual was used. Parallel leaching experiments in diluted and undiluted quinary brine indicated that a lower ionic strength resulted in a higher degree of reaction. Surface layers from both experiments were examined by SEM to confirm their similarity. Surface layers formed from diluted quinary brine were studied in more detail. A montmorillonite-like phyllosilicate is present. Comparison of Figures 4.7(b) and (c) and 4.11(a) and (b) indicates that there are differences in composition of the phyllosilicates formed in NaCl and Q brine.

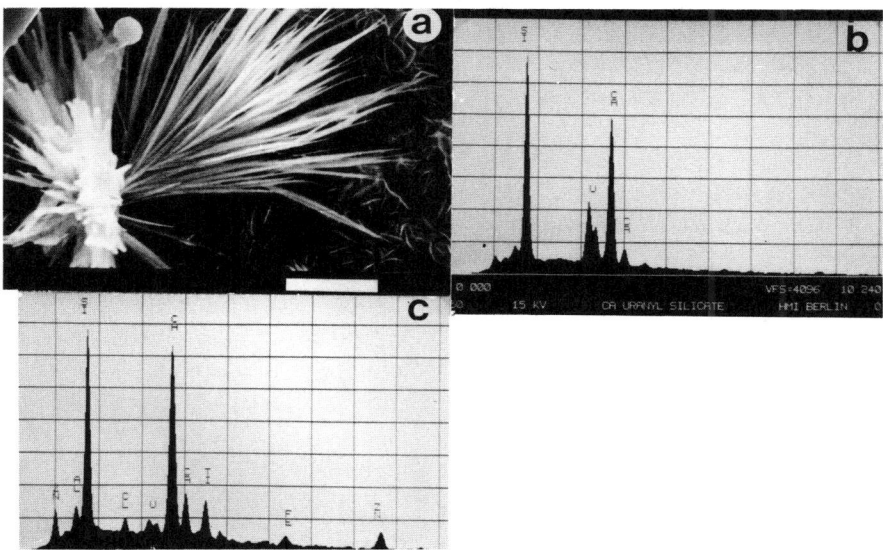

Fig. 4.13 A scanning electron micrograph of the specimen shown in Figure 4.7a. A cluster of needle shaped calcium uranyl silicate crystals and a nodule of the calcium silicate phase are shown in (a). (White bar equals 10 µm). The corresponding EDS spectra are given in (b) and (c).

A hydrotalcite phase is present. Metal ion ratios measured by semiquantitative EPMA suggest a composition of $Mg_{5.6}Al_{2.4}(OH)_{16}(SO_4)_{1.4} \cdot nH_2O$. This is in very good agreement with a composition of $Mg_{5.7}Al_{2.3}(OH)_{16})SO_4)_{1.4} \cdot nH_2O$ determined by

TABLE 4.8   An overview of leaching conditions and results for charges where C31-3 glass was leached with quinary brines and NaCl - 0.01 M MgCl$_2$

	quinary brine	quinary brine diluted	NaCl-brine 0.01M MgCl$_2$
S:V (cm^{-1})	3.6	3.6	3.6
Time (d)	21	10	21
Reaction progress (g·l^{-1})	0.7	2.0	2.5
Initial pH	5.1	4.6	9.3
Final pH	5.7	5.4	6.5
Si concentration (ppm)	15	15	50
Phases observed		14 Å clay hydrotalcite BaSO$_4$ BaKSrSO$_4$ RE molybdate RE arsenate	14 Å clay RE molybdate RE arsenate

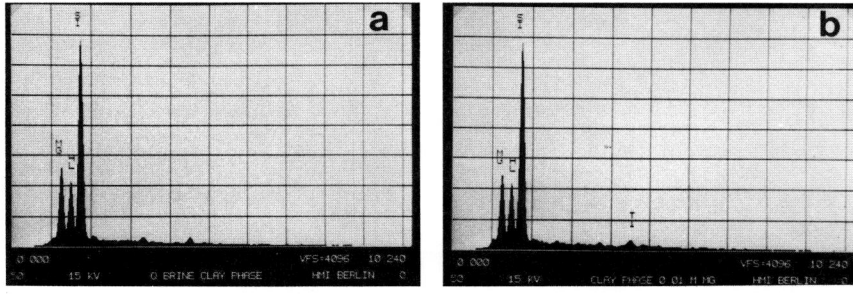

Fig. 4.14 EDS spectra for the phyllosilicate phase forming in quinary brine (a) and in NaCl with 0.01 M MgCl$_2$ (b). Both of these phyllosilicates have significantly less Zn than the phyllosilicates that formed from the reaction of C31-3 glass with NaCl brine, Fig. 4.7(b). Comparison of Figures 4.7(b), 4.7(c), 4.14(b) and 4.14(a) show that the Mg:Zn ratio increases as the total amount of Mg in the system increases. The Zn peak (see Figures 4.7b and c) is still indicated in Figure 4.14b but has vanished in Figure 4.14(a)

ICP analysis of the synthetic hydrotalcite produced by the hydrothermal reaction of quinary brine with aluminium hydroxide.

The mineral hydrotalcite has the ideal composition Mg$_6$Al$_2$(OH)$_{16}$CO$_3$ but many compositional variations are possible. Nitrate, chloride, perchlorate and sulphate analogues are known. The Mg:Al ratio is variable and numerous transition metal analogues are known [12]. Hydrotalcite has been previously reported as an alteration product arising from the reaction of sea water synthetic basaltic glass [13].

Barium sulphate is present in two distinct forms. Rounded grains always have some K, Sr and less Ba than normal. These grains may be a mixture of two or more phases. The second phase contains only Ba and sulphate, has well defined crystal faces and is presumed to be the barite phase identified by X-ray diffraction. An iron oxide phase, a TiO_2 phase and a U-containing phase are also observed by SEM-EDS analysis.

NaCl - 0.01 M $MgCl_2$ Brine

In the 21 days experiment involving this brine, not all of the Mg^{+2} is removed from solution and the pH remains acidic; near that of quinary brine (Table 4.4). Under these conditions zinc appears to leach congruently instead of entering a silicate or phyllosilicate, and analcime does not precipitate. Alkaline earth molybdates, chromates and arsenates are absent. Thus there are major differences in the phases and appearance of surface layers formed from NaCl and NaCl - 0.01 M $MgCl_2$ brines. The surface layer that forms on the glass with this brine resembles the surface layers formed in quinary brine in the following respects:

a. the 14Å phyllosilicate is nearly zinc free,

b. the rare earth elements form arsenates or molybdates, and

c. the surface layer is fragile.

The absence of $BaSO_4$, strongly absorbed chloride and sulphate are the most notable differences between quinary brine and NaCl - 0.01 M $MgCl_2$ type surface layers.

Effect of canister steel

The effect of canister steel on C31-3 glass surface layers was investigated and NaCl brine in quinary brine. Charges involving canister steel were reacted according to the same S:V, time and temperature parameters as described in Table 4.8 (3.6 cm^{-1}, 21 d, 200°C). The only difference being that the steel (as lathe turnings) was allowed to react with the solution for five days at 200°C before glass was added to the system. The relative weights of steel chips and glass were 0.32:1.00.

NaCl brine

Canister steel proved to be relatively unaffected by this solution. At the end of the leaching experiment, large amounts of fresh steel remained. It is unclear whether canister steel had any effect on the reaction rates since, with or without it, 21 days was long enough to digest almost all of the glass. ICP analyses of the completely reacted glass indicate that iron was not incorporated into the surface layer to any significant extent. No differences could be observed between surface layers of the glass formed with or without canister steel.

Quinary brine

Quinary brine was much more reactive toward canister steel than was NaCl brine. At the end of the leaching experiment most of the steel was decomposed. The addition of steel appeared to enhance the leach rate by a factor of approximately five. At the end of the experiment the solution had a turbid appearance and a yellow-brown precipitate developed over a period of weeks. SEM-EDS analysis indicated the presence of iron with some sulphur and silicon. β-$Fe_2O_3 \cdot H_2O$ was the only phase which could be detected by X-ray powder diffraction. Well crystallized hematite was detected in the diffraction pattern of the residual solid at the end of the experiment. As in the case of NaCl brine, incorporation of iron into surface layer phases was not observed.

Discussion and conclusions

The C31-3 glass surface layer has a high affinity for Mg and it may be that analcime only precipitates when insufficient Mg is present to form a clay as the dominant silicate. In every case, when some Mg remains in solution, acidic conditions prevail. It appears that under such conditions; zinc leaches congruently instead of entering the phyllosilicate phase or forming willemite. The alkaline earth molybdates, arsenate and chromate which precipitate from basic solutions are not observed in surface layers formed at pH 5. In their place a $NdAsO_4$ and a Nd-molybdate are observed. A montmorillonite-like phase appears in the acidic brines while a vermiculite-like phase is observed in NaCl brine. The surface layers formed at pH 5 are much more fragile than those formed in NaCl brine. It appears that the glass reacts more slowly with quinary brine than with NaCl. Mass balance among Mg, Al, Zn and Si [8].

In experiments involving C31-3 glass, the major crystalline silicate phases have been identified and their compositions have been estimated. After a high degree of reaction, in NaCl solution 10 g/l for example, the ratios Mg:Si, Al:Si and Zn:Si are nearly the same in the glass and in surface layers. Concentrations of measurements of these elements indicate that less than two percent are found in solution. Under these conditions the assumption that almost all of the Mg, Al and Zn are consumed by the crystallization of silicates permits a simple 'limiting case' mass balance of these elements to be calculated. The following subtraction sequence was used (Table 4.9).

a. the Mg-Zn-Al-phyllosilicate (assumed atomic proportions: 7Si:2Al:2Mg:3Zn) precipitates, consuming Mg, then

b. the zinc silicate phase (2Zn:Si) precipitates until the remaining zinc is consumed, and then

c. analcime can precipitate until the remaining Al is consumed (2.3Si:Al).

The sequence in which phases containing Mg, Al and Zn have been subtracted were chosen to satisfy the experimental observation that these elements are almost absent from solution after a high degree of reaction progress, Table 4.5. The actual precipitation sequence might be different. Table 4.9 represents a simplification

TABLE 4.9   Mass balance for C31-3 glass - NaCl brine: Mg, Al, Zn and Si

	Si	Mg	Al	Zn
Atom % in glass:	15.3	0.9	5.3	1.6
Phyllosilicate:	- 3.1	- 0.9	- 0.9	- 1.3
2:1 zinc silicate:	- 0.2	-	-	- 0.3
Analcime:	- 10.1	-	- 4.4	-
Residual:	1.9	0	0	0

and the exact numbers should not be taken too seriously. However, it appears that approximately 20 per cent of the Si is incorporated in the phyllosilicate, 1 per cent in the zinc silicate and 66 percent in the analcime phases respectively. According to this calculation, approximately 13 percent of the Si remains unaccounted for and may be assumed to partition among minor phases (the calcium uranyl silicate for example), the gel layer, solution and the walls of the teflon reaction vessel.

Time dependence derived from weight loss and solution analysis

The experiments were carried out with beads and chips. The surface area to volume ratio was 0.014 cm^{-1}. The weight losses were measured after drying the samples with surface layers on. The weight of the surface layer was estimated by weighing the sample after removal of the layer. The specific weight loss NL_G was derived from the surface area and the weight difference of the samples before leaching and after removing the layer.

The leachates were analyzed by ICP. Based on the behaviour of the components 'i' the normalized weight losses NL_i were calculated. The curves are given in Figure 4.15. The figure shows the time dependence of the specific weight loss of leaching in NaCl-brine. The measured (NL_G) and the normalized (NL_i) values show that the leaching process slows down with time. Up to 30 days the increase of NL_G is highest.

For the elements Al, Si, Mn the curves show a maximum. Their NL_i values decreased at about 30 to 60 days and at about 90 days NL_{Si} and NL_{Al} are constant or slightly increased in the long time. NL_{Mn} drops below detection limit. After 90 days the normalized specific weight losses of these elements and of Mg, Zn and Mo are considerably smaller than NL_G and the NL_i values of B, Li, Ca, Sr and Ba.

The leachates were also analyzed for Zr, Ti, Fe, Nd, Ni, Ce and U. Their concentrations were below the detection limits. Some semiquantitative analyses of the surface layers revealed enrichments of these elements and the elements whose NL_i values do not fit the NL_G curve.

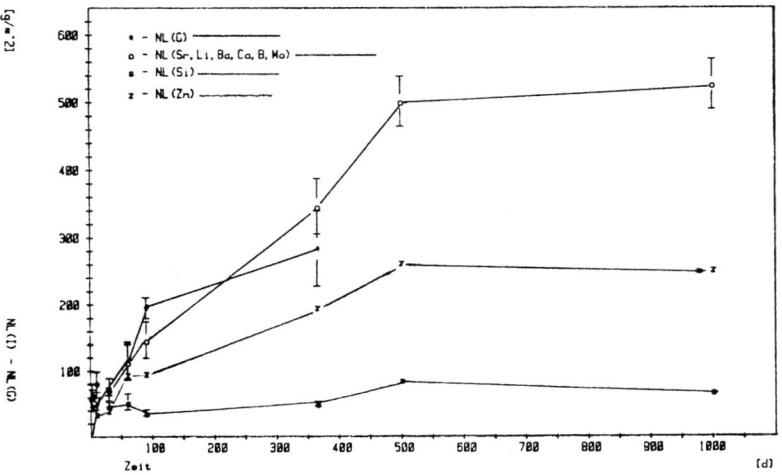

Fig. 4.15 Total and normalized elemental mass loss of leaching C-31-3-EC-SPC-Na glass in saturated NaCl solution at 200°C

Constant or decreasing NL_i- values mean constant or decreasing ion concentrations in the leachates, which are due to saturation and oversaturation effects respectively. In the leaching process solubility limits were obviously reached for some elements within 10 days. As a consequence concentrations of those elements are controlled by solid phases formed upon leaching. It has been shown that leachate concentrations for various elements can be interpreted with the help of thermodynamic data if one assumes the presence of appropriate solid phases in the leachate [6,14]. As discussed in the preceding chapters some solid phases have been detected which may control the concentration of some elements under static leaching conditions. The application of solubility limited concentrations to leaching kinetics will be discussed in the following section.

4.1.4 Leaching model in NaCl brine

The experimental results can be summarized by the following observations:

1. The silicon concentrations appears to be strongly correlated to the glass corrosion rate.

2. The disintegration of the glass is a congruent process (this is true at least within the depth resolution of X-ray line scans).

3. The leach rate is not necessarily a monotoneously decreasing function with time.

4. A number of crystalline phases precipitate from the leachate.

5. Most of the precipitates stay on the surface of the glass and form a more or less dense layer.

6. In most cases a gel layer exists between the reaction zone

of the glass and the precipitates.

Observation No. 1 indicates that silicon plays the major role in the mechanism of glass corrosion. Indeed, it is assumed [15] that the silicon concentration alone determines the reaction rate of the glass by a first order kinetic law:

$$\frac{dm_{Si}}{dt} = Sk_{Si}^+ (1 - a_{Si}/a_s) \qquad (1)$$

where k_{Si}^+ is the forward reaction rate ($C_{Si} = 0$), $a_{Si} = \gamma_{Si} C_{Si}$ the silicon activity in solution and a_s a theoretical equilibrium constant (zero-rate for $a_{Si} = a_s$). S is the surface area of the glass exposed to the leachate (see Figure 4.16). For reasons

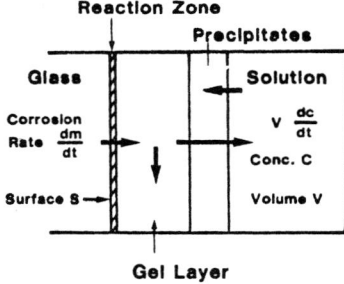

Fig. 4.16   Schematic view of a static leach experiment

discussed below it is not possible to measure a_s directly. In principle, k^+, a_s and the activity coefficient are temperature and pH dependent and the surface area decreases with increasing reaction progress. In leaching experiments the temperature is kept constant and for glass samples of typically several millimeters for the minimum dimension the change of the surface area can be neglected for most leaching times used. Furthermore, in brines of high ionic strength the pH is almost independent of the concentration of the corrosion products. Consequently, k^+, S, a_s and the activity coefficients will be kept constant for modelling purposes. Together with observation 2 equation (1) can be written as total glass reaction rate

$$\frac{dm}{dt} = Sk^+ (1-c_{Si}/c_s) \qquad (2)$$

where $m = m_{Si}/w_{Si}$, $k^+ = k^+_{Si}/w_{Si}$ and $c_s = a_s/\gamma_{Si}$, w_{Si} is the weight fraction of silicon in the matrix and the reaction rate of any element i is

$$\frac{dm_i}{dt} = w_i \frac{dm}{dt} \qquad (3)$$

If no surface layer is formed and no precipitation of phases containing silicon is considered equation (2) is easily integrated:

$$\frac{m}{V} = \frac{c_s}{w_{Si}} \left(1 - \exp\left(\frac{w_{Si} \cdot k^+ \cdot S}{c_s \cdot V} t\right)\right) \qquad (4)$$

where V is the volume of the leachate. As it is often observed in leaching experiments that silicon seems to be saturated in the leachate the simple exponential time dependence of the leach rate is frequently used in long term extrapolations.

However, leach experiments prove that the time dependence of the total mass loss can only be approximated over a short period of time by equation (4). In most cases the total mass loss rate does not approach a zero value, but (observation 3) might also stay constant or even increase. As will be shown in the following the precipitation of crystalline phases (observation 4) can explain most of the experimental results. If only silicon determines the total reaction rate only phases containing silicon have to be considered. Neglecting any kinetic effects in phase formation the silicon activity is determined by the equilibrium products

$$\pi_j a_i^{n_{ij}} = K_j \qquad (5)$$

of all solid phases which are in equilibrium with the solution. If again it is assumed that in brines of high ionic strength the activity coefficients and the pH are constant the activity in (5) can be replaced by the concentration with an appropriate constant, on the right site. It can be estimated from measured values of the concentrations.

In the following discussion only phases are considered containing silicon, one additional element of the glass matrix and possibly elements from the brine. However, the concentrations of the brine elements are kept constant due to their high initial value. Then the product (5) can be written as

$$c_i \cdot c_{Si}^{n_i} = K_i \tag{6}$$

i.e. the solid phases contain n_i silicon atoms per one atom of element i. Combination of equations (2), (3) and (6) leads to

$$\frac{dc_{Si}}{dt}(1+\sum_i n_i \frac{c \cdot M_{Si}}{M_i} \cdot \frac{K_i}{c_{Si}^{n_i+1}}) = \frac{w_{Si} k^+ S}{V}(1-\sum_i n_i \frac{w_i}{w_{Si}} \cdot \frac{M_{Si}}{M_i})(c_s - c_{Si}) \tag{7}$$

(the summation is over all elements participating in phase formation. The formation of other phases of element i without silicon is neglected). Equation (7) can easily be integrated [3] and the time dependence of the total mass loss derived.

Although only simple phases are considered equation (7) exhibits the main features of a leaching mechanism controlled by phase formation. The temperature or pH-dependence of the material constants k^+, c_s and K_i may affect the time scale, but not the sign of the change of the silicon concentration with time. It is only determined by the sign of

$$SC = 1 - \sum_i n_i \frac{w_i}{s_{Si}} \cdot \frac{M_{Si}}{M_i}$$

SC may be looked at as a stability criterion. Only for SC ⩾ 0 does the silicon concentration increase permanently and might approach the value c_s. For SC $\leq$ 0 the leach rate might finally return to its initial value. The criterion simply means that the supply of silicon by the corrosion process has to be higher than the amount of silicon consumed in phase formation. Otherwise it is principally impossible to stabilise the glass with respect to the solution. Indeed, both cases were observed for the glass C31-3, where the Si:Al-ratio is about 3. During leaching in NaCl the analcime phase (Si:Al = 2) forms, leading to an easily detectable depletion of aluminium in solution whilst Si stays almost constant. The total mass loss rate is approximately constant. During leaching in Q-brine a phyllosilicate precipitates due to the high Mg-concentration. This phase has a high Si:Al ratio (4:7) leading to a depletion of Si and consequently to a possible increase of the leach rate from the beginning of the phase formation. Even for the glass SON 68 the Si-concentration seems high enough to deplete the Si-concentraton in solution, again leading to an increase of the leach rate. This phase was - after some time - also observed during leaching of the glass C31-3 in NaCl where the Mg is only supplied from the glass. When the equilibrium is reached silicon might be depleted temporarily until the Mg in solution is consumed. Consequently, a maximum in the leach rate should be observed. Even though Fig. 4.15 shows a maximum for the leach rate between 30 and 90 days there might be another explanation. It is a common phenomenon in phase formation that the solution is oversaturated in the beginning and that the crystal growth rate depends on the amount already precipitated. Both phenomena lead to a higher reaction rate of phase formation than is calculated from the equilibrium product and therefore to a temporary depletion of silicon. Again, this causes a maximum in the leach rate.

Even in a system where the silicon concentration can increase permanently the value of c_s will not be reached as it is certainly higher than the solubility of silicon itself, i.e. SiO_2 precipitates (forms colloids or enters the gel layer) and keeps the silicon concentration constant. This results in an almost constant mass loss rate as long as the pH does not influence the values of k^+ and c_s and the surface is nearly constant. Actually, many leaching experiments show this behaviour with a final leach rate much smaller than k^+. However, close to the saturation value c_s, fluctuations of the silicon concentration will result in high changes of the relative mass loss rate. Finally, other observations (5 and 6) have to be considered. On all glasses a more or less structured system of layers is found consisting of the solid corrosion products of the glass, most of the precipitates from solution and a gel layer, positioned between the reaction zone of the glass and precipitates.

The gel layer normally contains some silicon. This can be modelled by the partitioning of the silicon fluxes:

$$V \frac{dc_{Si}}{dt} = a \frac{dm_{Si}}{dt} = a\, w_{Si} \cdot \frac{dm}{dt} \qquad (8)$$

where $(1-a)m_{Si}$ is the amount of silicon remaining in the gel layer. In the formula discussed so far this can be described by an effective silicon content $a \cdot w_{Si}$ in the matrix.

Additionally, the gel layer and/or the precipitates might form a diffusion barrier. Then the silicon concentration in solution will not be the same as the concentration in the reaction zone c'_{Si} determining the corrosion rate of the glass. In a simple diffusion model (one layer with constant diffusion coefficient D for silicon and no reaction of silicon in the layer) the flux to the solution will be

$$V \frac{dc_{Si}}{dt} = \frac{SD}{l} (c'_{Si} - c_{Si}) \qquad (9)$$

where l is the thickness of the layer. If the composition of the layer is uniform its thickness is proportional to the total mass loss

$$l = b \cdot m \qquad (10)$$

where $1/b$ is the specific density (mass/area) of the surface layer. Equation (9) can easily be integrated using (10) and (8) where the silicon concentration has to be c'_{Si} [16] and no phase precipitation is considered. The result is a more or less slower increase of the total mass loss depending on the value of D. However, the final concentration of silicon would be c_s which is not possible due to SiO_2 formation. Much more probable, the silicon concentration c'_{Si} is reached in solution all silicon not consumed in phase formation has to stay in the gel layer (a = 0, the silicon needed for phase formation from the solution will probably be supplied from the outer side of the gel layer). In consequence, a gel layer forms with composition very similar to the original glass, only depleted in elements which are able to move through the layer. As $c'_{Si} < c_s$ the glass will be continuously corroded, however the final rate might be very small and almost no elements are released to the solution.

Although the layers can clearly be identified (Figs. 4.1, 4.5) and some experiments show a decreasing mass loss rate in the long run, it is hard to prove whether the layer is a real diffusion barrier or essentially permeable to silicon species. Another possibility to explain a decreasing leach rate is seen in Fig. 4.11, where parts of the surface covered by precipitated crystals are less susceptible to corrosion than other parts of the surface. Therefore, the surface exposed to the leachant might be thought of as time dependent:

$$\frac{dm}{dt} = S(t)k^+(1-c_{Si}/c_s)$$

As in this case the amount of precipitated material is almost proportional to the total mass loss, a dependence of the type

$$S(t) = S_o(1-m/m_s)$$

might be applicable. It implies that the precipitate is
completely impermeable and that the surface exposed to the
leachate is finally zero. Only in a model of this type (for
application see [8]) is the final release rate of elements into
the solution zero. However, when the precipitated layer is
permeable to water the glass will continue to react and due to
swelling during the gel layer formation the barrier will not stay
undisturbed.

No fits of seal data are given here as too many phenomena and
parameters are involved and any mass loss curve can be explained
easily: the main emphasis of the discussion is to show that the
phase formation in the glass/brine system plays the major role in
determining the silicon concentration in solution and in the
reaction zone which in turn influences the total mass loss rate.

A temporary increase or decrease of the leach rate is easily
understood by phase formation rates calculated from equilibrium
products but may also be controlled by oversaturation and crystal
growth effects.

The protection of the surfaces by diffusion layers and solid
material is still a strong possibility keeping the leach rates low
over a long time, but it seems difficult to achieve a real proof
of its effectiveness in an experiment of relative short time. For
long time extrapolation it seems necessary to determine the leach
rate of a fresh glass in a solution which is in equilibrium with
the gel layer and other corrosion products and finally at least
the glass composition should prevent a possible formation.

4.1.5 Glass SON 68 17 L1 C2 A2 Z1 (Type R7 T7)

A summary of experimental conditions, results and lists of
observed phases for the reaction of SON 68 glass with various
brines is given in Table 4.10. Included in Table 4.11 are the
solution compositions after leaching as measured by ICP optical
spectroscopy.

TABLE 4.10 An overview of leaching conditions and results for the reaction of SON 68 glass with various brine solutions. $S:V = 3.6 \text{ cm}^{-1}$, 21 days, 200°C

	NaCl brine	quinary brine
Reaction progress (g/l)	2.6	1.7
Initial pH	9.1	5 - 5.1
Final pH	9.3	5.1
Final Si concentration (ppm)	91	11[a]
Phases	analcime	14 Å clay
	Ca-P-apatite	$BaSO_4$
		RE molybdate

a: This value is thought to be too low, perhaps affected by storage.

	NaCl brine– 0.002 M $MgCl_2$	NaCl brine– 0.01 M $MgCl_2$
Reaction progress	3.7	4.3
Initial pH	9.3	9.3
Final pH	8.6	6.7
Final Si concentration (ppm)	76	112
Phases	analcime	14 Å clay
	14 Å clay	RE molybdate

NaCl brine

No phyllosilicate reaction products were detected in SON 68 surface layers after reaction with NaCl brine at 200°C. However, a Zn montmorillonite or some other phyllosilicate might still be present. Since this glass has no Mg and only one-half as much Zn as C31-3 glass (Table 4.1) it would be expected that phyllosilicates were less prominent. Analcime is the major crystalline phase that is present in the surface layer. It has the dodecahedral crystal habit. EPMA indicates this phase has an unusually high Si:Al ratio compared to analcime from C31-3 glass, Table 4.7. A Ca-P phase, probably apatite, was also detected.

Quinary brine

A 14Å phyllosilicate is observed in SON 68 surface layers leached with quinary brine. Mg and Ca are supposedly the exchangeable ions, but attempts to convert the clay to the Mg saturated form

TABLE 4.11 Leached ion concentrations and relative release factors for the reaction of SON 68 glass with various solutions (21 days, 200°C, S:V = 3.6 cm^{-1})

Element	NaCl brine ppm	NaCl brine RR	Q brine ppm	Q brine RR
Si	91	0.16	11[a]	0.03
Al	0.9	0.01	0.7	0.01
Zn	--	--	35	1.00
Mo	32	1.08	19	1.00
Nd	--	--	16	0.67
B	115	1.00	73	1.00

Element	NaCl brine 0.002 M $MgCl_2$ ppm	NaCl brine 0.002 M $MgCl_2$ RR	NaCl brine 0.01 M $MgCl_2$ ppm	NaCl brine 0.01 M $MgCl_2$ RR
Si	76	0.09	112	0.12
Al	1.3	0.04	1.3	0.04
Zn	0.5	0.01	91	1.1
Mg	1.0	--	17	--
Mo	40	0.97	27	0.56
Nd	--	--	--	--
B	161	1.00	184	1.00

a: This Si value is unusually low; the solution may have lost Si while awaiting ICP analysis.

were unsuccessful. The surface layer contains large amounts of flocculant material so that it is all but impossible to prepare usable oriented diffraction mounts by sedimentation. Without knowing the exchangeable ion in the clay, it can be difficult to distinguish between montmorillonite and vermiculite. A RE molybdate and $BaSO_4$ were observed.

NaCl brine with 0.01 M and 0.00 M $MgCl_2$

The Mg concentration has a significant effect on phyllosilicate phase formation. In NaCl brine, adding small amounts of $MgCl_2$ causes a 14Å phyllosilicate (montmorillonite or swelling vermiculite) to appear. EDS spectra show this phase to be enriched in Mg with relatively little Al and Zn. As with C31-3 glass, Mg seems to have the effect of increasing the amount of phyllosilicate and decreasing the amount of analcime which is precipitated.

Time dependence derived from weight loss and solution analysis

Weight loss measurements were made after leaching glass chips up to 360 days in NaCl- and Q-brines saturated at 200°C. The leachates were analyzed by ICP optical spectroscopy. The results are shown in Fig. 4.17. NL_G in Fig. 4.17 are the curves obtained

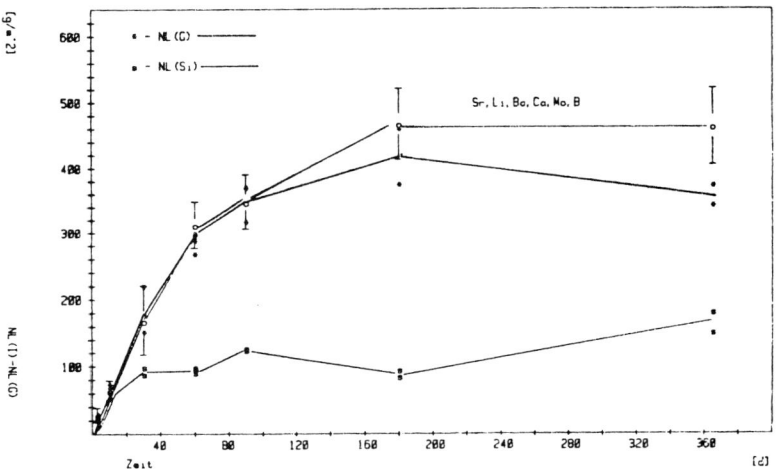

Fig. 4.17 Total and normalized elemental mass loss of leaching R7T7 glass in saturated NaCl- and Q-brines at 200°C

from weight loss measurements when the surface layers are scraped off, i.e. the total reacted glass mass, if concentration profiles in the glass beneath the surface layer are neglected. NL_i curves are normalized mass losses based on the behaviour of component 'i'. In case of a congruent dissolution of the glass and no phase formation on the surface the NL_i and NL_G curves fall together.

This is the case for elements like Sr, Li, Ba, Ca, Mo and B, whereas the mass losses based on Si are clearly below NL_G. As already discussed the apparently small specific weight loss based

- 106 -

on Si concentration in the leachate may be the consequence of the
formation of analcime in case of NaCl brine and a phyllosilicate
in case of Q-brine which were also detected on the surface of the
R7/T7 glass.

The NL_G- and NL_i- values show that the corrosion process slows
down with time. In NaCl-brine the mass losses up to one year are
considerably higher than in Q-brine. After 365 days the mass loss
values are the same in NaCl- and Q-brine. The NL_G-curves in NaCl
clearly lie below the NL_i curves. This may be an indication of a
progressive ion exchange of the mobile ions, but no experimental
evidence has been found as yet.

4.1.6 A comparison of C31-3 and SON 68 glasses

The effect of NaCl and quinary brines has been studied on C31-3
and SON 68 glasses. Superficially the surface layers resemble one
another but there are significant differences. In general the SON
68 glass was less reactive with NaCl brine than was C31-3 glass.
SON 68 also seemed to have a greater tendency toward amorphous
phase formation; for this reason it was the more difficult of the
two glasses to study and fewer phases were identified. This
difference may be due to the absence or low contents of silicate-
forming elements such as Al, Mg and Zn. The scarcity of these
elements in the SON 68 glass might mean that the removal of Si
from solution by formation of phyllosilicates, zeolites and zinc
silicates is a less important process in the overall glass -
solution reaction. The lower Al, Mg and Zn content of SON 68
glass as compared to C31-3 may be responsible for the different
behaviour of Si for these two glasses. SON 68 corrosion appears
to maintain about 100 ppm Si in solution while C31-3 maintains
about 25 ppm.

With both glasses, Mg seems to be an important element for the
formation of phyllosilicates. It appears that phyllosilicates
form in preference to analcime when sufficient Mg is present. The
amount of aluminium in the SON 68 phyllosilicate was much less
than in the C31-3 phillosilicate.

4.1.7 Investigation of high level waste glasses

High activity waste glass blocks were prepared in the Vulcain line by CEA in Marcoule [17]. The compositions of the glasses and their radioactive constituents are given in Tables 2.11 and 2.2.

SEM-EPMA-measurements

In order to perform SEM studies outside the hot cells, small pieces were taken from the blocks. Their weights were about 15 µg and their total β-, γ-activities were less than 2 µCi. The dose rate at the surface was about 40 mrem/h. Some pieces were cut with a low-speed saw and polished to get tiny low activity thin sections.

The as cast glasses

The as cast glasses have inclusions of Fe-Cr-Ni and Ru-Pd phases. Fig. 4.18(a) and (b) shows a piece of non-processed F-SON 641920 F3G3 glass. At higher magnification some inhomogeneities became visible as shown in Fig. 4.18(b). EPMA revealed noble metal Ru-Pd-phases, well known from the simulated inactive glasses [3]. Some phases were found in the radioactive sample, not detected in the inactive glasses. Fig. 4.19 shows an EDS spectrum of a phase containing Cr-Fe-Ni, probably a spinel. The spectrum in Fig. 4.20 belongs to an unidentified phase high in tungsten and Fig. 4.21 is the EDS spectrum of a phase which contains probably rare earth phosphates and some Cr, Fe, Ni. In Table 4.12 concentrations of some major constituents of the glasses and glass ceramic resp. are listed as measured by electron probe microanalysis.

Considering the oxide concentrations of the three waste forms in Table 4.12 when compared to theoretical values, positive or negative deviations can be seen for most of the oxides. However, there is no indication of a systematic impact on the EPMA measurements by self-irradiation of the samples. The smallest difference between theoretical and measured values in all products was found for SiO_2 and Fe_2O_3, whereas the highest deviation has been measured for Nd_2O_3. Taking into account the exceptional production technique of the glasses and the shortcomings of the energy dispersive microprobe analysis generaly the agreement between theoretical and measured concentrations is quite reasonable.

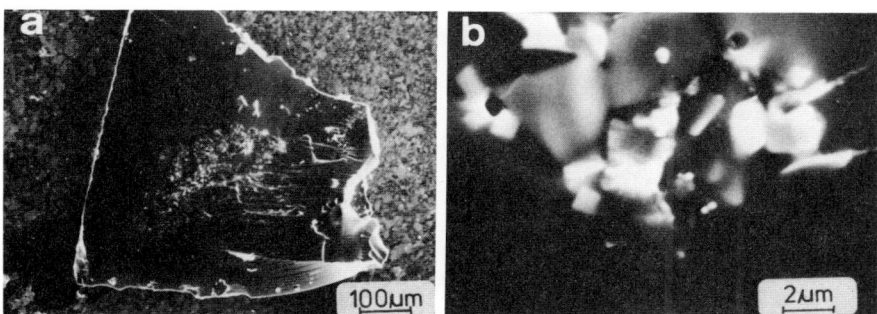

Fig. 4.18   SEM photomicrograph of radioactive glass
F-SON 64 19 20 F3G3. (a) at low (100x); (b) at
higher (5000x) magnification

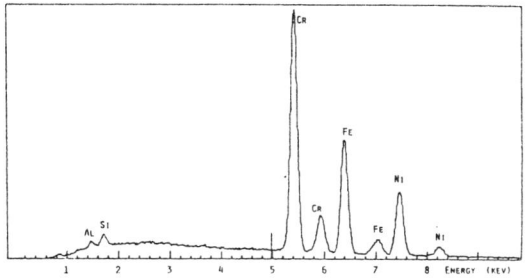

Fig. 4.19: EDX-spectrum of a phase shown in Fig. 4.18(b)

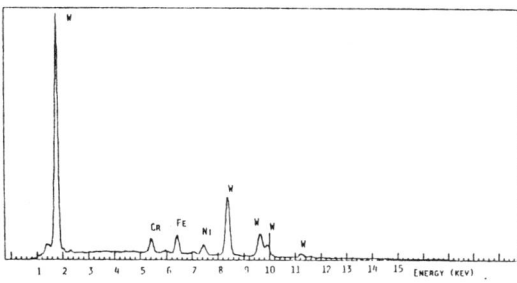

Fig. 4.20: EDX-spectrum of a phase shown in Fig. 4.18(b)

- 109 -

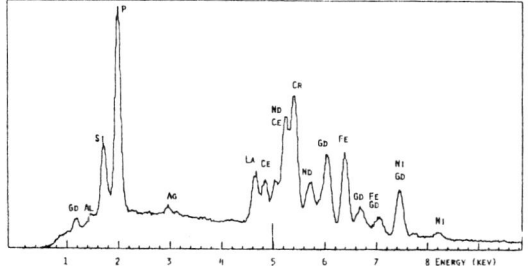

Fig. 4.21: EDX-spectrum of a phase shown in Fig. 4.18(b)

TABLE 4.12 Concentrations of oxides (wt.%) in the high activity glasses and glass ceramic:
1 = Theoretical values, 2 = EPMA measurements, number of measurements 10

Oxide	SON 64.19.20.F3G3		UK 209		C-31-3-EC glass ceramic	
	1	2	1	2	1	2
Na_2O	11.4	*6.7 ± 0.6	8.3	8.4 ± 0.9	3.3	5.5 ± 0.6
MgO			6.3	4.5 ± 0.3	1.4	1.5 ± 0.2
Al_2O_3	1.2	*4.1 ± 0.2	5.1	2.4 ± 0.1	10.3	10.5 ± 0.7
SiO_2	43.8	*43 ± 2	50.9	54 ± 1	34.8	38 ± 2
ZrO_2	1.9	*1.9 ± 0.1	1.4	1.0 ± 0.1	3.3	2.0 ± 0.3
MoO_2	2.1	*2.6 ± 0.3	1.8	3.3 ± 0.3	2.36	2.2 ± 0.2
CaO	0.01	0.06± 0.06			3.85	1.9 ± 0.2
Cs_2O	1.2	0.06± 0.07	0.3	n.m.	**1.86	0.5 ± 0.4
BaO	0.6	n.m.	0.4	n.m.	15.2	12.3 ± 0.6
CeO_2	1.3	0.5 ± 0.3	1.0	1.0 ± 0.3	1.31	0.2 ± 0.2
Nd_2O_3	1.8	*3.9 ± 0.5	1.8	5.1 ± 0.5	2.40	1.45± 0.5
Gd_2O_3	5.9	*5.6 ± 0.5				
Fe_2O_3	5.9	*6.1 ± 0.5	2.7	3.1 ± 0.2	1.47	1.6 ± 0.2
NiO	0.5	1.4 ± 0.2	0.4	n.m.		
Cr_2O_3	0.5	0.2 ± 0.1	0.6	n.m.		
MnO_2	0.6	0.4 ± 0.1	0.7	0.7 ± 0.1		
ZnO					4.91	4.3 ± 0.3

n.m. = not measured
 * = number of measurements = 20
 ** = Rb_2O, CsO_2

The annealed glasses

Pieces of the glass SON 64 19 20 F3G3 were annealed in the hot cells for about 5 days at 800°C. Fig. 4.22(a) and (m) are SEM micrographs of polished and non processed surfaces respectively. Various crystalline phases with different morphologies and compositions can be seen in the X-ray mappings in Figs. 4.22 b-i, n, o. Figs. 4.22 b, c and d-f indicate the existence of the already mentioned Fe-Cr-Ni phase. The X-ray mapping of Ca (Fig. 4.22 h), Sr (Fig. 4.22 n) and Mo (Fig. 4.22 o) suggest a (Ca,Sr)-molybdate phase is present.

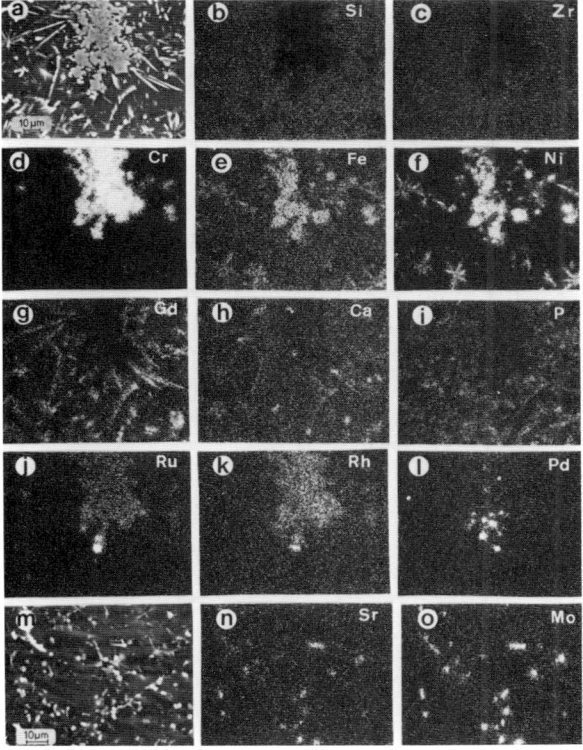

Fig. 4.22 a and m   SEM micrographs and b-l, n, o X-ray mappings of HLW glass SON 64 19 20 F3G3 annealed 5 days at 800°C

Pieces of the glass UK 209 were annealed for about 10 days at
700°C, Fig. 4.23 shows a SEM micrograph. Several crystal phases
have been formed. Some major constituents of the phases detected
by electron microprobe are shown in the EDX spectra in Figs. 4.24
to 4.27. Phase A in Fig. 4.23 shows a high concentration of Cr and
some Fe, Ni, Zn, Mg and Si (Fig. 4.24). Phase B seems to be a
magnesium iron nickel silicate (Fig. 4.25) identified in the simulated inactive glass [18]. Phase C contains mainly Si and Nd and
some Na, Mg, Al, Zr, Mo, La and Ce (Fig. 4.26). In Fig. 4.27 the
spectrum of the small bright needles (phase D) is shown, the major
constituents are Si and Mo and in minor concentrations Na, Mg, Al,
Zr, Ce, Nd and Mn. All phases excet phase A consist of very small
crystals, so that it is not possible to be certain if only the
pure phases were measured. Matrix effects cannot be excluded.

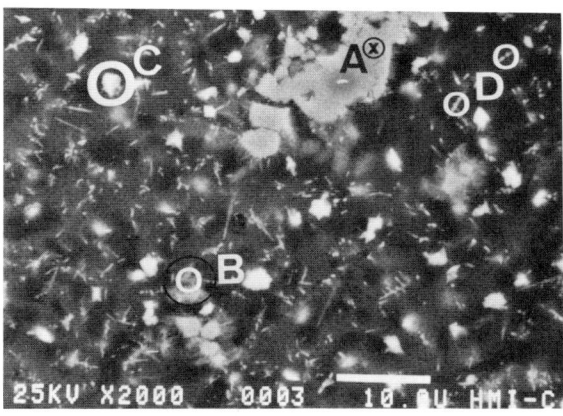

Fig. 4.23   SEM micrograph of HLW glass UK 209 annealed 10 days at
700°C

Leaching experiments

Experimental

For the leaching of highly radioactive glass samples the Soxhlet
apparatus described in [18,19] was modified for easy and
unbreakable remote handling in hot cells.

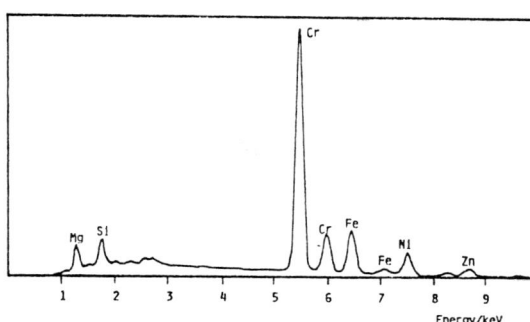

Fig. 4.24 EDX spectrum of phase A in Fig. 4.23

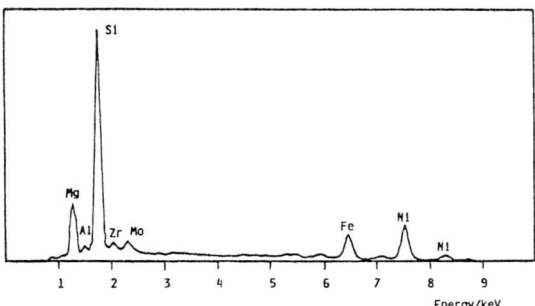

Fig. 4.25 EDX spectrum of phase B in Fig. 4.23

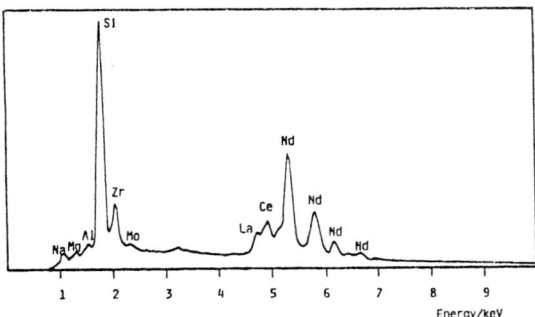

Fig. 4.26 EDX spectrum of phase C in Fig. 4.23

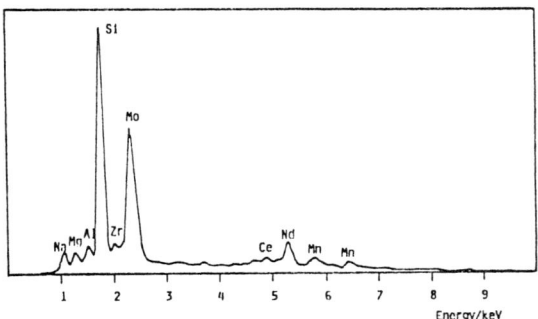

Fig. 4.27 EDX spectrum of phase D in Fig. 4.23

Leaching experiments were performed with the chips of 10 x 10 x 1 mm dimensions in bidistilled H_2O at 100°C, NaCl solutions and Q-brine at 200°C under static conditions. The water volumes were estimated by weighing. After finishing the experiments the samples were dried and their weight losses estimated by weighing in the hot cells.

Results

The glass samples of SON 64 19 20 F3G3 have been leached for 3, 10 and 30 days in the Soxhlet apparatus at 100°C. High temperature tests were carried out in autoclaves under static conditions at 200°C in bidistilled water and saturated NaCl solutions. The results can be seen in Fig. 4.28. In water the total mass loss increases from about 22 mg.cm^{-2} after 3 days to about 46 mg.cm^{-2} after 30 days, that is a factor of two, whereas the factor of time is ~ 10.

The curve in the middle of Fig. 4.28 shows the leaching results in saturated NaCl brine at 200°C under static conditions, and the lower curve gives the values for the leaching in the Soxhlet apparatus at 100°C.

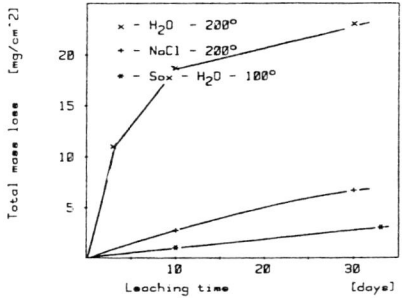

Fig. 4.28 Total mass losses of HAW glass SON 64 19 20 F3G3 in bidist. H_2O at 100°C and 200°C and in saturated NaCl solution at 200°C

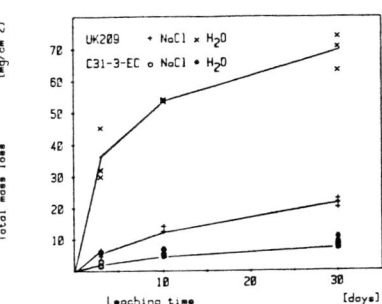

Fig. 4.29 Total mass losses of HAW glass UK 209 and glass ceramic C31-3-EC in bidist. H_2O and saturated NaCl solution at 200°C

Temperature has certainly a strong influence on the total mass loss as can be seen when comparing the values at 100°C and 200°C, whereas saturation of solution is less significant. With the Soxhlet method saturation of the leachant is impossible, therefore solubility limited effects, i.e. formation of secondary phases, can be excluded.

Fig. 4.29 shows the leaching results of the glass UK 209 and the glass ceramic C31-3-EC. As for the SON glass the highest total mass loss resulted when leaching in bidistilled H_2O at 200°C. Furthermore, the absolute values are as high as 70 mg/cm^2 which is about two times the SON glass and the glass ceramic. Also in saturated NaCl brine the mass loss of the UK 209 glass is more than two times higher than for the SON glass. A comparison between leaching in H_2O and NaCl solution revealed a mass loss in H_2O of about 4 times higher. When leaching the glass ceramic there is no significant difference between H_2O and NaCl solution. The mass loss of 10 mg/cm^2 is lower than that for the two glasses. The interpretation of the results at this time is not completely certain due to the lack of sufficient data. But, so far it may be concluded that H_2O is by far more corrosive than NaCl at 200°C under static conditions.

4.2 CONTRIBUTION OF NUCLEAR ANALYTICAL TECHNIQUES TO THE STUDY OF AQUEOUS CORROSION OF GLASS (Saclay)

4.2.1 Introduction

It is well known that the chemical analysis of leachates gives useful information concerning the durability of glasses. The characterisation of the hydrated layer which forms at the glass surface is also important. Correlation between the two physico-chemical approaches will contribute to a better understanding of the glass corrosion process.

Nuclear techniques based on the direct observation of resonant nuclear reactions and on the backscattering of ^{4}He ions seem to be

particularly useful to measure depth concentration profiles non-
destructively in the near-surface region of hydrated glasses, with
explorable depth of about one µm.

To characterise thicker alteration layers and the underlying
pristine glass the nuclear microprobe is used. This enables
complementary information to be acquired. However, in this case,
the examinations were made on polished sections of hydrated
glasses.

The experimental methods specially developed for this purpose are
described in Sections 4.2.2 and 4.2.3. Some examples of their
application to characterise leached samples are given in Section
4.2.4. These techniques were used in an extensive study of the
behaviour of some French nuclear glasses, including SON 68 18 17,
subjected to different aqueous corrosion tests. The role of two
parameters was specifically examined: the influence of the glass
composition on the development of the hydrated layer and the
evolution of the composition of this layer during the corrosion
test.

4.2.2 Basic principle and analytical capabilities of the experimental methods

Non destructive characterization of the near surface region
of glasses

Nuclear resonant reaction [20]

The principle of this method is based on the observation of
nuclear reactions whose excitation function exhibits narrow
isolated resonance, located at specific energy E_R (Fig. 4.30). As
the incident ion energy is progressively increased above E_R, so
the resonance penetrates deeper into the target medium, depending
on the stopping power of the material for the incident ions used.

The nuclear reactions considered for the determination of H, Na

and Al depth profiles in the near surface region of glasses are
listed in Table 4.13. Their main characteristics and the
performances allowed are specified.

The choice of the experimental configuration is important when
conducting quantitative measurements if secondary effects caused
by the intrinsic insulating nature of glass are to be avoided.
Glass is also a poor thermal conductor. Thus, to ensure adequate
dissipation of the electrical charges carried by the incident
beam, the sample surface is covered with a thin molybdenum grid
(wire diameter 50 µm, spacing between wires 500 µm). It is also
necessary to limit suitably both current density and total
integrated charge according to the values given in Table 4.14.

Elastic backscattering of ^{4}He ions [3,21]

The energy (E) of the backscattering ions in the direction θ is
fully determined by the three parameters m, M and E_0 according to
the following equation:

$$E = \left[\frac{m \cos\theta + (M^2 - m^2 \sin^2\theta)^{\frac{1}{2}}}{(m + M)}\right]^2 E_0$$

For given incident ions and observation direction θ (Figure 4.31),
the value of energy E is characteristic of the target atom M.

The capability of this technique to separate neighbouring elements
increases when E_0 is higher, the incident ions are heavier, θ is
nearer 180° and the target atoms M are lighter.

The sensitivity of the process depends on the differential
scattering cross section:

$$\left(\frac{d\sigma}{d\Omega}\right)_{E_0,\theta} = \left(\frac{z\,Ze^2}{2E_0}\right)^2 \frac{1}{\sin^4\theta} \left[1 + \frac{\cos\theta}{[1 - (\frac{m}{M}\sin\theta)^2]^{\frac{1}{2}}}\right]$$

which represents the probability of elastic scattering for a projectile (atomic number:z, mass:m, energy:E_0) by a target atom (atomic number: Z, mass: M).

It is clear that the sensitivity of this method will increase when m and M are higher and E_0 and θ are lower.

The number of detected ions is given by the formula:

$$N(E) = Q \cdot \left(\frac{d\sigma}{d\Omega}\right) E_0, \theta \cdot \varepsilon \cdot n \cdot \Omega$$

with   Q   number of incident ions;
$\left(\frac{d\sigma}{d\Omega}\right) E_0, \theta$ differential cross section;
        ε detector efficiency (about 100%);
        n   superficial concentration of target atoms (at/cm^2);
        Ω solid angle subtended by the detector.

In this expression we assume that the sample is so thin that it does not induce energy loss of the projectiles. In this case the differential cross section and the energy of the scattered particles are the same for all target atoms M. The backscattering spectrum is reduced to a unique ray located at energy E.

However, if the above condition is not satisfied, the incident ions, penetrating under the target surface, lose a small part of their energy before collision. At a depth x the energy becomes (Fig. 4.31)

$$E_{0x} = E_0 - \int_0^x \frac{dE}{dx} dx$$

where $\frac{dE}{dx}$ = stopping power of the target material.

The energy of the scattered ions can be calculated from:

$$E_x = K E_{0x} - \int_0^{x/\cos(\pi-\theta)} \frac{dE}{dx} dx$$

with $$K = \left[\frac{m \cos \theta + (M^2 - m^2\sin^2\theta)^{\frac{1}{2}}}{(m + M)} \right]^2$$

This yields a continuous spectrum whose upper limit corresponds to the energy of detected ions scattered at the surface of the sample by the heavier target component M.

There is a well-known relationship between the detected energy (E_x) and the interaction depth (x), from which quantitative values of surface composition and depth profiles may be calculated. A computer programme (VERDI) was written (in FORTRAN) to determine the concentration profiles of heavy elements ($Z > 40$, especially rare earths and actinides) in the near-surface region directly from the backscattering spectrum [22]. This program builds simulated spectra, which are iteratively compared to the experimental result by successive adjustments of parameters until a good fit is reached. The explorable depth is around 10000Å and the surface resolution can be estimated around 200Å.

Influence of the composition

The depth profile measurements, performed by direct observation of nuclear reaction or Rutherford backscattering spectrometry, require a very specific knowledge of the stopping power of the target material regarding the projectiles used. This physical parameter depends on the composition of the medium; therefore large changes in the chemical composition can affect significantly the shape of the profiles directly obtained.

A second computing program, entitled ANTIGEL, was written to estimate the amplitude of the distortions induced by this effect and to assess suitable corrections of the results [23,24]. It anticipates the effect of chemical composition or density changes versus depth on the concentration profiles determined in the near surface region of hydrated glasses. Then the correction of the direct profiles gives results in agreement with the real composition of the altered glass surface.

Nuclear microprobe analysis

The nuclear microprobe is a powerful analytical tool to examine polished cross sections of thick alteration layers (> 10 µm). Its basic principle is similar to the electron microprobe. In both, the sample is bombarded by a well focused incident beam (ϕ = 2 to 3 µm)) and the interaction products are detected (Fig. 4.32). However, in the case of the nuclear microprobe, the excitation mode is not unique because several types of charged particles can be accelerated: protons, deuterons, helions 3 or 4 etc.

The analytical applications are based either on the use of nuclear reactions, induced X-ray emission, or elastic backscattering spectrometry. The observation of nuclear reactions allows the specific determination of light isotopes (^{1}H, ^{2}H, ^{3}H, ^{6}Li, ^{7}Li, ^{11}B, etc.) while the interpretation of the induced X-ray emission permits the measurement of the distribution of some heavier elements (for example Si, Ca, Mn, Fe, Ni, Zn, Zr, La, Ce, Nd, Th, U etc.).

The scanning of the target surface by the well focussed ion beam and the simultaneous detection of the interaction products are used to build distribution maps of detected elements in the explored area. Detection thresholds are in the range 10^{-16} to 10^{-18} g (0.1 to 1 µg/g). They depend essentially on the nature of the detected element and on the chemical composition of the bombarded material.

It should be noted that analytical experiments can be performed in a non vacuum atmosphere (air, nitrogen, helium, etc.) using a specific device. This gives an important advantage to the nuclear microprobe with respect to the electron microprobe.

4.2.3 Experimental devices

(a)   Leaching apparatus

Leaching and sample preparation is done in two glove boxes linked

by a transfer tunnel and maintained at a slight negative pressure
(- 20 mm water gauge). One box contains a Soxhlet column in
stainless steel, Teflon coated on its inner surfaces, and a
thermostat bath of 500 ml capacity. The second box contains a
monocular microscope (x25, x75, x140) for sample surface
examination, an electronic comparator (TESATRONIC TTD 20) with
micrometer sample holder for corroded layer thickness measurement
and a sputter coating apparatus (HUMMER VI). The sputter coater
is used to deposit thin (100 - 200Å) copper films on samples
before analysis. The film ensures surface conductivity of the
target and acts as a contamination barrier during irradiation. It
does not affect the detected signal.

(b)   Irradiation equipment

The set up includes the following main parts (Fig. 4.33):

—   the beam line equipped with an acoustical delay line, two
fast aperture valves and a vacuum system (200 l/s);

—   the irradiation chamber enclosed in a glove box
($V \simeq 0.5$ m^3);

—   a second vacuum system (200 l/s) coupled with the chamber,
installed in a second glove box ($V \simeq 1$ m^3).

During the bombardment of the samples, these two boxes are
maintained in depression (- 15 mm of water). The air inspired and
the vapours from the vacuum system are filtered and rejected out
of the laboratory.

The acoustic delay line, comprising a series of 40 aluminium
diaphragms (5 mm dia. x 2 mm thick), is located between the
deflecting magnet of the 2 MV Van der Graaff accelerator and the
irradiation chamber. Its purpose is to prevent contamination of
the accelerator in the event of an air leak in the equipment.

A fast (200 ms) aperture valve is fitted near the deflecting
magnet. This valve is controlled by two gauges, one in the
irradiation chamber, the other between the chamber and the delay
line. The pressure threshold is set at 5×10^{-5} torr. A second
fast aperture valve (800 ms) located near the target chamber is
closed during the sample positioning in the chamber to avoid the
alteration of the vacuum into the beam line.

The cylindrical irradiation chamber (diameter: 300 mm - height:
200 mm) contains two pyrex observation windows (diameter: 300 and
100 mm respectively), and an eight-target sample holder. The
chamber is equipped with a surface barrier detector (active area:
25 mm^2 - sensitive depth: 100 µm - intrinsic resolution: 13 keV)
whose orientation θ regarding the incident beam direction can vary
between 130 and 170°. The distance between the target and the
detector is in the range 50 - 120 mm; its detection area is
limited by a tantalum collimator (diameter of the aperture: 1 mm)
to reduce pulse pile up.

Another surface barrier detector (active area: 500 mm^2, sensitive
depth: 100 µm, intrinsic resolution: 17 keV) is located in the
chamber where it is impossible to receive backscattered ions. This
detector which can be decontaminated, is used to control the
pollution level of alpha emitters (Pu, Am, etc.).

Finally, the chamber lid can carry a gamma detector (NaI (Tl) or
Ge HP), to record prompt gamma rays emitted during nuclear
reactions and which can be used to determine concentration
profiles of sodium and aluminium.

The irradiation area is defined by a graphite diaphragm (diameter:
5 mm) inserted in the beam line near the chamber.

Thirty minutes are needed to obtain a good vacuum in the whole
device (10^{-5} torr); extending the pumping time, a vacuum of around

5.10^{-7} torr can easily be reached.

(c) Nuclear microprobe

The instrument is installed near a 4 MV Van de Graaff located in the Research Centre of Bruyeres le Chatel [25].

After the production and the acceleration steps the incident charged particles pass through several devices dispersed along the beam line to obtain a final focused spot near the 'object aperture box' of the microprobe. The microfocusing of the beam on the target is achieved by four electromagnetic quadrupole lenses.

An electrical deflection device comprising two pairs of metallic plates suitably polarized by four HT power supplies controlled from the acquisition system, ensures the scanning of the target surface ($\simeq 1$ mm^2) and enables the construction of element distribution maps.

The target chamber, made of stainless steel (diameter: 30 cm - height: 35 cm), is equipped with the following set up:

- a micromanipulator (XYZ translator and rotary drive system) for sample changing and positioning along the beam axis;

- a sample holder with 5 positions;

- a binocular microscope (X 200 and X 400) for the final focusing of the beam and the observation of the sample;

- a Si(Li) detector for X-ray analysis;

- surface barrier detectors used to collect charged particles emitted during nuclear reactions or scattered by elastic collision with the components of the target.

A TN 4000 computer-based (LSI - 11/23) multichannel analyser digitally controls the beam position and processes the signals detected during the projectile-target interaction.

The 'imaging' subroutine is able to reconstitute either multicoloured elemental distribution maps or pictures in which each element is represented by a specific colour.

4.2.4 Comparative Study of the Hydrated Layer of Some Nuclear Glasses

Five different glass compositions, studied in the frame of the French high level waste solidification programme [26,27] and two commercial glasses were leached according to the following scheme:

- temperature: 90°C;
- duration of the test from 7 days to 12 months;
- bidistilled water (initial pH = 6.7 at 22°C);
- static mode with $S/V = 0.5$ cm^{-1};
- container and sample holder made of teflon.

The glass compositions are given in Table 4.15.

In addition, a glass sample (composition SON 68 18 17) was leached in dynamic conditions (Soxhlet test 100°C) for 28 days and specially prepared for nuclear microprobe examination.

All the leaching tests were conducted by the High Level Waste Service (DRDD/SDHA/SEMC) in Marcoule. Leachate analysis, weight loss measurements, pH assessment and preparation of the samples were also performed in Marcoule. The leaching procedure is schematically described in Fig. 4.34: teflon containers and an oven were mainly used.

Each glass block (25 mm x 25 mm x 7 mm) was polished before leaching with SiC 600 paper on each face. For each glass the test was duplicated to supply one sample for 'macro' nuclear technique analysis and another for nuclear microprobe examination. Leached samples were air dried and weighed.

The pH of the leachates was determined at 90°C and also at room temperature (22°C). These solutions were then acidified and analysed by Atomic Absorption and ICP spectroscopy.

Leached samples for the nuclear microprobe experiments were prepared according to the following scheme:

- embedding in an epoxy resin (room temperature and atmospheric pressure);

- sawing under flowing ethylene glycol with a slow diamond saw;

- polishing using SiC 400 and 800, and thin diamond paste of grain size from 1 to 8 µm.

Direct observation of the resonant reaction $^{23}Np(p,\alpha\gamma)^{20}Ne$ in the energy range 1000 - 1080 keV and Rutherford backscattering spectrometry ($^{4}He^{+}$ ions - Eo = 1.5 MeV - θ = 160°) were employed to determine the concentration depth profile of sodium (explorable depth 1.5 µm - surface resolution 500Å) and both rare earth and actinide profiles (analysed depth 1 µm - surface resolution 200Å) in the hydrated layer developed at the glass surface.

Nuclear microprobe analysis combined with X-ray spectrometry (proton energy - 2 MeV, beam intensity < 1 nA, microbeam diameter ≃ 3 µm) enables the elemental distribution of most of the glass matrix components (Na, Si, Ca, Fe, Mn, Ni, Zn, Zr, La, Nd, Th and U) in both the hydrated layer and the underlying pristine glass to be determined.

Results

Table 4.16 shows the elemental analysis, mass losses and final pH values for the 3 months at 90° tests.

Figure 4.35 indicates the 'equivalent altered glass thicknesses' for the same tests. The equivalent altered glass thickness corresponds to the thickness of glass which would need to be dissolved to find in the leachate the concentration C [28].

$$E_i = \frac{C_i}{\rho\, p_i\, S/V}$$

where ρ = glass density
p_i = weight fraction of element i
C_i = concentration in leachate.

If an element is totally absent from the hydrated layer, its equivalent thickness is then equal to the altered glass thickness.

The data concerning the study of the evolution of the corroded layer of the SON 68 18 17 glass versus the leaching duration (7 days up to 12 months) are reported in Tables 4.17, 4.18 and 4.19 (leachate analysis, equivalent thickness, pH).

Figs. 4.36 to 4.38 show the sodium concentration depth profile determined by displacement of the 1012 keV resonance of the nuclear reaction $^{23}Na(p,\alpha\gamma)^{20}Ne$, in the near surface region of the leached glasses. In Fig. 4.37 the seven glass compositions are compared. Figs. 4.38 and 4.39 illustrate the progressive sodium depletion in SON 68 18 17 induced by aqueous corrosion.

Similarly, Figs. 4.39 to 4.42 show the progressive changes in the rare earth and actinide concentration profiles. Similar measurements were performed on the six other glass compositions [29,30].

All these profiles have been corrected to take into account the distortions caused by the variation of the depth composition, according to the procedure developed in 'ANTIGEL' computing program.

Finally, many elemental distribution measurements were conducted on the transverse cross sections of the seven leached glass composition using the nuclear microprobe [26,27]. However, taking into account the very thin hydrated layer induced by 90°C static tests and because of problems encountered during the sample preparation, most of these results are not viable.

To illustrate the analytical capabilities of the nuclear microprobe, some results obtained, from examination of the SON 68 18 17 sample leached in a Soxhlet column for 28 days, are given in Figs. 4.43 to 4.49 [31]. The upper pictures show elemental distribution maps (shape 60 μm x 60 μm), below these concentration profiles determined along the horizontal scanning axis at the level marked with an arrow on the maps are given.

Lithium and boron distribution and profiles in the same sample have been recorded elsewhere, preliminary results clearly show the strong depletion of these two light and mobile elements in the hydrated layer, particularly for the Li [32].

Discussion

The experimental results presented here (leachate analysis and surface characterizations) enable a classification of the seven glass composition studied to be made in increasing order of chemical durability:

> SON 64 < WINDOW < FRITTE SON 68 < SAN 55
> < ASA < SON 68 < PYREX

This classification confirms the excellent behaviour of the French reference glass SON 68 18 17. In the case of this composition, the glass matrix components fall into two distinct groups:

— elements which are noticeably released: Na, Si, Ca, Sr, Mo, Cs;

- the elements which contribute to the development of the hydrated layer: Al, Mn, Fe, Ni, Zn, Zr, La, Ce, Nd, Th, U.

Finally, the maps and concentration profiles (Figs. 4.43 to 4.49) of the glass sample leached for 28 days at 100°C in a Soxhlet column (dynamic mode) exhibit several different regions in the hydrated layer. This experimental result was first described by MALOW [4] for another glass composition leached under different conditions. Some mobile elements (Na - Si - Ca) appear to concentrate at the interface between the hydrated layer and the pristine glass, which might indicate the formation of a diffusion barrier according to KENNA's observations [33].

This sample was also examined using electron microprobe analysis [34]. The depletion of Na, Ca, Si and the enrichment of Fe, Zr and rare earths in the hydrated layer were also shown. Electron microprobe demonstrates strong iron enrichment in the external part of the hydrated layer according to Fig. 4.46. However, enrichment of some mobile components at the hydrated layer – pristine glass interface was not found which seems very surprising. Therefore some complementary experiments are required to confirm this structure.

4.2.5 Contribution to the Study of the Formation Mechanism of the Hydrated Layer Developed at the Glass Surface

The results published by GRAMBOW in 1982 [14] showed that the formation of metallic hydroxides like $Fe(OH)_3$, $Zn(OH)_2$, $Nd(OH)_3$ and $Ce(OH)_4$ seemed to control the release of these elements during the glass corrosion process. The chemical equilibrium representing these reactions can be expressed by the following equation:

$$M^{n+} + nH_2O \underset{K'}{\overset{K_H}{\rightleftarrows}} M(OH)_n + nH^+$$

Assuming that ionic activities are equal to concentrations, the following relationship between the hydrolysis constant pK_H, the pH of the aqueous medium and the concentrations, is satisfied:

$$pK_H = npH + \log (M^{n+})$$

Moreover, AHRLAND et al. [35] have shown that ionic exchange between silica gel and metallic ion solutions occurs. STANTON et al. established the mathematical formation to describe this chemisorption phenomenon for cations M_{n+} with $1 < n < 4$ [36].

Taking the acid-base equilibrium of the silanol group:

$$SiOH \underset{K_a}{\rightleftharpoons} SiO^- + H^+$$

with
$$K_a = \frac{(H^+)(SiO^-)}{(SiOH)}$$

and choosing a parameter θ, where

$$\theta = \frac{(SiOH)}{(SiO^-)(SiOH)}$$

the chemisorption equilibrium

$$M^{n+} + m(SiOH) \underset{}{\overset{Kc}{\rightleftharpoons}} M(SiO)_m^{(n-m)+} + mH^+$$

is described by the relationship

$$pKc = \left(\frac{1-\theta}{m\theta^m}\right) \left(\theta + \frac{1-\theta}{m}\right)^{m-1} \frac{(H^+)^n}{(M^+)^n}$$

Generally $n = m$ and this equation becomes:
$$pKc = npH + \log (M^{n+}) - A(\theta,n)$$

with
$$A(\theta,n) = \log \frac{1-\theta}{n\theta^n} \left(\theta + \frac{1-\theta}{n}\right)^{n-1}$$

The analytical results for leaching tests covering a very large range of variation for t, T, pH and nature of the leaching solution, S/V ratio and glass composition, were used to assess both hydrolysis and chemisorption constants defined here above in the case of following metallic cations M^{n+} ($1 \leq n \leq 4$): Li^+, Na^+, Cs^+, Zn^{2+}, Sr^{2+}, Ba^{2+} UO_2^{2+}, Mn^{2+}, Cr^{2+}, Co^{2+}, Fe^{2+}, Al^{3+}, Fe^{3+}, Co^{3+}, Cr^{3+}, Mn^{3+}, La^{3+}, Ce^{4+}, Zr^{4+}, Mn^{4+}, Th^{4+}.

The details of these calculations are fully described elsewhere [37,38]. The results obtained for the values of pK_H and pK_C lead to the following conclusions:

1 - for n = 1, in other words for the alkali cations (Li^+, Na^+, Cs^+) the values of the hydrolysis and of the chemisorption constants do not explain the release of these elements. It is well known that the behaviour of the alkali ions during glass leaching is governed by the leachant pH;

2 - for n = 2, especially with Zn^{2+} and Zr^{2+}, hydrolysis is more probable than chemisorption in contrast to Cr^{2+}, Mn^{2+}, Co^{2+}, Fe^{2+}, Ba^{2+} and UO_2^{2+};

3 - for trivalent cations mainly Cr^{3+} and Mn^{3+} there is a competition between hydrolysis and chemisorption; for Al^{3+}, Co^{3+}, Fe^{3+}, La^{3+}, Ce^{3+} chemisorption seems more probable;

4 - Finally, when n = 4 the ionic species Mn^{4+}, Ce^{4+}, Zr^{4+}, obey the chemisorption process.

4.2.5 Conclusions

The surface examination performed using the specially developed nuclear techniques described and analysis of the leachates showed that the glass composition SON 68 18 17 is particularly resistant to aqueous corrosion compared with some other nuclear glass compositions.

The hydrated layer developed at the SON 68 glass surface as a consequence of the leaching process reaches only a few micrometres even after corrosion duration up to twelve months. Nuclear microprobe analysis of a sample leached in a Soxhlet column for 28 days reveals a corroded layer with a thickness around 15 μm.

The results obtained clearly demonstrate the differences in the behaviour of mobile elements (Li, B, Na, Si, Mo, etc.) and other metallic glass components (Fe, Zn, Ni, Zr, rare earths and actinides) enriched in the hydrated layer.

In addition, regarding the release mechanisms of the glass constituents in the leaching solutions, hydrolysis and chemisorption assumptions were studied using numerous leachate analysis results from a very wide range of variation in the main leaching parameters (t, T, S/V, glass composition, pH and nature of the leachant). Chemisorption of the metallic ions by the silanol groups is more probable when the charge state of the cation is higher than 2.

An experimental study of thermal diffusion of actinides in SON 68 18 17 glass is described in Section 4.3. Additional experiments are in progress.

The analyses performed using the nuclear microprobe, which were the first of their kind undertaken with this new method, show the very powerful capabilities of this instrument to characterise thick hydrated layer and the underlying pristine glass. Using a specific suitable experimental device, this apparatus makes possible the examination of highly radioactive glasses, ceramics, metals, rocks etc. at atmospheric pressure, in air or other gas. As a result the analytical observations can be performed behind a thin water film so that the deleterious effects of dehydration (cracking, swelling etc.) can be avoided. The 'atmospheric pressure' operation greatly reduces the experimental difficulties in the case of highly radioactive material.

So, the access to such an analytical tool to characterise and control candidate materials for the immobilization of radioactive waste would hopefully be of interest.

TABLE 4.13 Main characteristics of the nuclear reactions used to determine hydrogen, sodium and aluminium depth profiles

Profiled element	H	Na	Al
Nuclear reaction	$^{1}H(^{15}N,\alpha\gamma)^{12}C$	$^{23}Na(p,\alpha\gamma)^{20}Ne$	$^{27}Al(p,\gamma)^{28}Si$
Resonance Energy E_R (MeV)	6.385	1.012	0.992
Energy of the γ ray emitted (MeV)	4.439	1.634	1.778
Half width of the resonance (keV)	13	3	0.1
Resonance Cross Section (mb)	200	45	900
Energy range (MeV)	6.385 à 10	1.012 à 1.080	0.992 à 1.010
Explorable depth (μm)	2	1.3	0.4
Depth resolution ($\overset{\circ}{A}$) $x = 0$	60	400	200
Depth resolution ($\overset{\circ}{A}$) $x = 1\,\mu m$	500	2 000	-
Detection Limits ponderal ($\mu g/g$)	200	2 000	4 000
Detection Limits superficial (g/cm^2)*	3.10^{-5}	3.10^{-3}	$1.5.10^{-4}$

* For a layer with a thickness of the same order of magnitude than the resonance width.

TABLE 4.14   Experimental conditions required to determine the concentration depth profiles of H, Na and Al at the surface of leached glasses

Element	Maximum incident current density* ($\mu A/cm^2$)	Total integrated charge* (μC)
H	0.2 **	200
Na	0.5 ***	1 200
Al	2 ***	2 000

*   *without apparent damage.*

**   *for an irradiated area about 0.1 cm^2.*

*** *for an irradiated area about 0.2 cm^2.*

TABLE 4.15   Compositions of the glasses studied*

OXYDE	Concentration (weight %)						
	SAN 55 20 20	SON 64 19 20	SON 68 18 17	ASA 15-7	PYREX	WINDOW	FRITTE SON 6
SiO_2	39	47.6	45.48	67	78.78	70.89	58.84
Al_2O_3	13.8	-	4.91	4.9	1.79	0.4	4.28
B_2O_3	17.3	18.6	14.02	-	13.16	-	18.15
Na_2O	17.3	12.4	9.86	15.9	3.49	12.59	7
CaO	-	-	4.04	-	-	8.06	5.23
MgO	5	-	-	-	-	3.93	-
Li_2O	-	-	1.98	-	-	-	2.56
Fe_2O_3	2.6	6.4	2.91	6.20	-	0.09	-
ZnO	-	-	2.50	-	-	-	3.24
MnO_2	0.23	0.67	0.72	0.27	-	-	-
NiO	0.12	0.36	0.74	0.14	-	-	-
SrO	0.15	0.44	0.33	0.18	-	-	-
ZrO_2	0.69	2.09	2.65	0.84	-	-	0.7
MoO_3	0.81	2.45	1.70	0.98	-	-	-
Nd_2O_3	0.59	1.78	1.59	0.71	-	-	-
La_2O_3	-	-	0.90	-	-	-	-
Ce_2O_3	-	-	0.93	-	-	-	-
Cs_2O	0.4	1.18	1.42	0.48	-	-	-
UO_2	-	-	0.52	-	-	-	-
ThO_2	0.31	0.91	0.33	0.37	-	-	-
K_2O	-	-	-	-	2.25	0.2	-

* Among the other oxides present in the composition of these glasses, there are :

Cr_2O_3 - CoO - P_2O_5 - Y_2O_3 - Ag_2O - BaO - SnO_2 - CdO_2 - Sb_2O_3 - TeO_2 - Rb_2O.

TABLE 4.16 Analysis of the leachates from the 3 month tests at 90°C

	Elementary concentration in the leachate.						
	SAN 55	SON 64	SON 68	ASA	PYREX	WINDOW	FRITTE
Si	34.8	194.8	32.8	32.8	8.0	104.8	79.8
Al	1.25	-	0.27	4.43	0.12	0.43	0.88
B	12.2	117	7.45	-	2.15	-	25.45
Na	35.3	204.8	15	20.3	1.5	43.8	22.3
Zr	0.14	0.05	0.003	0.19	0.001	-	0.06
Fe	1,25	0.75	0.1	2.65	-	-	-
Mo	1	1	1.85	0.65	-	-	-
Sr	0.08	0.04	0.26	0.2	-	-	-
Cs	0.65	6.35	2.45	0.42	-	-	-
Th	0.16	0.002	0.002	0.26	-	-	-
U	-	-	0.09	-	-	-	-
Ce	-	-	0.005	-	-	-	-
Zn	-	-	0.23	-	-	-	1.77
Li	-	-	1.9	-	-	-	5.6
Mg	2,95	-	-	-	-	1.62	-
Mass loss (mg)	8	35.8	4.7	4.6	1.7	13.4	11.2
pH (90° C)	8.42	8.81	8.48	8.55	6.26	9.14	8.70

TABLE 4.17 Analysis of the leachates from glass SON 68 18 17: Corrosion duration 7 days to 12 months

	Elementary concentration in the leachate.							
	7 days	14 days	1 month	2 months	3 months	6 months	9 months	12 months
Si	17.8	20.8	20.8	29.8	32.8	27.3	23.0	50.35
Al	1.4	1.56	1.69	2.29	1.43	1.78	2.02	0.87
B	4.62	5.52	5.72	10.92	7.5	6.22	4.80	34.9
Na	6.09	7.29	7.99	14.89	15.0	11.19	12.0	39.9
U	0.035	0.050	0.055	0.065	0.10	0.095	0.09	0.22
Sr	0.15	0.17	0.15	0.11	0.26	0.18	0.121	0.107
Mo	0.65	0.87	0.87	1.28	1.85	1.05	1.1	6.75
Zn	0.07	0.09	0.17	0.25	0.23	0.33	0.25	1.15
Fe	*	*	0.15	0.20	0.05	0.12	0.17	0.40
Li	0.85	1	1.1	1.9	1.9	1.5	1.5	6.1
Cs	1.2	1.4	1.6	2.4	2.5	1.8	1.9	4.5
Mass loss (mg)	1.6	3.1	2.6	3.4	4.7	3.1	3.1	6.0

*   The obtained value is of the same order of magnitude than the detection limit allowed by ICP for this element : 5×10^{-3} mg/l.

** For Th C < 0.002 mg/l ⟶ 9 months and < 0.005 for 12 months.
     Ce C < 0.005 mg/l ⟶ 6 months and < 0.010 for 9 months and 12 months.
     Zr C < 0.008 mg/l ⟶ 9 months and = 0.025 for 12 months.

TABLE 4.18   Equivalent altered glass thicknesses for glass SON 18 17 leached for 7 days to 12 months

	Equivalent tickness (μm)							
	7 days	14 days	1 month	2 months	3 months	6 months	9 months	12 months
Si	0.6	0.8	0.8	1.0	1.1	0.9	0.7	1.6
Al	0.4	0.5	0.5	0.7	0.4	0.5	0.5	0.2
B	0.8	1.0	1.0	1.9	1.3	1.0	0.7	5.4
Na	0.6	0.8	0.8	1.5	1.5	1.1	1.1	3.7
Zr	< 0.05	< 0.05	< 0.05	< 0.05	< 0.05	< 0.05	< 0.05	< 0.05
U	0.1	0.1	0.1	0.1	0.2	0.1	0.1	0.3
Th	< 0.05	< 0.05	< 0.05	< 0.05	< 0.05	< 0.05	< 0.05	< 0.05
Ce	< 0.05	< 0.05	< 0.05	< 0.05	< 0.05	< 0.05	< 0.05	< 0.05
Sr	0.4	0.5	0.4	0.3	0.7	0.5	0.3	0.3
Mo	0.4	0.6	0.6	0.8	1.2	0.7	0.7	4.0
Zn	0.03	0.03	0.06	0.09	0.09	0.12	0.08	0.39
Fe	u	u	0.06	0.07	< 0.05	0.04	0.06	0.13
Li	0.7	0.8	0.9	1.5	1.5	1.2	1.1	4.5
Cs	0.7	0.8	0.9	1.3	1.4	1.0	1.0	2.3
Mass loss	0.3	0.6	0.5	0.7	0.9	0.6	0.6	1.1

u : undetermined

TABLE 4.19   Evolution of the pH value vs leaching time for glass SON 68 18 17

Leach duration	pH at 90° C	pH at 22° C
7 days	8.15	8.96
14 days	8.15	8.93
1 month	8.24	8.90
2 months	8.40	8.96
3 months	8.48	9.05
6 months	8.10	8.49
9 months	8.57	8.59
12 months	8.40	8.90

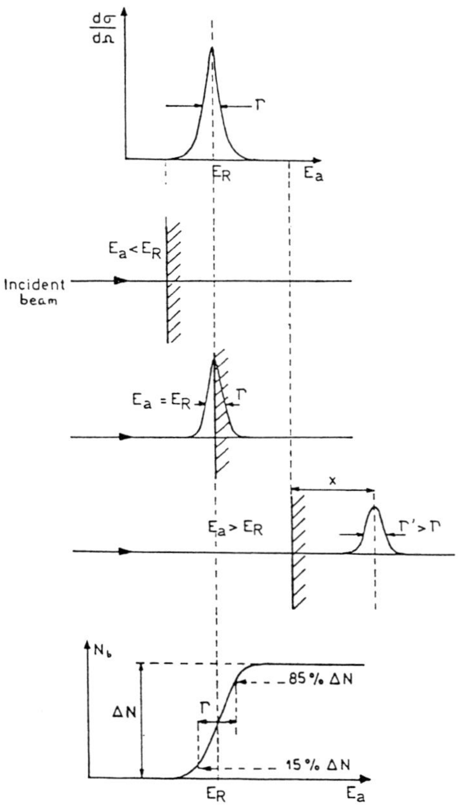

Fig. 4.30 Basic principle of concentration depth profile measurement using resonant nuclear reaction

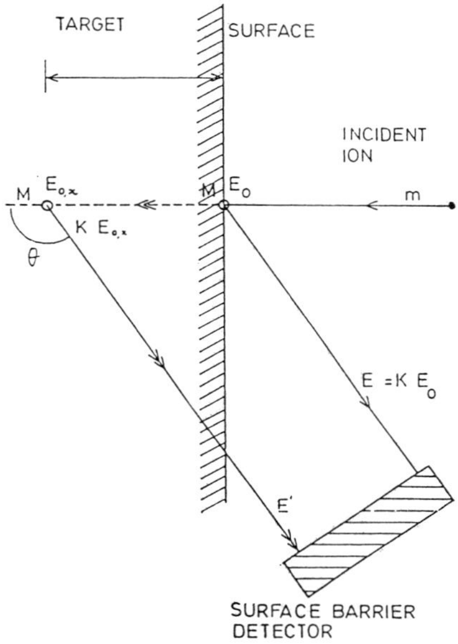

Fig. 4.31 Basic principle of surface analysis using Rutherford backscattering spectrometry

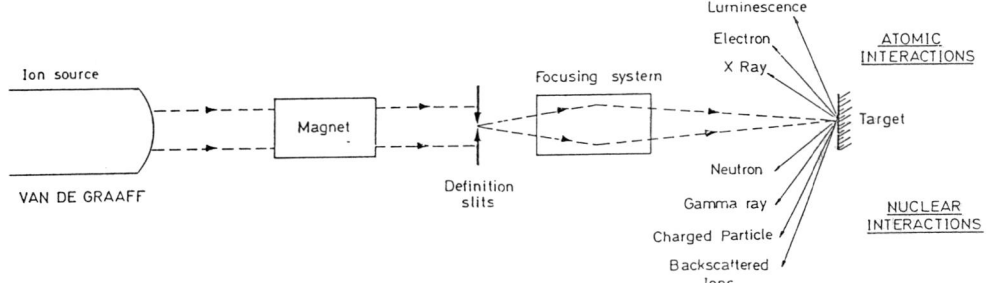

Fig. 4.32 Schematic diagram of the nuclear microprobe system

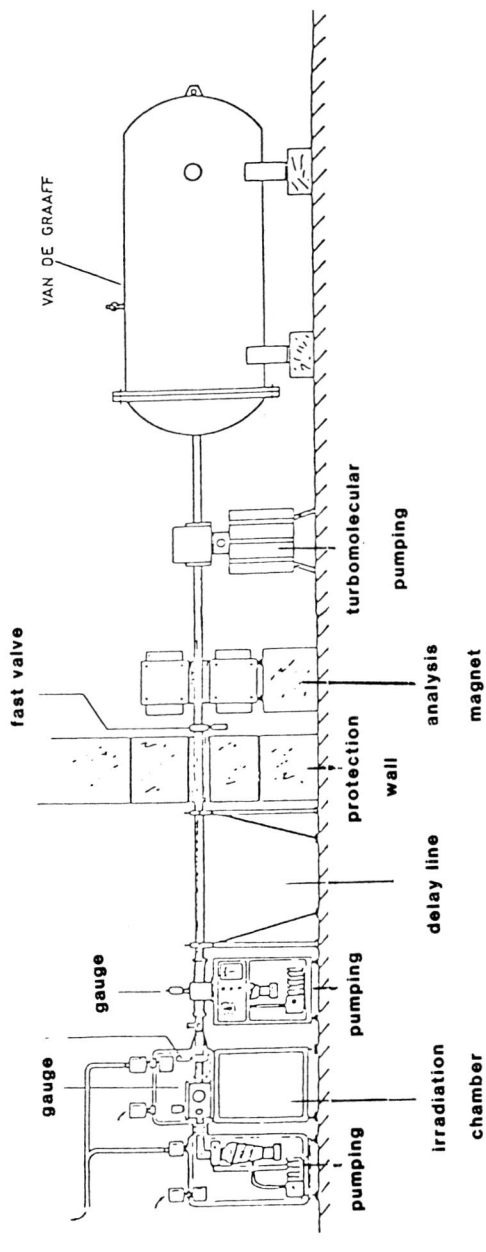

Fig. 4.33 Apparatus for the examination of the near surface region of glasses containing actinides using resonant nuclear reaction analysis and Rutherford backscattering spectrometry

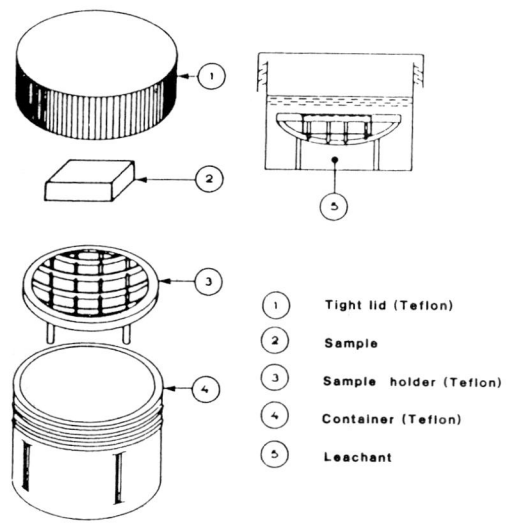

Fig. 4.34 Leaching vessel for the 90°C tests

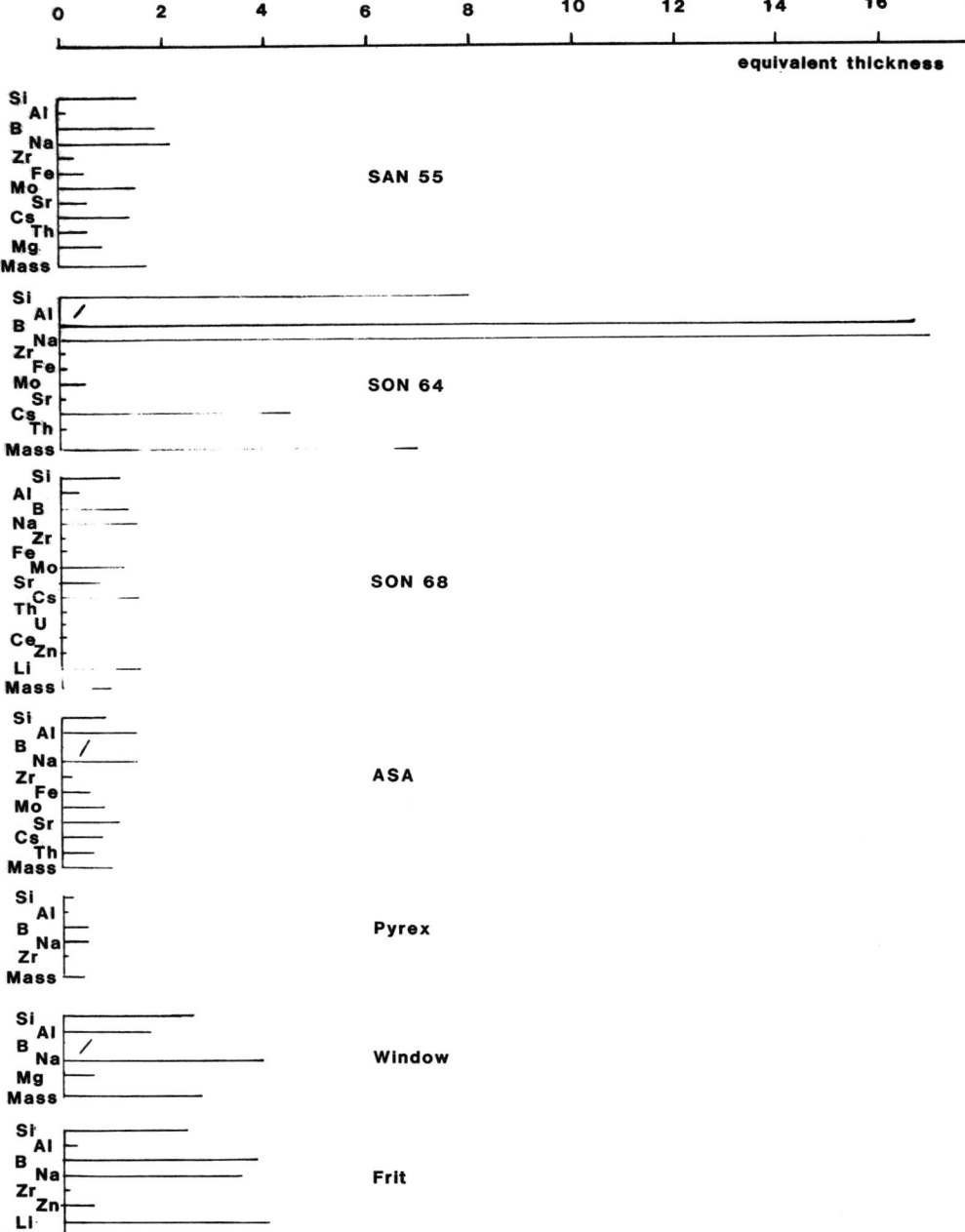

Fig. 4.35 Equivalent thickness results for three month tests and 90°C for seven glasses

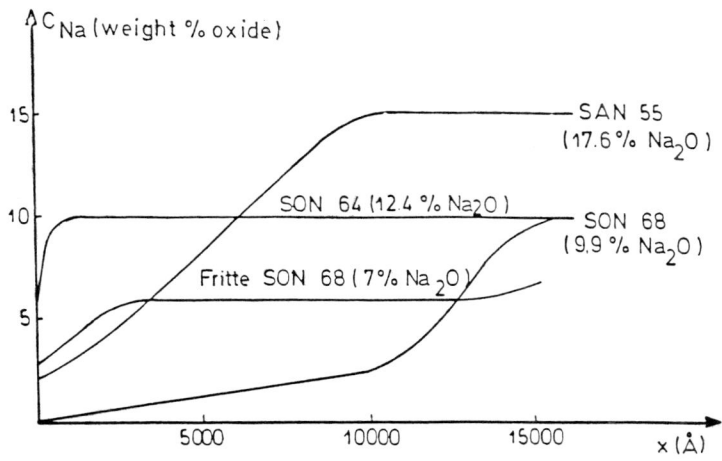

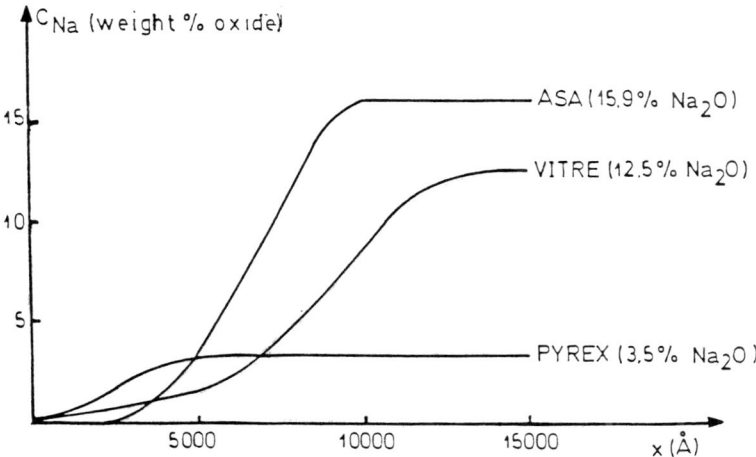

Fig. 4.36 Sodium concentration depth profiles in the near surface region after 3 months leaching at 90°C for the seven glasses

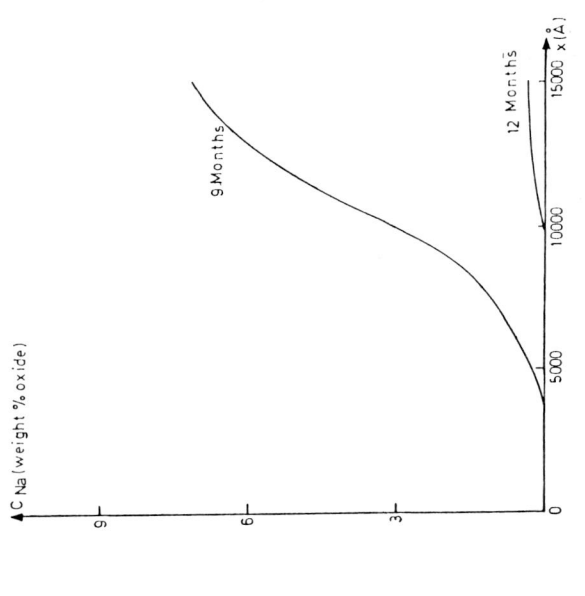

Fig. 4.38 Sodium concentration depth profiles for SON 68 leached at 90°C for 9 and 12 months

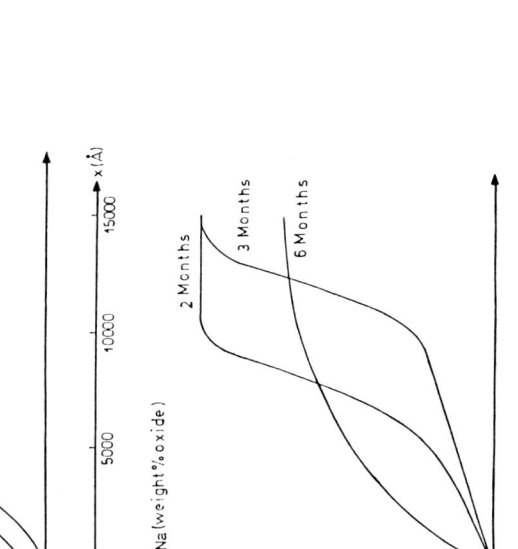

Fig. 4.37 Sodium concentration depth profiles for SON 68 leached at 90°C from 7 days to 6 months

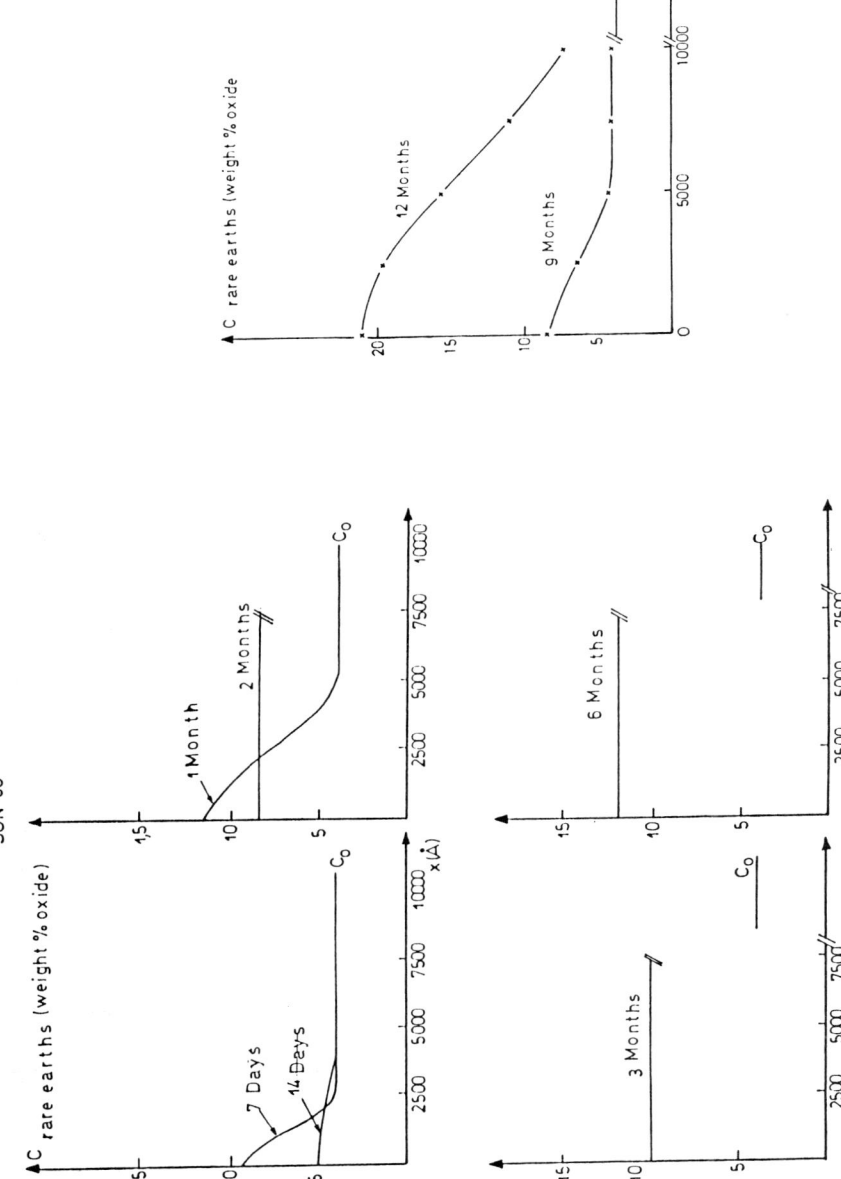

Fig. 4.39 Rare earth concentration depth profiles for SON 68 leached at 90°C from 7 days to 6 months (C_o = concentration in unleached glass)

Fig. 4.40 Rare earth concentration depth profiles for SON 68 leached at 90°C for 9 and 12 months (C_o = concentration in unleached glass)

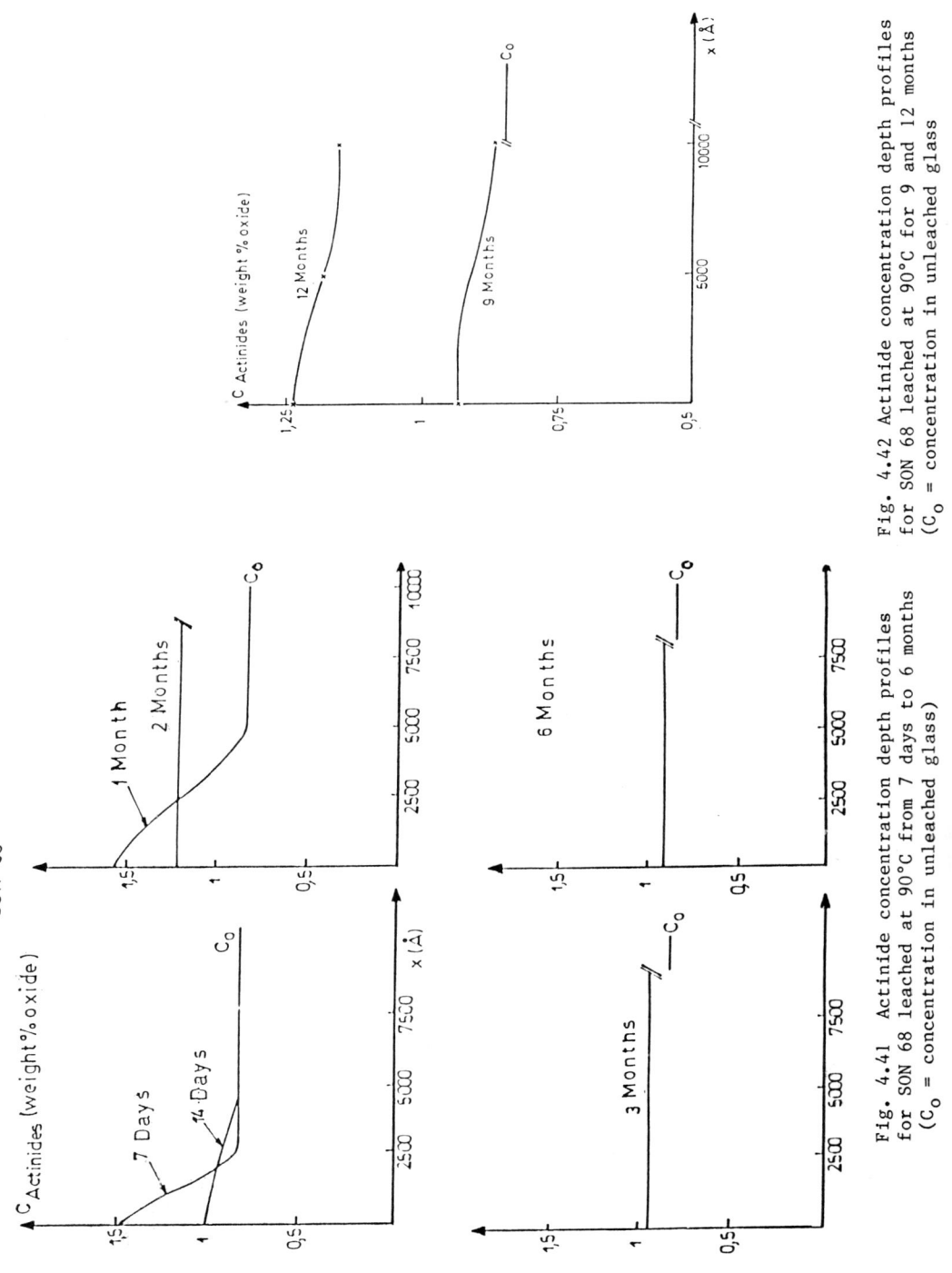

Fig. 4.41 Actinide concentration depth profiles for SON 68 leached at 90°C from 7 days to 6 months (C_o = concentration in unleached glass)

Fig. 4.42 Actinide concentration depth profiles for SON 68 leached at 90°C for 9 and 12 months (C_o = concentration in unleached glass)

Figs. 4.43 to 4.49 Maps of the element redistribution in the near surface region of SON 68 after 28 day's 100°C Soxhlet leaching

Fig. 4.43   Na
Fig. 4.44   Si
Fig. 4.45   Ca
Fig. 4.46   Fe
Fig. 4.47   La
Fig. 4.48   Nd
Fig. 4.49   Th + U

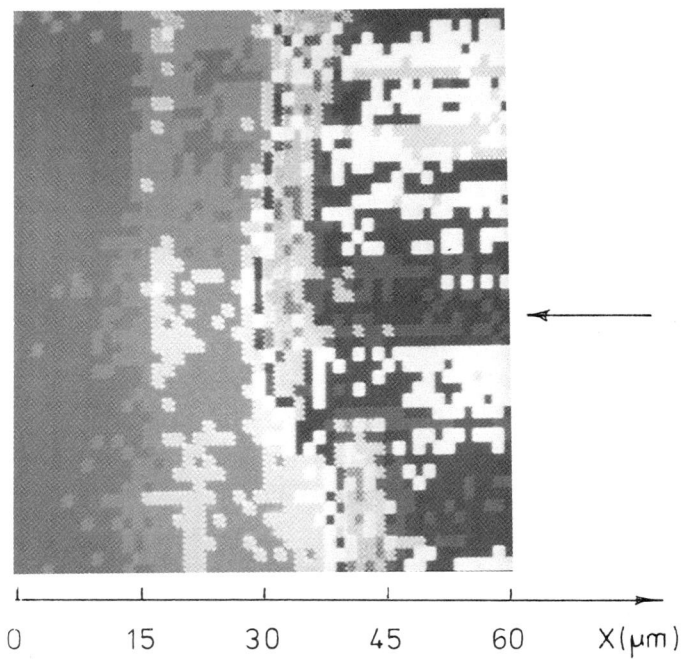

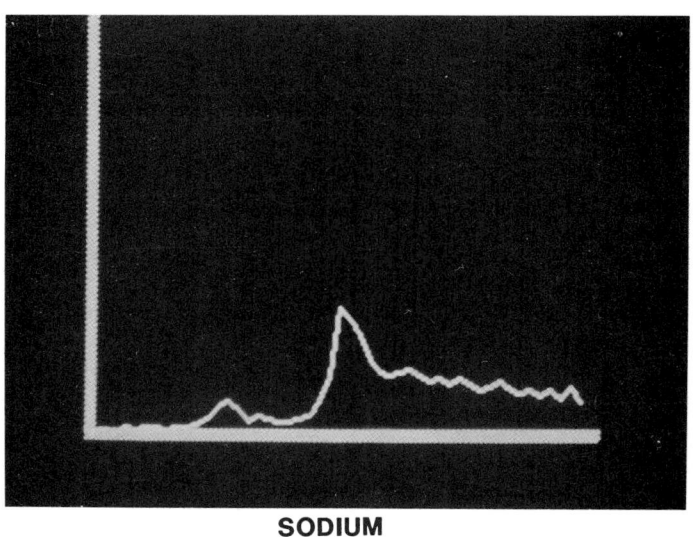

SODIUM

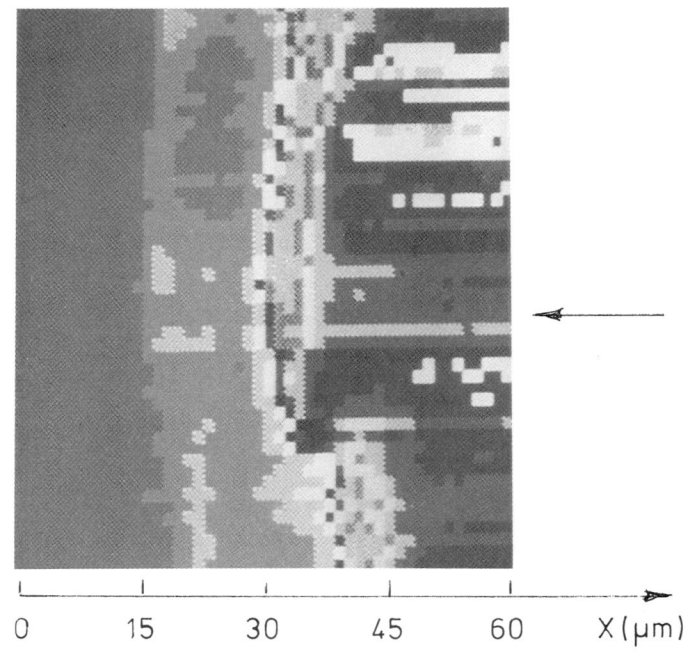

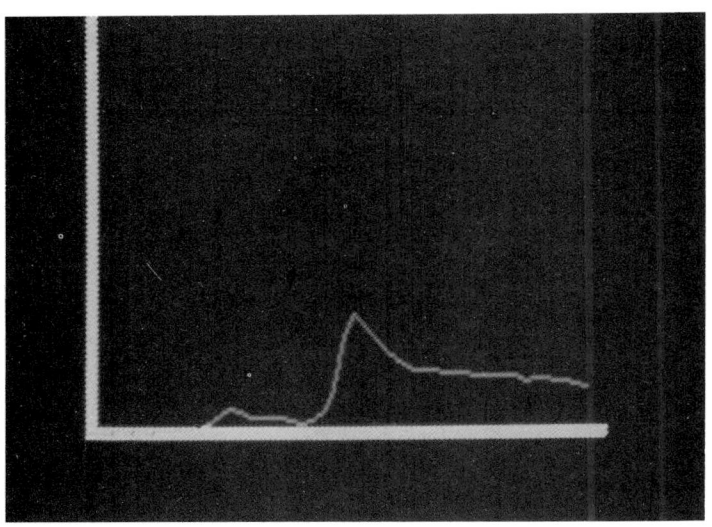

SILICON

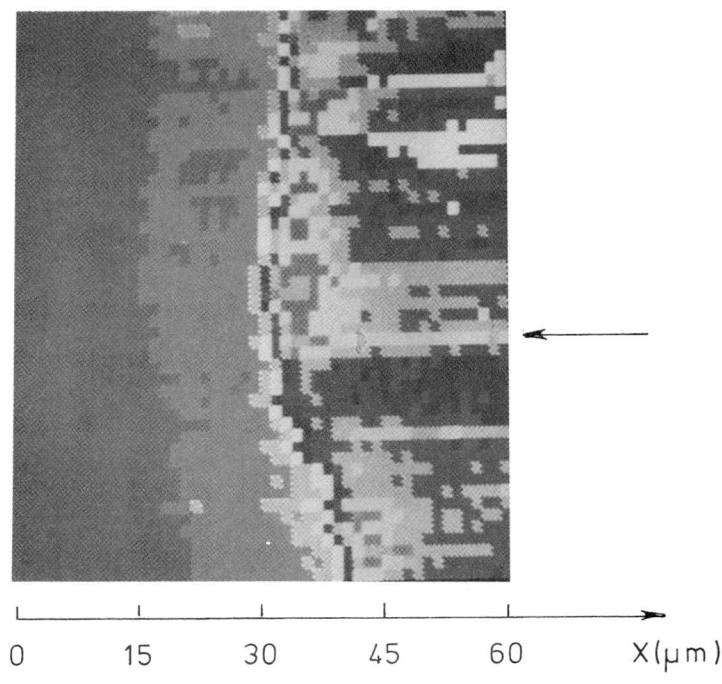

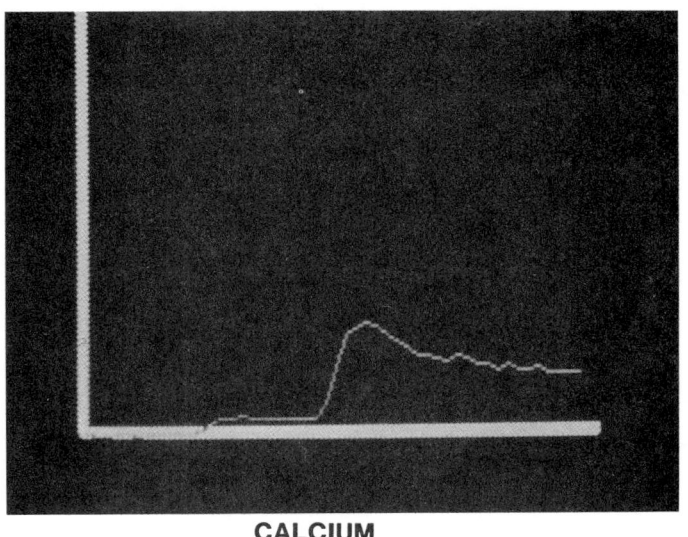

CALCIUM

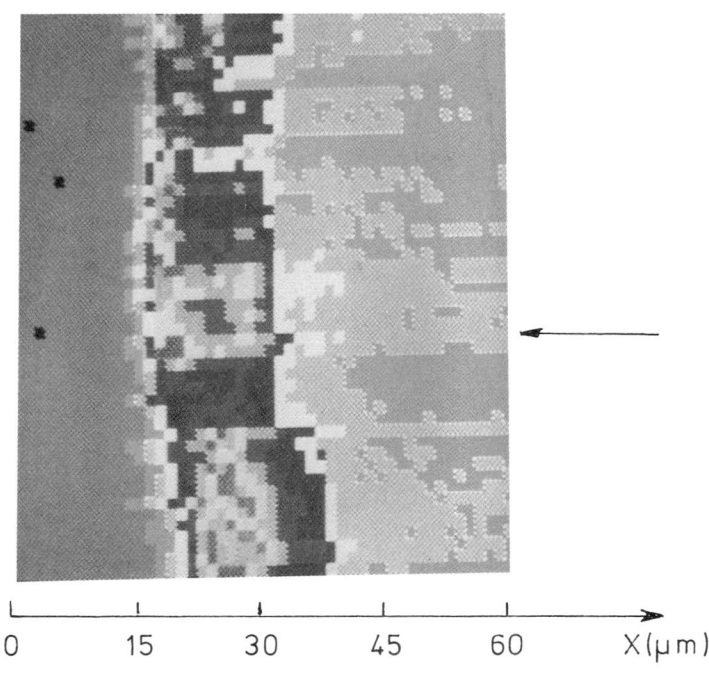

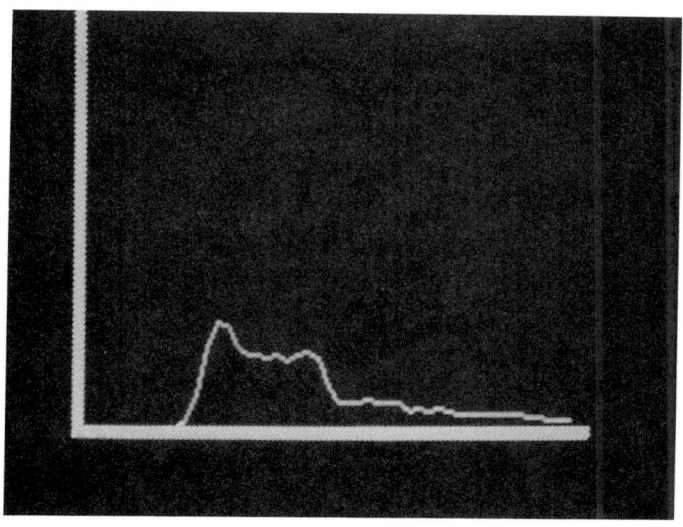

IRON

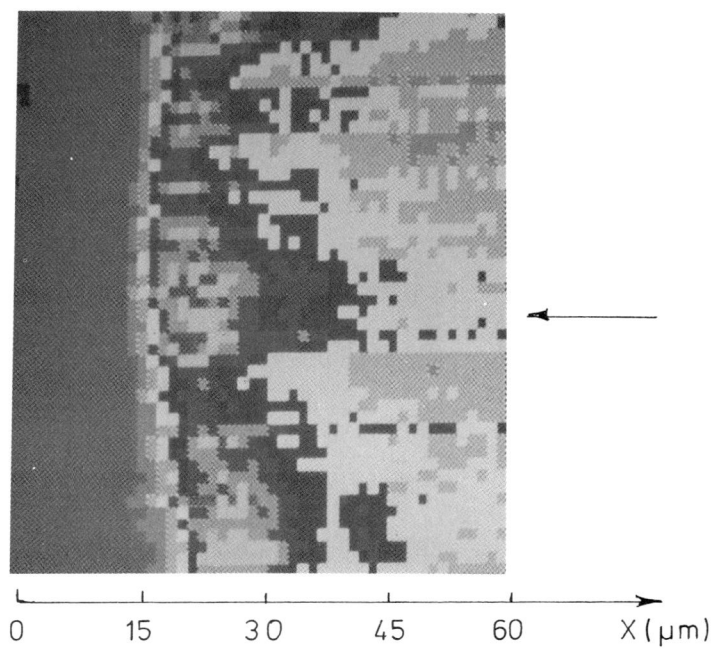

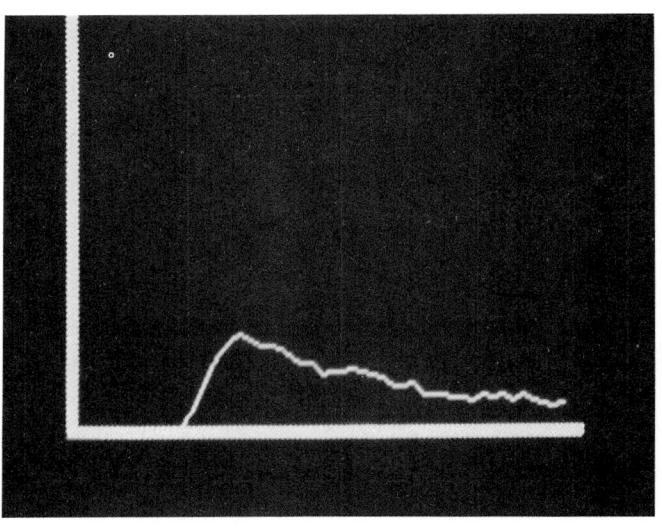

LANTHANUM

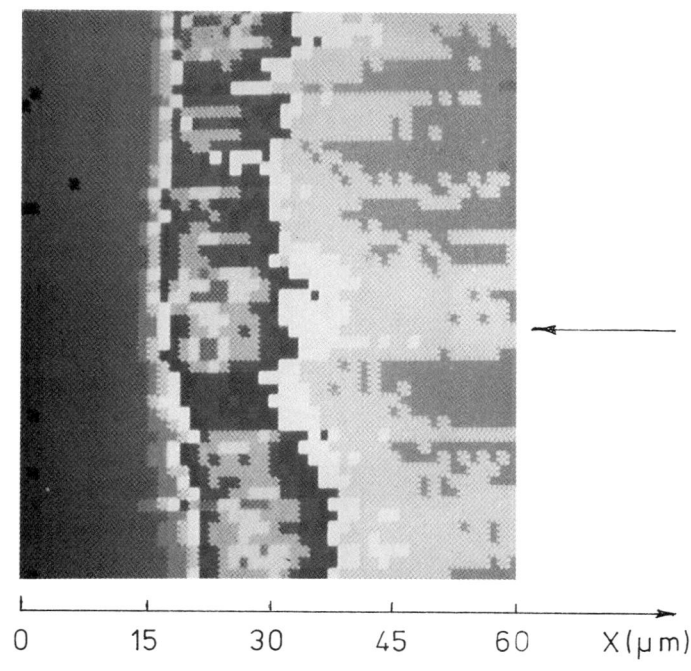

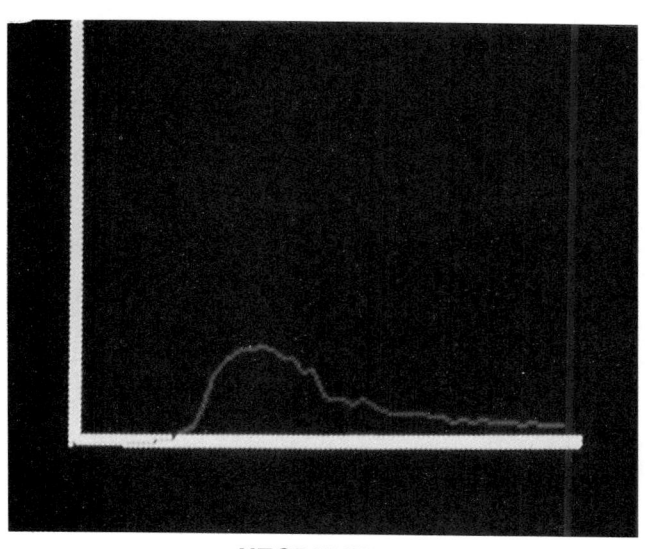

NEODYMIUM

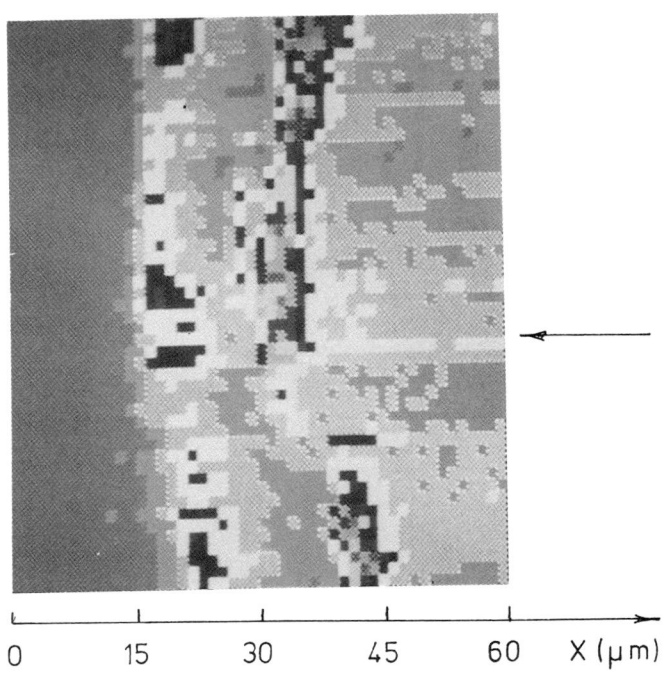

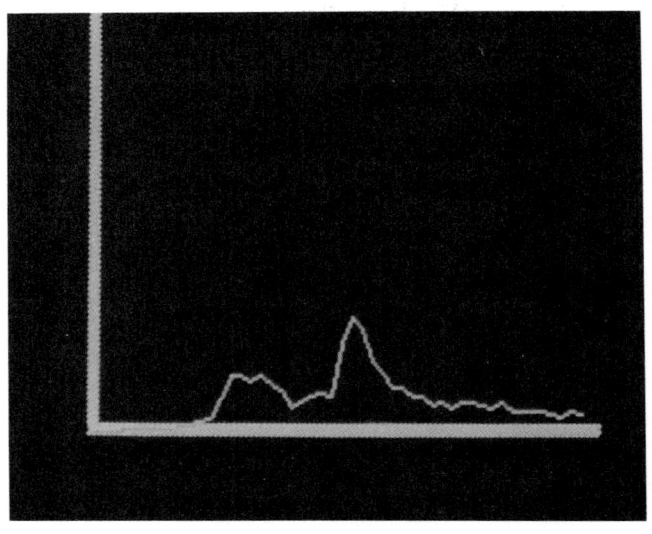

ACTINIDES

4.3 THE USE OF NUCLEAR TECHNIQUES IN THE STUDY OF THERMAL DIFFUSION OF ACTINIDES AND THE CHARACTERISATION OF NUCLEAR GLASS SURFACES (Saclay)

4.3.1 Introduction

Extension of the research contract WASF-326-83-55 for six months (1st January to 30th June 1985) has enabled the study of thermal diffusion of thorium and uranium in the reference glass SON 68 18 17 to be pursued [38].

A new technique of sample preparation has been successfully tested for uranium specimens.

Analyses of active glass by ^{4}He Rutherford backscattering have been made for the first time, on a sample of SON 68 18 17 containing 0.85 wt.% ^{237}NpO$_2$ leached by the Section d'Etude des Materiaux de Confinement (DRDD/SDHA-CEN VALRHO) (SEMC) using a specially-developed experimental device [39].

The concentration profiles of several light element constituents (H, Li, Na, Al) of the vitreous matrix have been determined by direct observation of resonant nuclear reactions in the surface regions of glasses leached by the SEMC for 28 days at 90°C in static mode and at 100°C in continuously flowing mode.

4.3.2 Study of Thermal Diffusion of Actinides in Glass SON 68 18 17

4.3.2.1 (a) Sample preparation and experimental procedure

Thin targets of natural thoria and natural uranium oxide (40 to 50 µg/cm^2) were prepared by painting on coupons of glass SON 68 18 17. The targets were then heated to 300°C for the required time and the following test procedures applied:

- cleaning off the remaining deposit with 11 N HCl;

- analysis of the near-surface region by ^{4}He Rutherford back-

scattering (E_0 = 2 MeV, θ = 160°, i ⩽ 30 nA, Q ⩽ 150 μC) [40].
- determination of the actinide concentration profile by means of the simulation programme VERDI [36];

- iterative calculation of the diffusion coefficient by the graphical Boltzman-Matano method [41].

This method is applicable to a concentration-dependent diffusion coefficient. We merely recall that starting from Fick's second law:

$$\frac{\partial c}{\partial t} = \frac{\partial}{\partial x} \left(D \frac{\partial c}{\partial x} \right)$$

then by means of the transformation λ x/√t and by integration the Boltzman-Matano equation is obtained, from which D may be calculated.

$$D(c) = \frac{-1}{2t} \frac{\int_{c1}^{c} x \cdot dc}{(dc/dx)_c}$$

where c1 represents one of the extremities of the diffusion couple.

The denominator of this expression corresponds to the slope of the tangent of the concentration profile at point c and the numerator is the area bounded by the concentration profile, the Matano plane of symmetry and the horizontal axes at c_1 and c_2, as shown in Fig. 4.50 reproduced from reference [41].

(b)   Results previously obtained

Figures 4.51 to 4.53 illustrate the reconstruction of the Rutherford backscattering experimental spectra for glass SON 68 18 17 without deposit and untreated, and for the thorium glass after 8 months and 13 months at 300°C using the VERDI simulation programme. Fig. 4.54 shows the thorium concentration profiles for the two tests.

The thorium diffusion coefficients from these profiles at the Matano interface are 5.08 and 7.83 x 10^{-20} cm^2 s^{-1} respectively ($\Delta D/D = 20\%$).

This variation of D with time is another illustration of concentration dependence which could be attributed to a phenomenon of interdiffusion with a mobile species (e.g. alkali or alkali earth) as in the case of the hydration of a glass

$$(Na^+ \text{ glass} + H_2O \quad H^+ \text{ glass} + Na^+ + OH^-).$$

The method of cleaning the sample before analysis with 11 N HCl was found to be unsuitable for samples coated with a thin film of uranium oxide.

4.3.2.2   Calculation of diffusion coefficients for thorium and uranium after 21 months' treatment at 300°C

A similar procedure applied to a thorium-coated glass sample after heat treatment for 21 months at 300°C (see Figs. 4.55 and 4.57) gave a value of $(1.43 \pm 0.3) \times 10^{-19}$ cm^2 s^{-1}.

For uranium oxide coatings, the remaining deposits were successfully removed with 6 N nitric acid. Figures 4.57 and 4.58 show the experimental spectrum and concentration profile after 21 months heat treatment. The diffusion coefficient value obtained from the uranium concentration profile (Fig. 4.57) by the Boltzman Matano method is $(1.65 \pm 0.33) \times 10^{-20}$ cm^2 s^{-1}.

Some thin thoria and urania deposits are currently being heat treated at 300°C. Thin oxide deposits of ^{237}Np and ^{239}Pu on glass SON 68 18 17 are also being heat treated at 300°C; measurements will be made at the end of 1985.

4.3.3 Characterisation of the surface region of leached glasses by nuclear techniques

4.3.3.1   Inactive glasses

To assist in the evaluation of different physico-chemical surface analysis techniques, the SEMC have prepared two specimens of glass SON 68 18 17 leached for 28 days in de-ionised water, one at 90°C in static leachant, the other at 100°C in continuously flowing leachant. These samples were divided into several portions and sent to various laboratories. In addition to the 90° and 100°C samples we were supplied with a reference sample of unleached glass.

Figs. 4.58 to 4.61 illustrate the use of resonant nuclear reactions in the determination of the changing composition profiles of a glass as leaching proceeds.

- Hydrogen profile by $^1H(^{15}N,\alpha\gamma)^{12}C$.
- Lithium profile by $^7Li(p,\gamma)^8Be$.
- Sodium profile by $^{23}Na(p,\alpha\gamma)^{20}Ne$
- Aluminium profile by $^{27}Al(p,\gamma)^{28}Si$.

Examination of these figures gives rise to the following comments

- the glass leached at 90°C shows compositional changes to a depth of ca 1 µm, whereas that leached at 100°C exhibits a hydrated layer of at least 5 or 6 µm thick;

- The hypothesis of ion exchange Na^+ H^+ (H_2O) is in agreement with the shapes and amplitudes of the 90°C sodium and hydrogen profiles, on the other hand this mechanism does not govern the mechanism at 100°C.

- The lithium concentration profile for the 90°C sample, which extends well beyond 1 µm (nominal hydrolysed layer thickness) suggests the existence of concentration gradients at the interface.

- The aluminium concentration profiles in Fig. 4.61 confirm the complex behaviour of this element during leaching (surface depletion for 90°C, very limited zone of relative enrichment at 100°C).

4.3.3.2   Active glasses

Two samples of SON 68 18 17 containing 0.85 wt.% of ^{237}Np oxide and Pu oxide (mostly ^{238}Pu) respectively were leached by the SEMC Marcoule for 15 days at ambient temperature with SA/V of 0.15 cm^{-1}, daily leachant renewal, were received for analysis. Only the ^{237}Np specimen could be examined by ^{4}He Rutherford backscattering (E_0 = 2 MeV, θ = 160°), following a method already well described [39]. Figs. 4.62 and 4.63 show the high energy portion of the backscattering spectrum, which shows a near-surface enrichment in neptunium. However, the backscattering spectrum shows that in general the surface concentration of rare earths is the same as for an unleached sample.

4.3.4 Conclusions

^{4}He Rutherford backscattering is shown to be a tool which is well suited for the measurement of actinide diffusion coefficients, particularly Th and U, in glass SON 68 18 17. The values obtained at 300°C for uranium and thorium range from 1.65 to 14.3 x 10^{-20} cm^2 s^{-1}. For a heat treatment period of 21 months thorium diffuses more rapidly than uranium, moreover the values determined are close to those previously obtained by other authors [42]. This study is being extended to the transuranics, particularly ^{237}Np and ^{239}Pu.

The direct observation of several resonant nuclear reactions has enabled the changing surface composition of the alkali borosilicate glass SON 68 18 17 to be illustrated during leaching, in static conditions at 90°C or flowing water at 100°C. The persistence of concentration gradients beyond the hydrated layer

and more precisely at the level of the front of attack
(particularly in the case of lithium) appears to be verified.

For the first time it has been possible to analyse a leached
active (doped with $^{237}NpO_2$) glass sample by the Rutherford
backscattering method. A significant surface enrichment in
neptunium is clearly shown; this enrichment affects a zone of less
than 1 µm thick.

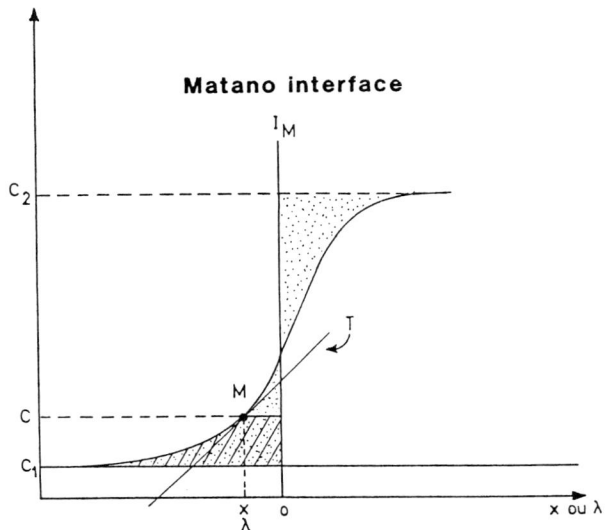

area//// = numerator of the equation

T   tangent at M where the slope is equal to the the denominator of the equation

Fig. 4.50 Illustration of the graphical Boltzman-Matano method

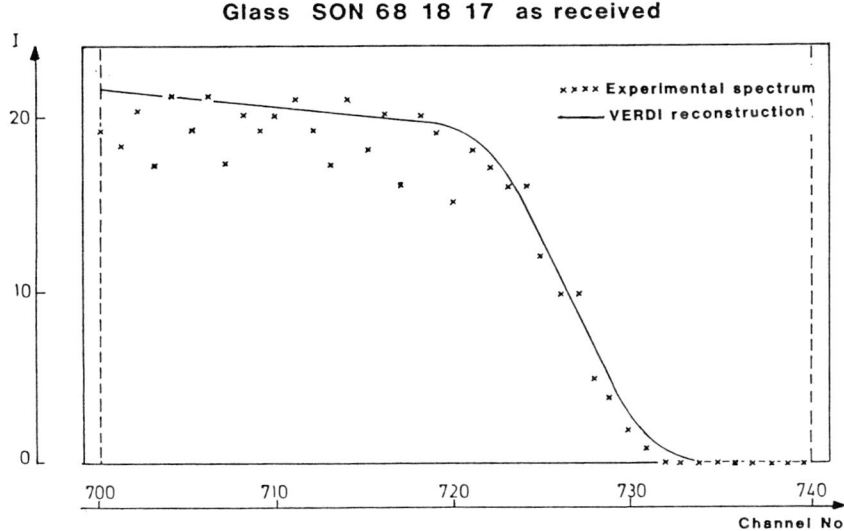

Fig. 4.51 Rutherford backscattering spectrum of pristine glass using the VERDI programme

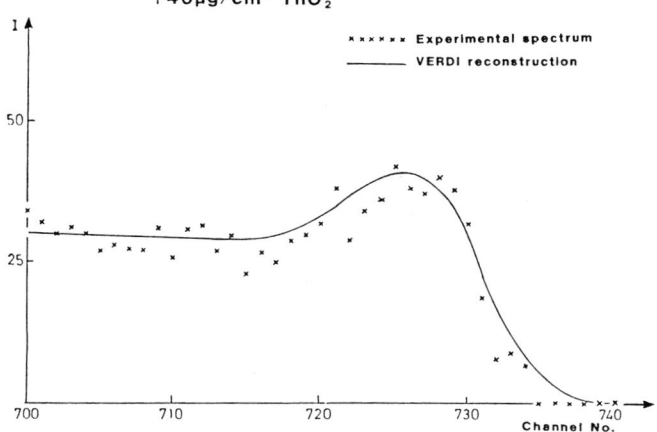

Fig. 4.52 Rutherford backscattering spectrum for a "thorium sample" heat-treated for 8 months at 300°C

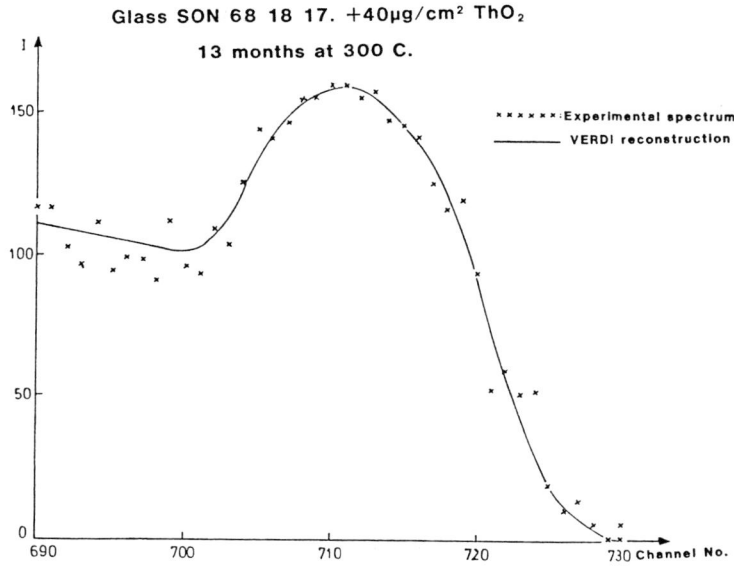

Fig. 4.53 Rutherford backscattering spectrum for a "thorium sample" heat-treated for 13 months at 300°C

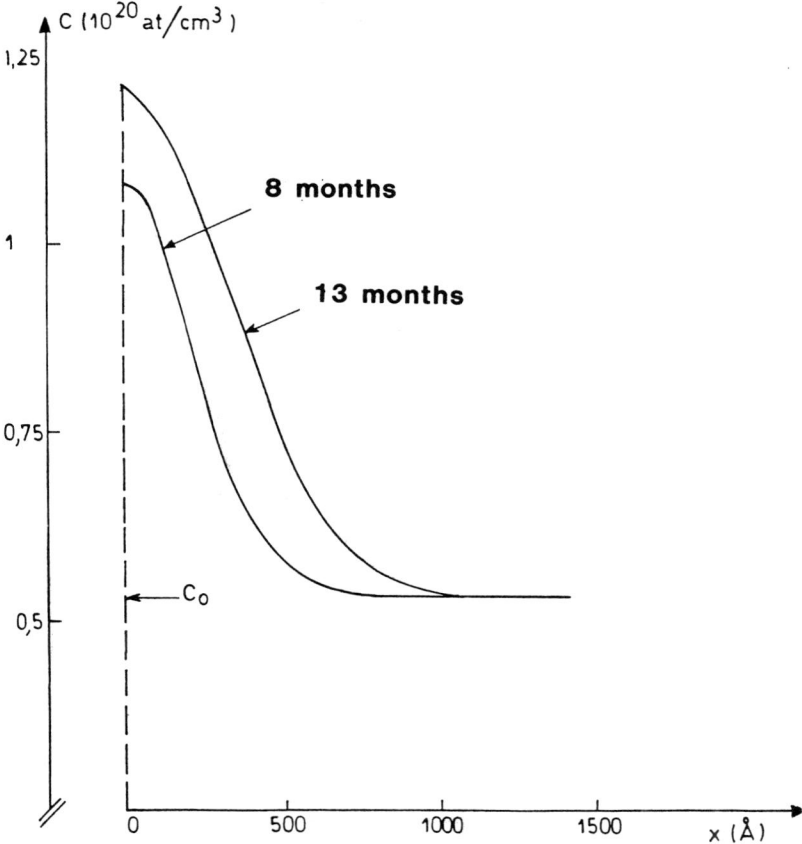

Fig. 4.54 Thorium concentration profiles for SON 68 18 17 after heating for 8 and 13 months at 300°C

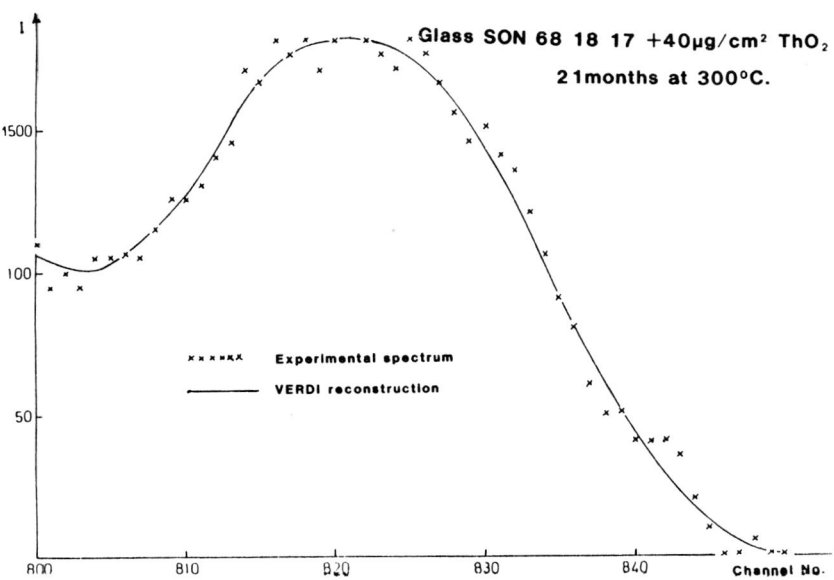

Fig. 4.55 Rutherford backscattering spectrum for a "thorium sample" heat-treated for 21 months at 300°C

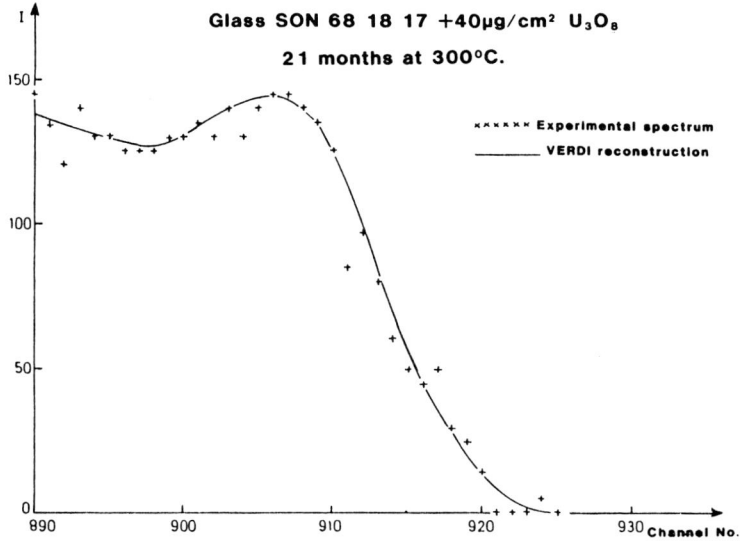

Fig. 4.56 Rutherford backscattering spectrum for an "uranium sample" heat-treated for 21 months at 300°C

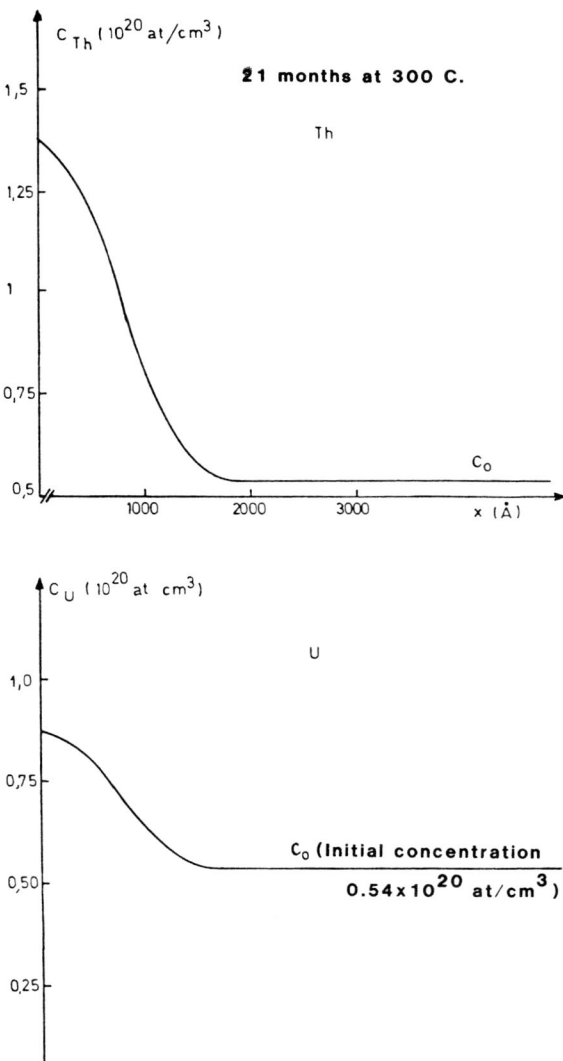

Fig. 4.57 Thorium and uranium concentration profiles for SON 68 18 17 after heating for 21 months at 300°C

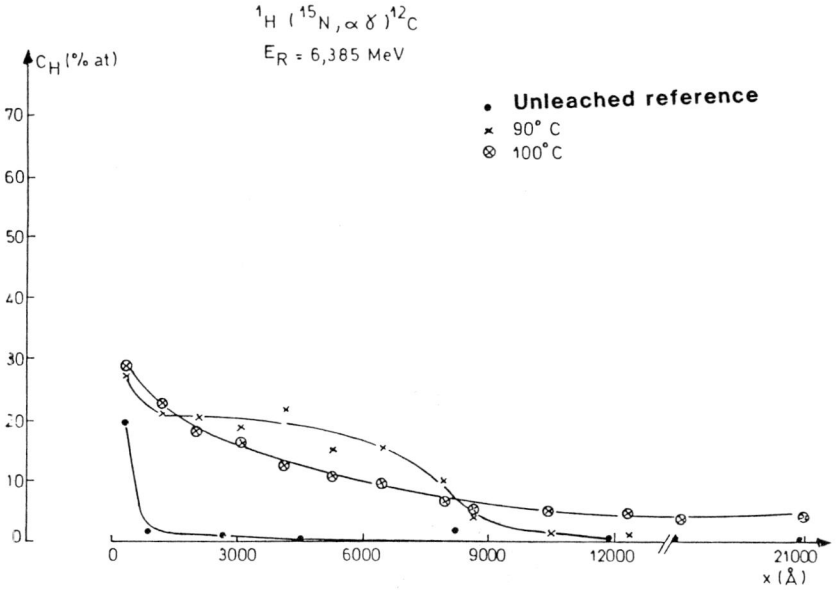

Fig. 4.58 Hydrogen concentration profiles by the $^1H(^{15}N,\alpha\gamma)^{12}$ reaction for SON 68 18 17

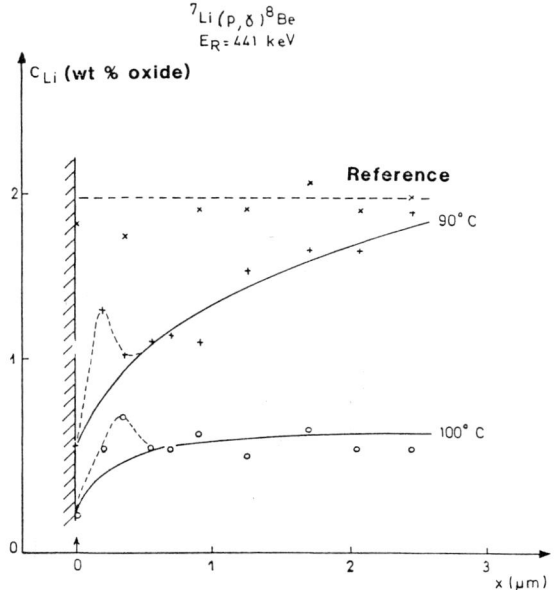

Fig. 4.59 Lithium concentration profiles by the $^7Li(p,\gamma)^8Be$ reaction for SON 68 18 17. (The positions of the first three points show a secondary ion migration effect as a consequence of the high incident current used: 5 µA/cm^2)

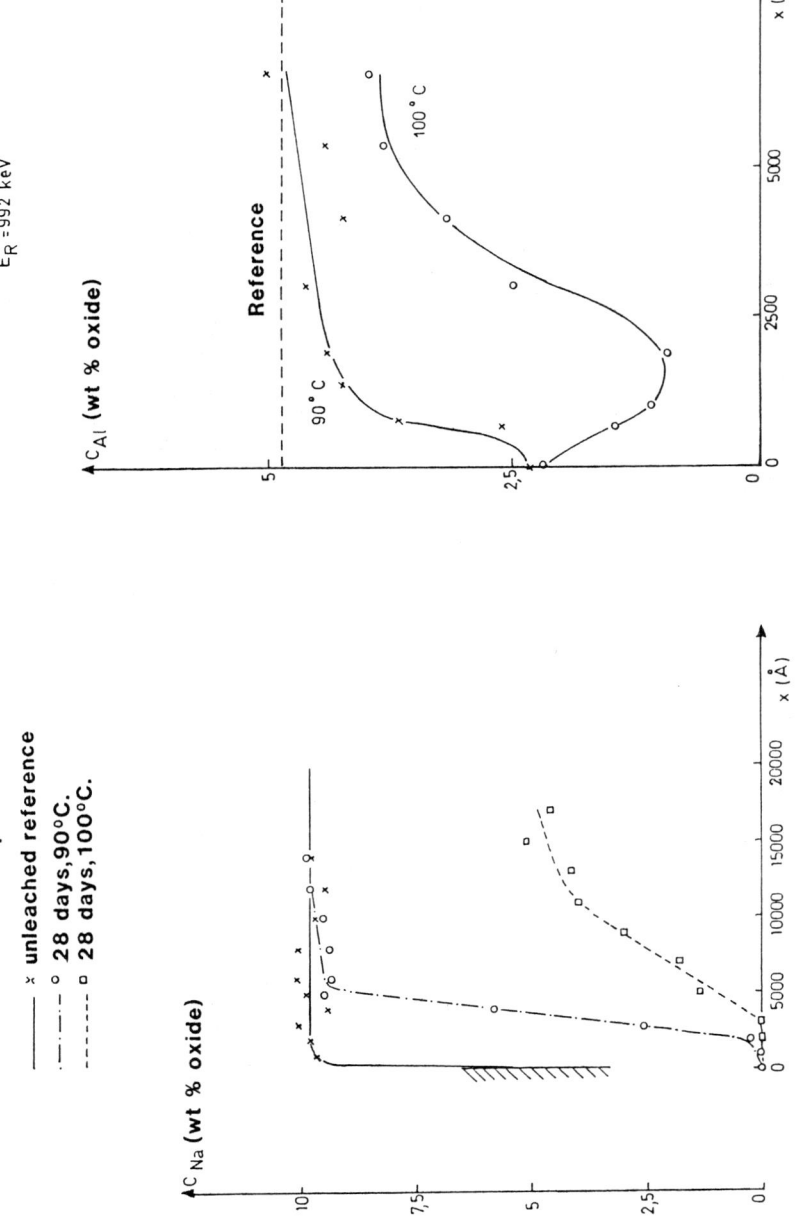

Fig. 4.60 Sodium concentration profiles by the $^{23}Na(p,\alpha\gamma)^{20}Ne$ reaction, for glass SON 68 18 17

Fig. 4.61 Aluminium concentration profiles by the $^{27}Al(p,\gamma)^{28}Si$ reaction, for glass SON 68 18 17

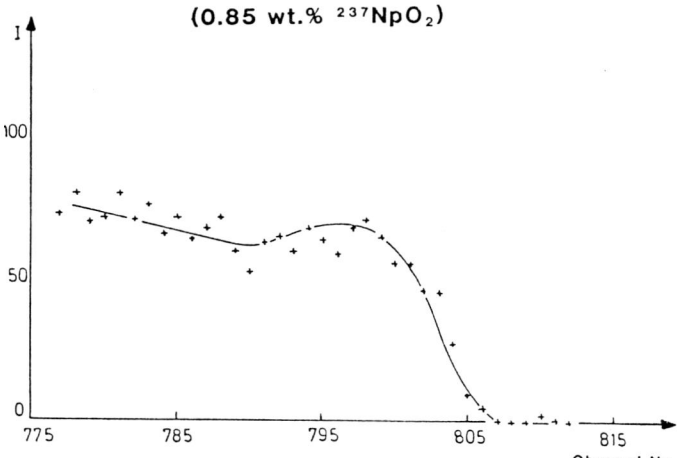

Fig. 4.62 Rutherford backscattering spectrum (^{4}He ions at 2 MeV) for SON 68 18 17 doped with 0.85 wt.% of $^{237}NpO_2$ (and containing no Th or U), leached for 15 days at ambient temperature

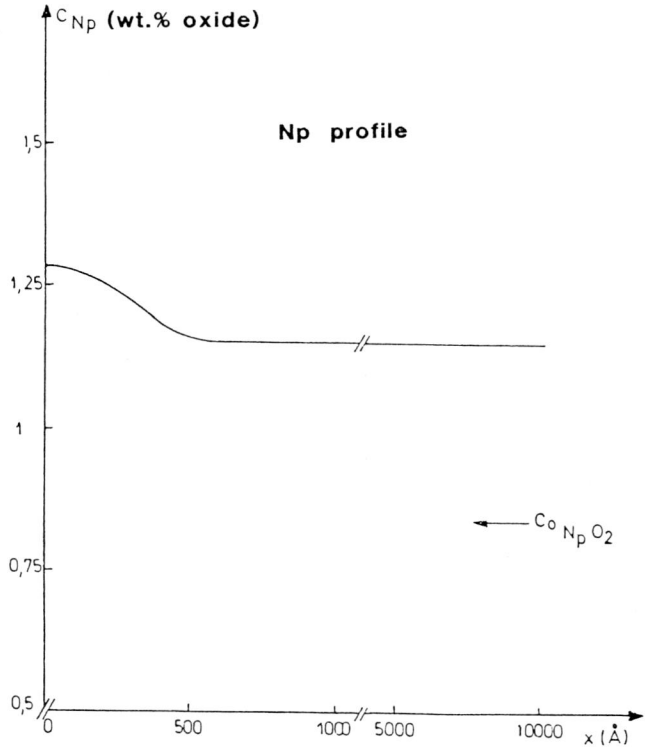

Fig. 4.63 Neptunium concentration profile

References – Chapter 4

[1] W. Heimerl, A Usable Method for the Management of High-level Waste Form from Nuclear Fuel Reprocessing, Annals of Nuclear Energy 4 (1977) 273-277.

[2] A.G. Herrmann, Composition of Salt Solutions for Leaching and Corrosion Experiments with Solidified Radioactive Waste, unpublished.

[3] de Batiste, R. et. al., Testing and Evaluation of Solidified High-level Waste Forms, EUR 8424 EN (1983).

[4] G. Malow, The mechanisms for hydrothermal leaching of nuclear waste glasses: Properties and evaluation of waste glasses, Scientific Basis for Nuclear Waste Management V, W. Lutze Ed., North-Holland, New York (1982) 25-36.

[5] F.K. Altenhein, W. Lutze, G. Malow, The Mechanism for Hydrothermal Leaching of Glass and Glass-Ceramic Waste Forms, Scientific Basis for Nuclear Waste Management, Vol. 3, J.G. Moore Ed., Plenum Press, New York (1981) 363-370.

[6] W. Lutze, G. Malow, H. Rabe, T.J. Headley, Surface layer formation on a nuclear waste glass, Scientific Basis for Nuclear Waste Management VI, D.G. Brockins Ed., North-Holland, New York (1983) 37-45.

[7] G. Brown, The X-ray identification and crystal structures of clay minerals, Mineralogical Society London, Jarrold and Sons Ltd., Norwich (1961), 143-207 and 297-324.

[8] R.F. Haaker, G. Malow and P. Offermann, Effect of phase formation on glass leaching, Scientific Basis of Nuclear Waste Management, VIII, C.M. Jantzen, J.A. Stone, R.C. Ewing Ed., North-Holland, New York (1985), 121-128.

[9]   P. Saha, The system NaAlSiO$_4$ (nepheline)-NaAlSi$_3$O$_8$ (albite) - H$_2$O, Am. Mineral, 46 (1961), 859-885.

[10] P. Cerny, The present status of the analcime-pollucite series, Can. Mineral., 12, (1974), 334-341.

[11] D. Roy and F. Mumpton, Stability of minerals in the system ZnO-SiO$_2$-H$_2$O, Econ. Geol., 51 (1956), 432-443.

[12] G. Brindley and S. Kikkawa, A crystal chemical study of Mg, Al and Ni, Al hydroxy-perchlorates and hydroxy-carbonates, Am. Mineral., 64 (1979), 836-843.

[13] J. Thomassin and J.C. Touray, L'Hydrotalcite, an hydroxycarbonate transitorie precocement forme lors de l'interaction verre basaltique-eau de mer, Bull. Mineral., 105, (1982), 312-319.

[14] B. Grambow, The role of metal ion solubility in leaching of nuclear waste glasses, Scientific Basis for Nuclear Waste Management, Vol. 11, Lutze, W. Ed., Elsevier North-Holland, New York (1982), 93-102.

[15] B. Grambow, A general rate equation for nuclear waste glass corrosion, Scientific Basis for Nuclear Waste Management, VIII, C.M. Jantzen, J.A. Stone, R.C. Ewing Ed., North Holland, New York 1985, 15-27.

[16] R.M. Wallace, G.G. Wicks, compiler J.E. Mendel, Report PNL-4382 UC 70 (1982).

[17] J.A.C. Marples et. al., Testing and evaluation of the properties of various potential materials for immobilizing high activity waste, European Appl. Res. Rept.-Nucl. Sci. Technol (1981) EUR 7138 EN.

[18] G. Malow et. al., Testing and evaluation of the properties of various potential materials for immobilizing high activity waste, European Appl. Res.Rept.-Nucl. Sci. Technol. (1980) 453-515, EUR 6617 EN.

[19] G. Malow et. al., Testing and evaluation of the properties of various potential materials for immobilizing high activity waste, First annual report 1977, EUR 6213 EN, 1978.

[20] P. Trocellier, B. Nens, Ch. Engelmann, Measurements of the hydrogen, sodium and aluminium concentration versus depth in the near surface region of glasses by resonant nuclear reactions. Nucl. Inst. Meth. 197 (1982) 15.

[21] P. Trocellier, B. Nens, Ch. Engelmann, Utilization of charged particle backscattering to study the near surface region of glasses. Application to depth profiling of lanthanum, cerium, thorium and uranium induced by aqueous leaching. Scientific Basis for Nuclear Waste Management, Ed. W. Lutze, North-Holland (1982).

[22] B. Nens, P. Trocellier, Traitement informatique des spectres de rétrodiffusion élastique de particules chargées. Application à la détermination des gradients de concentration d'éléments lourds dans des substrats légers. Note Technique DCAEA/SEA No. 82.437 (1982).

[23] P. Trocellier, Correction des distorsions engendrées par les variations de composition superficielle sur les profils de concentration déterminés à la surface d'un solide par des techniques nucléaires d'analyse. Application d'un programme informatique de simulation dans le cas de l'altération aqueuse de borosilicates alcalins. Note Technique DCAEA/SEA No. 83.519 (1983).

[24] Ch. Engelmann et. al., Testing and evaluation of solidified high-level waste forms. Report EUR 9268 EN (1984).

[25] Ch. Engelmann, J. Bardy, The French AEC nuclear microprobe: description and first application examples, Nucl. Inst. Meth. Phys. Res. 218 (1983) 209.

[26] J.L. Nogues, E. Vernaz, N. Jacquet-Francillon, Etude de la lixiviation des verres nucléaires liée à la sûreté du stockage. Note Programme SSD/GEL No. 82.01 (1982).

[27] J.L. Nogues, Les mécanismes de corrosion des verres de confinement des produits de fission. Thèse Docteur Ingénieur, Montpellier II (1984)

[28] J.L. Nogues, E. Vernaz, N. Jacquet-Francillon, J.P. Mestre, J.P. Morlevat, J. Jolivet, M. Chaudet. Altérabilité par l'eau de différentes compositions de verre en mode statique à 90°C. Etude de la couche hydrolysée. Rapport Technique SDHA/SEMC No. 83.03 (1983).

[29] P. Trocellier, Ch. Engelmann, J.L. Nogues, E. Vernaz. Etude comparative de la couche hydrolysée de verres nucléaires au moyen de methodes nucléaires d'analyse. Rapport Technique DCAEA/SEA No. 84.565 (1984).

[30] P. Trocellier, Ch. Engelmann, J.L. Nogues, E. Vernaz, N. Jacquet-Francillon. Etude comparative de la couche hydrolysée de verres nucléaires au moyen de méthodes nucléaires d'analyse. Rapport Technique DCAEA/SEA No. 84.573 (1984).

[31] P. Trocellier, J. Gosset, Ch. Engelmann. Exemple d'application de la microsonde nucléaire à l'etude de la corrosion aqueuse des verres. Rapport Technique DCAE/SEA No. 84.576 (1984).

[32] Ch. Engelmann, P. Trocellier. Sur les possibilités d'emploi de la microsonde nucléaire pour déterminer les éléments légers à la surface des verres. Rapport Technique DCAEA/SEA No. 84.600 (1984).

[33]   B.T. Kenna, Analysis of long term Soxhlet tests. Nuclear and Chemical Waste Management 3 (1982) 69.

[34]   J.L. Nogues, A. Terki, E. Vernaz, N. Jacquet-Francillon, J.P. Morlevat, J. Jolivet, M. Chaudet, Effet du mode de lixiviation sur l'altérabilité du verre "eau légère" R7T7 (SON 68 18 17 L1C2A2Z1) par l'eau à 100°C. Rapport Technique SDHA/SEMC No. 83.04 (1983).

[35]   S. Ahrland, I. Grenthe, B. Noren, The ion exchange properties of silica gel. Part I: The sorption of Na^+, Ca^{2+}, Ba^{2+}, UO_2^{2+}, Gd^{3+}, Zr(IV) + Nb, U(VI) and Pu(IV). Acta Chem. Scand. 14 (1960) 1059; Part II: Separation of plutonium and fission products from irradiated uranium. Acta Chem. Scand. 14 (1960) 1077.

[36]   J.N. Stanton et. al., The reaction between aqueous uranyl ion and the surface of silica gel. J. Coll. Sci. 18 (1963) 132.
Thorium interaction with the silanol group, water and nitrate in the pores of silica gel. J. Coll. Sci. 18 (1963) 878.
The exchange of twenty metal ions with the weakly acidic silinol group of silica gel. J. Phys. Chem. 68 (1964) 757.

[37]   P. Trocellier, Approche théorique de la formation de la couche hydrolysée d'un verre altéré en milieu aqueux. Rapport Technique DCAEA/SEA No. 84.597 (1984).

[38]   P. Trocellier, Contribution à l'étude de la corrosion aqueuse des verres et de la diffusion des actinides dans ces derniers. Rapports Technique DCAEA/SEA No. 84.603 (1984).

[39]   Ch. Engelmann, J. Gosset, B. Nens, P. Trocellier, Contribution to 1981 Annual Report for Action No. 5. SEAIN 82-435 (1982).

[40]   Ch. Engelmann et al, Technical Note SEAIN 81-394 (1981).

[41] Y. Adda, J. Philibert, 'La Diffusion dans les Solides". Presses Universitaires de France, Paris (1966).

[42] Hj Matzke, in "Thermodynamics of Nuclear Materials 1979". IAEA-SM 236/05 (1979) 311.

5. PARAMETRIC STUDY OF CORROSION STABILITY

5.1 CHARACTERIZATION OF SIMULATED HIGHLY ACTIVE WASTE FORMS (Mol)

5.1.1 Study of the corrosion mechanisms in static distilled water at 90°C

5.1.1.1 Experimental

Static corrosion tests have been performed on glasses SAN60, SM58, UK209, SON58 and SON64, glass-ceramic C31, and partially devitrified glasses SAN60 and SM58, for periods ranging between a few minutes and two years. As for the Soxhlet tests (see 5.1.7.2), and for the other corrosion tests discussed later, thin, polished (600 grit paper) samples were used. The samples, with a surface area of about 5×10^{-4} m^2 were enclosed in teflon cups, filled with the appropriate amount of leachant. Two surface area to solution volume ratios ($SA.V^{-1}$) were investigated: 10 m^{-1} (identical to the MCC1 conditions and 100 m^{-1} (more closely related to the expected conditions in a highly impermeable clay repository, and considered as the reference value in this programme). All experiments were performed twofold.

The experiments were interpreted in terms of:

- Specific mass loss (ML): the samples were dried at 110°C and surface layers, if formed, were not removed before weighing.

- Leachate analysis: the leachates were analysed in terms of pH (measured at room temperature) and cation composition (atomic absorption spectrometry (AAS), spectrophotometry (for Si at $SA.V^{-1}$ = 100 m^{-1}), or inductively coupled plasma analysis (ICP)). The elemental leaching data was transformed into normalized mass losses (NL_i) using the following formula:

$$NL_i = \frac{q}{Q/100} \frac{1}{S}$$

where q is the amount of element i in the leachate (in g), Q the weight percentage of the element in the sample and S the geometrical surface area of the sample (in m^2).

- Surface analysis: when relevant, the sample surface was analysed by infrared reflection spectroscopy (IRRS), scanning electron microscopy coupled with energy dispersive X-ray analysis (SEM-EDAX). Auger electron spectroscopy coupled with argon ion milling (AES: for the outer 1 µm zone) or electron microprobe analysis (EMPA: for the outer 100 µm zone).

Especially for the longer corrosion times (max. one year), important water losses from the teflon vessels were observed. When water losses were in excess of 20% the results were discarded.

5.1.1.2 Corrosion at 100 m^{-1}

Mass loss and leachate pH evolution

Graphs showing mass loss and pH variation versus time are plotted in Fig. 5.1. For the glass samples, the overall slopes of the mass loss curves (log ML vs log t), between 2 minutes and 8 months of corrosion, vary between 0.20 (SAN60) and 0.35 (SON58, SM58), corresponding to steeply decreasing corrosion rates. The ML's for glass ceramic C31 are even less time dependent: they appear to have reached a constant value after only about 40 minutes of corrosion. No attempt has been made to analyse more carefully the ML time dependence, because the ML values initially barely exceed the sensitivity of the balance (10^{-4} g). The mass losses after 8 months are between 1 (C31) and 8 (SON58, SON64, SM58) gm^{-2}.

After 2 minutes of corrosion, pH for all waste forms has already reached values between 7.7 and 7.9 (from an initial value of about 6). For glass SAN60, the equilibrium pH value of about 8.7 is reached after a few days, whereas for the other waste forms, pH apparently increases continuously, and reaches values between 8.9

TABLE 5.1   Slopes in the elemental ($\log NL_i$ vs $\log t$) plots during corrosion in distilled water at 90°C, $SA.V^{-1} = 100$ m^{-1}

	SAN60	UK209	SM58	SON58	SON64	C31
Si	0.5 (2'→4d)	0.5 (2'→80d)	0.5 (2'→27d)	0.65 (10'→1d)	0.65 (10'→1d)	0.2 (2'→240d)
Na	0.3 (2'→3d)	0.5 (2'→1d)	0.5 (2'→1d)	0.65 (10'→27d)	0.65 (10'→27d)	0.2 (2'→240d)
		1.0 (1d→9d)	1.0 (1→27d)			
Li	0.5 (2'→240d)	0.5 (2'→1d)	0.5 (2'→1d)	-	-	-
		1.0 (1d→27d)	1.0 (1→27d)			
B	0.5 (8h→4d)	0.5 (8h→1d)	0.5 (8h→1d)	0.65 (40'→9d)	0.65 (40'→9d)	0.2 (2h→240d)
		1.0 (1→80d)	1.0 (1→27d)			
Mg	-	no increase (≥ 2h)	no increase (≥ 2h)	-	-	-
Ca	0.5 (2h→1d)	-	0.5 (2h→3d)	-	-	0.20 (2h→240d)
Cs	-	0.8 (2'→9d)	-	0.5 (2'→8h)	0.5 (2'→1d)	0.35 (2'→240d)
Sr	-	0.3 (2'→1d)	-	0.5 (2'→40')	0.5 (2'→40')	0.4 (2'→2h)
						0.15 (2h 240d)
Al	0.5 (8h→4d)	0.5 (10'→9d)		...	...	...
Fe	-	-	n.d.	-	-	-
U	n.d.	n.d.	-	0.65 (2'→1d)	0.65 (40'→3d)	...
Ba	n.m.	n.m.	n.m.	n.m.	n.m.	0.4 (2'→2h)

n.m. : not measured
n.d. : not detectable

and 9.7 after 243 days of corrosion.

Leaching of Si, B, Na and Li

One may distinguish four types of leaching behaviour for these matrix constituents.

Glass SAN602519L$_3$C$_2$

The slopes in the (log NL_i vs log t) - plot yield fairly constant values over a large time span for either Si, B, Na or Li (see Fig 5.2a and Table 5.1). Diffusion phenomena, yielding a time dependence $NL_i \cong t^{0.50}$ [2] seem to control their leaching. After 4d of corrosion the amounts of Si, Na and B leached apparently have reached maximum (saturation) concentrations (see also Table 5.2). After periods of one to several months NL_{Si}, NL_{Na} and NL_B increase again, however. For Li no saturation effects are observed.

Glasses UK209 (Fig. 5.2b) and SM58LW11 [3]

Si and (Na, Li, B) leach differently from both glasses. Whereas Si yields a fairly uniform time dependence, suggesting diffusion as the rate controlling process, Na, Li and B seem to show a two-step leaching behaviour, consisting of a diffusion controlled stage followed by a matrix dissolution stage (with a time dependence $NL_i \cong t^{1.0}$ [3]. Apparently maximum concentrations for Si, Na, Li and B are reached in many cases.

Glasses SON583020U$_2$ (Fig 5.3a) and SON641920F$_3$ [3]

The leaching time dependence for Si, Na and B from both glasses is the same, yielding a slope 0.65 in the (log NL vs log t) plot. It is believed that both diffusion and matrix dissolution processes determine the leaching of these elements. The amounts of Si, Na and B leached reach maximum values after a period ranging from 1 to 27 days.

Glass-ceramic C31.3EC (Fig 5.3b)

Again the leaching time dependence for Si, Na and B is the same. The slope in the ($\log NL_i$ vs $\log t$) plot, however, is very low (about 0.20). No saturation effects are observed until 240d of leaching.

Leaching of Mg and Ca (all waste forms)

After a first interval ($\leq$ 2 h), during which the NL_{Ca} and NL_{Mg} are anomalously large [3], the leaching of Ca shows the same time dependence as for Si, for the waste forms SAN60, SM58 and C31. The dissolved amounts of both Ca and Mg however reach saturation values even earlier in the case of the waste glasses SAN60, SM58 and UK209 (e.g. 8h for Mg).

Leaching of Cs, Sr and Ba (all waste forms)

Until maximum 9d the leaching of Cs and Sr from glasses SON58 and SON64 seems to be diffusion controlled. The leaching of Cs and Sr from glass UK209 apparently occurs differently (see Table 5.1). For these three glasses however the amounts of Cs and Sr dissolved in the leachate have reached maximum concentrations after $\leq$ 9d. For glass SAN60 not sufficient data are available to conclude about a time dependence in the pre-saturation stage.

The leaching of Cs and Sr from glass-ceramic C31 is more complicated. No saturation effects in the leachate are occurring yet. However, Sr leaches congruently with Ba from C31 initially. Ba, however, quickly saturates in the leachate.

Leaching of Al, Fe, U (all waste forms)

The leaching of Al (UK209, SAN60 after 8h) and U (SON58, SON64) in the pre-saturation stage exhibits the same time dependence as Si. It is hard to distinguish a single time dependence for the leaching of Al from SM58, SON58, SON64 and C31 and of U from C31.

In most cases (except for Al from SAN60 and U from C31) Al and U are found to dissolve up to 10 times slower than Si. Also the amounts of Al or U dissolved apparently reach saturation concentrations in most cases (except for U from C31). After 27 d NL_{Al} for SAN60 again increases with time – see also the behaviour of Si, Na, B. Fe generally could not be measured in the leachates. With progressing corrosion Fe was sometimes detected (see glasses UK209, SON58 and SON64).

Discussion

Two major conclusions can be drawn, as far as the vitreous waste forms are concerned (SAN60, SM58, UK209, SON58, SON64).

1. Diffusion processes control the leaching of the glass matrix components Si, Na, Li, B, Ca, Al. It has been argued elsewhere [4] that these diffusion processes correspond with diffusion in bulk glass (D_{Na} for glasses UK209 and SM58, calculated from the leaching data, was 2 and 5 x 10^{-18} m^2 s^{-1}, respectively).

2. Saturation of many glass components, including Si and even very soluble elements such as B, Na and Li apparently occurs after different corrosion times and at various concentrations (depending on the glass composition – see Table 5.2). For the high alumina glass SAN60 (18 wt% Al_2O_3) relatively low 'equilibrium' concentrations are found for Si, Na, B, Cs, Sr. The formation of solubility controlling alumino-silicate complexes in the leachate has been suggested [5,6].

In contrast with the other glasses, however, the 'saturation' stage for SAN60 appears to be an intermediate phase prior to a resumption of the leaching of the various glass components (see Fig 5.2a, including data up to 2 years of corrosion).

TABLE 5.2 Observed elemental saturation concentrations in the leachate (in mg l^{-1}) during corrosion in distilled water (90°C, $SA.V^{-1} = 100$ m^{-1})

	SAN60	UK209	SM58	SON58	SON64	C31
Si	16	160	150	90	100	n.s.
Na	10	n.s.	160	350	300	n.s.
Li	n.s.	130	70	–	–	–
B	7	n.s.	110	200	120	n.s.
Mg	–	1.5	1	–	–	–
Ca	3	–	4	–	–	n.s.
Cs	0.1	1	–	2	1	n.s.
Sr	0.02	0.1	–	0.3	0.2	n.s.
Al	7	2	0.2	0.1	0.2	0.3
Fe	–	n.s.	n.d.	n.s.	n.s.	–
U	n.d.	n.d.	–	1	0.5	n.s.
Ba	n.m.	n.m.	n.m.	n.m.	n.m.	7

n.d. : not detected
n.m. : not measured
n.s. : not saturated

Data on surface analysis, using either IRRS, SEM-EDAX or AES
(coupled with argon milling) are given elsewhere [3,7]. They
indicate that surface layers up to 5 µm have been formed after
240d of corrosion, whose composition clearly differs from that of
the non corroded glass. IRRS analysis revealed a different compositional evolution with time for the layers formed on either glass
SAN60 (as well as the glass-ceramic C31) or the other glasses.

There may be some evidence for matrix dissolution processes
occurring in the pre-saturation stage (see SON58, SON64, where
slopes of about 0.65 were measured in the (log NL_i vs log t) plots
for Si; see also the Si leaching from glasses SM58 and SAN60
between 8h and 9d and between 2 and 8h respectively).

The leaching behaviour of the glass ceramic C31 is strongly
different from that of the glasses. The elemental releases occur
even slower.

Some additional observations can be made:

*   Generally Al and U leach much slower than Si in the pre-
saturation stage (exceptions: Al from SAN60 and SM58; U from C31).
This might indicate some trapping, yielding an enrichment of these
elements in the waste form surfaces.

*   Li leaches slower than Na in the diffusion controlled stage;
Ca and Mg leach congruently with each other, however (see SM58).

*   The elemental amounts of Mg, Cs, Sr and Al dissolved in the
leachate decrease as a function of the corrosion time, once the
elemental saturation concentrations are reached.

5.1.1.3   Corrosion at 10 m^{-1}

Mass losses and leachate pH results

The evolution of the mass losses and pH is shown in Fig 5.4.

Three stages can be discerned in the (log ML vs log t) plot. Between 2' and 8h the mass losses do not increase yet. A second stage lasts from 8h to about 28d, and shows a linear time dependence for the ML's, corresponding with a constant corrosion rate due to matrix dissolution for glasses SON58 and SON64. For the other waste forms, the available data do not allow to conclude about the ML time dependence. After one month of corrosion there is some evidence for the onset of saturation of the leachate in case of waste forms SAN60, SON58 and C31.

The leachate pH is about 7.0 during the first 8h, suggesting the samples are not heavily corroded in the initial time period. In the matrix dissolution stage leachate pH increases from 7.0 ($\leq$ 8h) to 8.5 - 9.0 after 28d.

Elemental leaching

The elemental leaching data, expressed as NL_i, are plotted in Fig 5.5a-b (UK209 and SON583020U$_2$; data for the other waste forms are found in [8], and reveal a similar time dependence as the mass loss data. As the glass leaching data indicate similar phenomena occurring for the five waste glasses, a global discussion is presented.

(a) During the first 2h $NL_{Na,Mg,Ca}$ stay rather constant; Si, Al,... could not be detected, however. No further attempt was made to investigate this time interval.

(b) Si, B, Na, Li, Mg and Ca leach rather congruently after 8h, and the slope in the (log NL_i vs log t) plots is about 1. This suggests matrix dissolution as the rate determining corrosion mechanism. This second stage lasts till about 10 (SAN60, UK209, SM58), 30 (SON58) or 100d (SON64).

The leaching of Sr (UK209, SON58, SON64) seems to be

diffusion controlled between 1 and 10d. Insufficient data are available for Cs to conclude about its leaching behaviour. The leaching of U and Al is mainly characterized by their very low (U) or low (Al) NL's, indicating that both elements leach much slower than the other glass components.

(c) Beyond the second stage, the leaching behaviour of Si, B, Na and Li strongly differ depending on glass composition (see Fig 5.6). The leaching of these main matrix components is halted (SON58, probably SON64 as well), proceeds at a decreasing rate (SAN60) or is stopped temporarily to resume after one month (UK209, SM58). In the latter case the leach rates for B, Na and Li seem to be constant with time in this third stage, but the leach rate for Si tends to decrease.

The apparent saturation concentrations are listed in Table 5.3. Smaller Si saturation concentrations are found for the high Al_2O_3 glass SAN60, confirming the observation from the similar experiments at $SA.V^{-1} = 100$ m^{-1} (see 5.1.1.2). Less soluble elements such as Mg, Ca, Cs and Sr have reached their saturation concentrations already much sooner; once these maximum concentrations are reached, the respective concentrations decrease with progressing corrosion. U apparently has reached a maximum concentration in the leachate in case of SON58, but not yet in case of SON64.

The leaching data for glass-ceramic C31 suggest a rather similar time evolution as for glass SAN60 [8]. During the first 10 days matrix dissolution apparently controls leaching of Si, B, Na,... Thereafter leaching proceeds at a decreasing rate. As for the glasses, U and Al leach (much) slower than eg Si, Na, B.

Surface Analysis

SEM-EDXA surface and EMPA cross-section analyses were performed on the waste form samples corroded during a few days or 8 months (see Figs 5.7-8).

TABLE 5.3　　Elemental saturation concentrations (in mg l^{-1}) observed during the experiments in distilled water, 90°C $SA.V^{-1} = 10\ m^{-1}$

	SAN60	UK209	SM58	SON58	SON64	C31
Si	18	n.s.	n.s.	130	(150)	n.s.
Na	n.s.	n.s.	n.s.	100	n.s.	n.s.
Li	n.s.	n.s.	n.s.	–	–	–
B	n.s.	n.s.	n.s.	85	n.s.	n.s.
Mg	–	2	0.4	–	–	–
Ca	2	–	6	–	–	(5)
Cs	<D.L.	(0.7)	–	(2)	(2)	(1.5)
Sr	0.03	0.2	–	0.3	0.3	0.6
Al	7	1.5	0.4	0.5	0.8	1
U	–	–	–	0.6	n.s.	0.05
Ba	–	–	–	–	–	5

D.L. = detection limit

n.s. = not (yet) saturated

( ) = the value given might correspond with the onset of saturation, but evidence is not sufficient.

The main observations are:

*   During the first 3 or 9d of corrosion no 'gel-like' surface layer has formed on glasses SM58, UK209 or SON58 (no analyses after these corrosion times were performed for the other waste forms) [8]. The Na content in the outer 1 μm of the glass surface (corresponding with the penetration depth of the SEM) is still quite high. Both observations are consistent with the matrix dissolution leaching mechanism as discussed in (b) above.

*   Thick, double surface layers have formed after 8 months of corrosion (e.g. UK209: 35 μm; SM58: 60 μm; SON64: 100 - 150 μm). These layers cracked upon drying. The upper, thinner layer appears to consist of crystalline precipitates (see e.g. Fig 5.7), and contains mainly Mg (SM58, UK209) or Mn, Ni and Mg (SON64). The inner layer is depleted in Mg, Na, Mo (except for a 40 μm thick zone near the glass bulk in case of SON64), and contains large amounts of Si. The layers might be enriched (w.r.t the bulk) in Fe, Al, Zr, Ti, Ca and the rare earths (Nd, Ce, Gd, La). No such thick surface layer is observed on glass SAN60; after 8 months of corrosion, its surface is covered by a layer, about 8 μm thick and enriched in Ca [8]. As to the glass-ceramic C31, it may also have developed a double surface layer after 8 months of corrosion. The outer part of the layer, whose overall thickness is about 25 μm, consists mainly of Zn and Mg, while the inner part contains Al, Zr, Nd, Ti, Ba, but no Mg or Zn. The underlying bulk has considerably roughened.

Discussion

By performing the 'standard' MCC1 test, which was introduced in the present work to elucidate leaching mechanisms in the absence of leachate saturation, some interesting features were revealed:

*   For $SA.V^{-1} = 100$ m^{-1}, the leaching of the main glass components (Si, B, Na) is generally diffusion controlled before the onset of saturation of the leachate for these and other glass components. In larger amounts of water (SA.V^{-1} = 10 m^{-1} in the present tests), the corrosion is dominated by matrix dissolution (yielding constant corrosion rates) initially. But, more important is the divergent long term (t > 10d) behaviour - see Fig 5.6 - yielding a steady state leach rate for B-Na-Li in case SM58, UK209 but a halt in the Si-B-Na leaching in case of SON58 (probably also SON64). SAN60 and C31 present an intermediate behaviour.

* Intensive surface layer formation occurs at SA.V^{-1} = 10 m^{-1}, following the matrix dissolution stage. At least two processes may contribute to layer formation; removal (by diffusion ?) of e.g. Mg, Mo, Na and retention of high-valence elements such as Fe, Al and the rare earths in a Si rich zone. In view of its high Si content, this layer does not seem to have formed by precipitation. Precipitation of insolubles from the leachate apparently led to the formation of the outer layer, which might be partially crystalline as well.

Intercomparison of different waste forms, based on the Si and Na release after 28d in the MCC1 test in DW leads to the following ranking of the chemical stability:

decreasing chemical stability ↓
| C31, SAN60, WG123*, WG124*, FLK HP*
| UK209
| SON64, WG119*, WG122A*, PNL76-68**, SRL131**
| SM58
| SON58

* WG119-124 are base or reference silicate glasses for the incinerator FLK alpha waste product (FLK HP) [9].

** PNL76-68 and SRL131 are US reference glasses for commercial and defense high level waste concentrates, respectively [10].

The difference in elemental release rates from the best (e.g. SAN60) and worst (SON58) waste form is about a factor of ten, and cannot be explained easily on the basis of glass compositional considerations. The glass compositions differ in many respects, indeed. Distinction between the contents of SiO_2, ($Na_2O + B_2O_3$) and the remaining glass constituents, as is often done (e.g. [11]) is not applicable in the present case (see: SiO_2 and $Na_2O + B_2O_3$ concentrations of SAN60 and SON58 are almost the same, but their corrosion stability strongly differs).

It should also be stressed that a qualification of various waste forms on the basis of their relatively short term (28d) performance does not necessarily fit with the long term performance (see e.g. Fig 5.6). Finally, the leaching data obtained at both 100 and 10 m^{-1} show that the retention capacity of the different waste forms for the various radionuclides does not necessarily follow the same tendency as the Si-B-Na (which represent the glass matrix) leaching behaviour.

5.1.1.4 Influence of partial crystallization (glasses SAN60 and SM58)

Glasses SAN60 and SM58 were subjected to corrosion in DW, 90°C, at both $SA.V^{-1}$ of 100 and 10 m^{-1}, after being partially crystallized (SAN60: 10d at 700°C; SM58: 10d at 800°C). The experimental details are similar to those described in 5.1.1.1 for the parent glasses. No surface analyses were performed, however, and the corrosion times at 10 m^{-1} were limited (2h, 1, 9, 27, 80, 720d). At both 100 and 10 m^{-1} the two years experiment is still underway.

Based on the data for the first 240 (100 m^{-1}) or 80 days (10 m^{-1}), the following general tendencies can be derived:

— For SAN60, partial crystallization does not markedly influence the corrosion. Mass losses, elemental leaching data and leachate pH values are roughly similar whether or not the glass is partially crystallized (see e.g. Table 5.4).

— For SM58 partial crystallization results in a slight (2 to 5 times) increase of both mass losses and elemental leaching (see e.g. Table 5.4). Equally the slope in the (log NL_i vs log t) plots at

TABLE 5.4   Comparison of the leaching data for the parent and partially crystallized glasses SAN60 and SM58) (DW, 90°C, 100 m^{-1})

	AMORPHOUS			PARTIALLY CRYSTALLINE		
time (days)	1	9	80	1	9	80
SAN60						
ML (g m^{-2})	1.23	1.01	1.68	0.23	0.25	- 0.47
Si (mg l^{-1})	11.0	16.0	20.0	9.2	13.0	15.0
Na (mg l^{-1})	6.5	11.0	14.0	4.8	7.3	8.6
B (mg l^{-1})	3.6	6.5	10.5	3.3	5.2	5.7
pH	7.10	7.93	7.95	8.96	8.56	8.88
SM58						
ML (g m^{-2})	1.53	3.42	8.17	0.66	8.5	15.8
Si (mg l^{-1})	35.0	118	153	62.0	160	265
Na (mg l^{-1})	10.5	49.6	161	30.0	152	310
B (mg l^{-1})	5.3	33.0	110	19.6	95	236
pH	7.32	9.22	9.78	9.64	9.75	10.44

100 m^{-1} increases somewhat (about 0.7 instead of 0.5 for e.g. Si). At 10 m^{-1}, insufficient data are available to conclude about a time dependence for the leaching.

The observations for both glasses correspond with those made after Soxhlet corrosion testing. Also, the very typical saturation phenomena for glass SAN60 (see 5.1.1.2 and 5.1.1.3) are still occurring after partial crystallization of the samples.

5.1.2 Influence of temperature (40 – 200°C) upon corrosion in static distilled water (SA.V^{-1} = 100 m^{-1})

5.1.2.1 Experimental

The temperature interval 40 – 200°C was screened by performing two types of experiments. First, at 40 and 70°C tests were performed for similar time intervals as for 90°C, with total corrosion time of one year. But, at each intermediate time the specimens were retrieved from the teflon corrosion cups, dried, weighed and placed again in the teflon cup. Also the leachate pH was measured. No further analyses were performed.

Secondly, experiments at temperatures in excess of 100°C (120, 150 and 200°C) were performed by heating up several teflon cups (containing the waste form samples) in an autoclave for times of 1, 7 or 28d. Upon heating, the pressure inside the teflon cups reaches the saturation value (0.2 MPa at 120°C, 0.7 MPa at 150°C and 1.7 MPa at 200°C). In this case, the experiments were interpreted in terms of specimen mass loss, leachate pH, cation composition (see 5.1.1.1) and specimen surface analysis by SEM-EDXA (for the 200°C) 28d).

At 200°C, glasses SAN60 and SM58 were also tested in their partially crystalline state (after annealing for 10d at 700 or 800°C, respectively). These tests were interpreted only in terms of mass loss and leachate pH.

All tests were performed triply, using either thin waste form plates (all temperatures) or 6 mm sized cubes (T $\geq$ 120°C).

5.1.2.2 Corrosion at 40 and 70°C

The results in terms of mass losses and leachate pH evolution with progressing corrosion are given in [3,7]. The main indications were:

* The time dependence of the mass losses is quite similar to the 90°C tests: slopes of ≤ 0.50 were obtained in (log ML vs log t) plots. The increase of the ML's between 40 and 90°C yields on the average a factor x 4, independent of time in most cases (1 or 12 months).

* The leachate pH increase with time occurs much slower at lower temperatures; e.g. compare pH values after one month corrosion for UK209; 7.0 (40°C), 8.0 (70°C) and 9.3 (90°C).

No further interpretation was given to these experimental data, since in many cases the mass losses measured were of the same order of magnitude as the sensitivity of the balance used (10^{-4} g).

5.1.2.3 Corrosion at 120-150-200°C

The mass losses measured during autoclave exposure are given in [7] and the 200°C results are repeated in Table 5.5 (but, averaging for three samples per test condition). The most important feature is that, with the exception of SAN60 at 200°C, all corrosion rates – calculated on the basis of the mass losses – decrease with time. In a few cases, the increase of mass loss with time was even negligible (e.g. SON64; besides, no removable surface layer was formed on this glass). Generally, glass SAN60 and glass-ceramic C31 performed the best, glasses SM58, SON58 and SON64 the worst – like at 90°C (see 5.1.1.2). But at 200°C excessively large mass losses, and surface alteration were recorded for SAN60 and SON58 (see further).

Partial devitrification of glasses SAN60 and SM58 does not influence the corrosion stability. Almost identical mass losses and leachate pH values, with similar time dependence, were recorded; this includes

TABLE 5.5   Mass losses (in gm^{-2}) after autoclave corrosion at 200°C, 100 m^{-1} (after removal of weakly bounded surface layers)

Corrosion medium	Distilled water			Clay-water mixture			Wet clay		
Waste form \ Corrosion time (d)	1	7	28	1	7	28	1	7	28
SM 58 LW 11	7.14	6.46	13.41	17.12	25.31	19.0	16.94	52.7	121.39
UK 209	6.87	10.32	13.15	12.88	15.06	12.74	15.59	24.30	48.27
SON 58 30 20 U$_2$	12.39	36.69	85.91	22.26	33.20	25.94	62.18	158.04	390.0
SON 64 19 20 F$_2$	10.12	11.30	11.09	18.10	35.96	32.29	29.34	49.48	204.05
SAN 60 25 19 L$_3$C$_2$	4.69	10.40	249.03	21.37	42.29	199.11	25.78	146.27	154.20
C31 - 3EC	1.90	3.11	3.24	6.23	8.55	6.74	6.94	22.34	57.45

almost no increase in mass loss for SM58 beyond one day, and a tremendous increase of the mass loss between 7 and 28d for SAN60. The conclusion for SAN60 is similar with that made upon corrosion at 90°C (see 5.1.1.4); for SM58 however an increased corrosion was observed at 90°C, upon partial devitrification.

The chemical analysis of the various leachates revealed further interesting features. The slowly increasing mass losses correspond with slowly increasing releases of the main glass components such as Si, B, Na, Li. Generally, the concentrations in the leachate of the less soluble elements such as Ca, Mg, Al, Sr, U, and also Cs, are constant during the whole corrosion sequence (1 to 28d) or even decrease with time. Two peculiarities, consistent with the observations at 90°C (see 5.1.1.2 - 5.1.1.3), stress the different corrosion behaviour of the various glasses:

*   Si, B, Na and Li leach congruently from SAN60, either in case of (apparent) saturation of the leachate for Si (120°C), or in the absence of saturation (150 - 200°C). For the other glasses, B, Na and Li continue to leach even when Si has ceased to do so (and again the Si saturation concentration for SAN60 is much lower than for the other glasses).

*   The leaching behaviour of Cs, Sr and U is not similar to the overall waste form performance. At 120 and 150°C, maximal concentrations for Cs and Sr in the leachate were about 0.2 (SAN60 and SON58 - with the worst overall stability), or up to 5 mg l^{-1} (UK209, SON64, C31). The maximum U concentrations were 0.01 (SON58), 0.7 (C31) or 5 mg l^{-1} (SON64).

Mass balances fit reasonably well. It was found that - contrary to the experiments reported earlier [7] - the sum of the elements leached (as oxides) differed generally by less than 20% from the mass loss. As an exception, for SON58, the elemental leaching data in some cases suggested much smaller glass corrosion (up to 10 times!) than measured from the mass loss. No explanation can be proposed as yet.

The temperature dependence of the corrosion is rather complex. Only rarely an Arrhenius type behaviour was observed for either corrosion time (1, 7 or 28d), considering the mass losses [7]. In most cases there is a kink at 90°C, above which the slopes (and, therefore, the activation energies) in the log ML vs 1/T plots decrease. Also, the slopes are corrosion-time dependent, and decrease with progressing corrosion. These phenomena may be due to different cross-over times between corrosion mechanisms (e.g. ion exchange, dissolution, precipitation,....) or to different mechanisms occurring at different temperatures.

The formation of crystals at the glass-water interface might be one of these 'different' mechanisms. After 28d corrosion at 200°C, various kinds of crystals, with dimensions between 10 and 100 μm were observed on top of e.g. SAN60, SM58, C31 [3] (see also 5.1.3.2 for SAN60). It has been suggested that this surface crystallization may speed up the corrosion in the actual corrosion conditions (occurrence of saturation for some elements), because precipitation of crystallites from the liquid phase should allow more of the glass components (e.g. Al, Si, Na) to be dissolved in the leachate (see also 5.1.3). Additionally this phenomenon, at least for SAN60 and SON58, induced a strong increase of the pH (e.g.: 10.6 for SON58, 9.5 for SAN60 after 28d, 200°C) which will strongly enhance the solubility of eg Si,... (very large Si, Na, Ca, Cs concentrations were measured for SAN60 and SON58). Finally, one may observe that the crystalline surface layer does not protect the underlying glass from dissolving.

As a conclusion, corrosion times in excess of one month should be applied at temperatures in excess of 100°C as well, to enable to fully understand the eventual role of surface crystallization at temperatures below 200°C. Also, it is doubtful whether the saturation phenomena mentioned (e.g. for Si at 120 - 150°C) really correspond to an equilibrium situation, since the corresponding leachate pH is still far beyond the corresponding 'saturation leachate pH' at 90°C.

5.1.3 Corrosion of glass SAN602519L$_3$C$_2$ in near saturation conditions

5.1.3.1 Introduction

To obtain better confirmation of the saturation effects in the leachate which are typical for glass SAN60 (see 5.1.1.2), some corrosion tests were designed in conditions expected to enhance saturation.

First, powdered (fraction between 43 and 123 μm) samples of SAN60 were corroded in pure, distilled water, with SA.V^{-1} values of about 20, 700, 4200, 7000 and 11500 m^{-1}. Test duration was 28d and test temperature 90, 120, 150 or 190°C. Thereafter the leachates generated during corrosion at SA.V^{-1} = 4200 m^{-1} were used as corrosion media for fresh, monolithic SAN60 samples at the corresponding temperature.

Complete experimental results can be found [3,8].

5.1.3.2 Corrosion in distilled water

The corrosion tests in pure, distilled water revealed increased elemental releases with increasing SA.V^{-1} (at either temperature) for matrix components B, Na and Li, and to some extent also for Ca. On the other hand, Si and Al concentrations have reached maximum values in the leachate for SA.V$^{-1} \geq 700$ m^{-1} (see Table 5.6; at 150 and 190°C, the Al concentrations for SA.V$^{-1} \geq 700$ m^{-1} are even lower than at 20 m^{-1}). For these SA.V^{-1} conditions leachate pH was almost constant. Comparison of the B, Na, Li and Si, Al leaching data suggests that saturation of the leachate for Si and Al occurred. Comparison of the Si, Al data with the data of 5.1.1.2 for SA.V^{-1} = 100 m^{-1} (see Table 5.1; pH ~ 8.75) also confirms the dependence of the Si-Al solubility upon the leachate pH [5].

The different pH values measured at the 'equilibrium' stage (saturation of the main matrix elements) for different SA.V^{-1} (at 90°C pH $\simeq$ 8.75 at 100 m^{-1}; pH $\simeq$ 10.0 at ≥ 700 m^{-1}) indicates that the use of

$SA.V^{-1}$ as an accelerating parameter [2] is limited. Apparently increasing $SA.V^{-1}$ might not only accelerate the corrosion by accelerating the accumulation of corrosion products, but also alter the corrosion processes.

In general, for a single $SA.V^{-1}$ value, the elemental amounts dissolved increase with increasing temperature until 150°C. Between 150 and 190°C these amounts sometimes stay constant or even decrease.

Extensive surface crystallization has been observed at 150 and 190°C (see Fig 5.9). Powdered samples were even completely altered at 190°C. The crystalline phases generated at both temperatures are not identical. At 150°C cubic analcime ($NaAlSi_2O_6.H_2O$) and a tetragonal zeolite ($Na_3Al_3Si_5O_{16}.6H_2O$) were observed; at 190°C three crystal types could be detected: analcime, orthorhombic eucryptite ($LiAlSiO_4.2H_2O$), and a needle-shaped, presently undetermined Ca, Si based crystal.

No surface crystallization was observed at either 90 or 120°C.

5.1.3.3 Corrosion in pre-concentrated water

The corrosion of fresh, monolithic specimens in pre-concentrated (an saturated in e.g. Si and Al) water largely depends on temperature. Glass samples corroded in preconcentrated solutions (resulting from 28d interaction of powdered SAN60 with distilled water at $SA.V^{-1} = 4200\ m^{-1}$) were in fact less corroded than in non-preconcentrated solutions, at both 90 and 120°C. At 150 and 190°C about 25 x larger weight losses (excluding the surface layer) were measured (between 2 and $4 \times 10^{-2}\ gm^2$). The leached species were partly present as precipitates in the teflon cell (about 50% at 150°C, 20% at 190°C); and B, Na and Li concentrations in the leachate strongly increased (almost independent of the degree of preconcentration) with respect to the as-prepared preconcentrated solution, the Si and Ca concentrations stayed much the same, but the Al concentration was even reduced. Besides this large glass dissolution, extensive laye

TABLE 5.6   Maximum Si and Al concentrations (in mg l^{-1}) in the leachates upon corrosion of glass SAN60 for 28d in various $SA.V^{-1}$ conditions (> 700 m^{-1})

°C	Si	Al	pH
90	70	27	10.0
120	105	43	10.1
150	350	20	10.4
190	340	35	10.0

TABLE 5.7   Normalised leach rates for Si and mass losses upon corrosion at a flow rate of 2 cm^3 d^{-1} (90°C, distilled water)

	$LR_{Si}(g.m^{-2}.d^{-1})$	$ML(g.m^{-2})$
SAN 60	0.25	29.4
SM 58	0.40	44.7
UK 209	0.20	12.2
C 31	0.25	14.9
SON 58	0.80	81.6
SON 64	0.60	77.2

formation was observed (see Fig. 5.10).

Obviously this divergent corrosion behaviour depending on the temperature (almost no corrosion at 90 - 120°C, enhanced corrosion at 150 - 190°C) may be related with the onset of surface crystallization at 150 - 190°C.

We suggest that crystallization can occur at 150 - 190°C when certain elemental saturation concentrations are reached (e.g. Si, Al), thereby allowing not only their own further leaching, but also that of B, Na, Li. At 90°C where no surface crystallization was observed, the amounts of B, Na and Li indeed apparently saturated congruently with Si, Ca, Al. Surface crystallization in this case ($\geq$ 150°C in pre-concentrated solutions) therefore acts as a catalyst for further corrosion. One may also conclude that in these experiments the surface layer did not act as protecting layer at all.

5.1.4.1 Corrosion at constant flow rate (90°C, distilled water)

A screening test was performed to investigate the influence of the flow rate of the corrosion medium. The test was conducted at 90°C at a constant flow rate of 2 $ml.d^{-1}$, and was stopped after 132 days. Samples were taken daily. Chemical analysis of the leachates was performed on the samples taken after 3, 7, 14, 28, 42 and 80 days.

Two major observations can be made:

1. The normalized leach rates for Si, Na, Li, B, Ca and Cs for all waste forms are fairly constant with time. The average leach rates based on Si, are given in Table 5.7.

The ranking between the glasses based on the leach rate for Si agrees with a ranking based on the specific mass losses. The differences between the leach rates based on the following elements: Si, Na, Li, B, Ca, Cs are within a factor two for a specific waste form, except for glass SON58, where LR_{Cs} is lower than $LR_{Na,B}$ by one order of magnitude. These constant elemental leach rates suggest that matrix dissolution is the process governing the corrosion.

2. The elemental leach rates for Mg, Al, Fe and U in all waste forms are markedly smaller than the corresponding leach rates for e.g. Si. As in many cases $LR_{Al,Fe,U}$ decreases with time, these high valence elements can be supposed to be trapped within the surface layer. The leaching behaviour of Sr is not clear, since in two cases (UK209 and C31) its leach rate corresponds quite well with the leach rate of Si, while in glasses SON58 and SON64 its leach rate is lower by more than one order of magnitude than the leach rate for Si.

Although constant leach rates were observed both in this test, at a flow rate of 2 ml d^{-1}, and in the Soxhlet test at about 1500 ml d^{-1}, the value of the leach rate is about 10 - 50 times greater in the Soxhlet test, (see Section 5.1.7.2). The pH of the leachate is about the same in both tests ($\simeq$ 6.0) and there is no evidence of significant saturation effects in either experiment. It may therefore be deduced that the flow rate itself is the important parameter that affects the leach rate.

5.1.5 Influence of the pH of the leachate

Corrosion tests at a constant leachate pH were performed to evaluate the specific influence of the leachate pH upon the corrosion of high level waste forms. Data on the corrosion at different leachate pH values are useful, since the long term corrosion tests in clay media resulted in rather divergent pH values: see e.g. in CWM after one year: 3.5 - 4.0 (40°C) and 7.4 - 9.0 (90°C).

Two types of chemicals were used to adjust pH. Initially, the ISO/DIS 6961 standards were used to obtain pH values of 5.7 (0.02 M glycine) or 9.7 (0.05 M THAM) [7]. The corrosion test conditions were 90°C, $SA.V^{-1} = 10$ m^{-1}. Despite frequent renewals of the leachate, pH excursions up to 2.0 units were measured (pH 7.6 instead of 5.7). New tests were started therefore, using piperazine, $C_4H_{10}N_2 \cdot 6H_2O$ as buffer, and HCl for adjusting the pH at 5.7 or 9.5.

With weekly renewals of the leachates, pH excursions were limited to 0.5 unit.

The results for the mass losses (measured without removal of the surface layer; the same specimens were further corroded after weighing) are plotted in Fig 5.11. The following observations can be made:

* At pH 5.7 the corrosion rates for all six waste forms, except for glass SON64, are almost constant with time (see the slope equal to 1 in the log ML vs log t plot). For glass SON64 a slope of 0.80 is obtained. In case of SON58 the corrosion might be diffusion controlled in the initial stage. The corrosion behaviour of glasses SON58, SAN60, SM58 and UK209 is almost identical. Surprisingly, glass SON64 corrodes much faster than the others. The glass-ceramic C31 is relatively less stable, too (in contrast with its good performance in distilled water in either Soxhlet or static conditions (see 5.1.7 and 5.1.1).

* At pH 9.5 the corrosion rates decrease with progressing corrosion time. The slope is about 0.75 (all waste forms except SAN60) and suggests that both diffusion and matrix dissolution might determine the corrosion. In case of SAN60 an almost constant corrosion rate is obtained. In the basic conditions of pH 9.5 all waste forms, except glass SON64, are corroded more heavily than in the almost neutral conditions of pH 5.7.

5.1.6 Corrosion in clay related media

5.1.6.1 Experimental

Various clay* related media were defined: wet clay (clay rendered plastic by addition of a small amount of water), a mixture of

---

*The mineralogical composition of the Boom clay, considered in Belgium, is (in wt %): 25 illite, 20 smectite, 30 vermiculite, 15 illite-montmorillonite, 10 chlorite-vermiculite

distilled water equilibrated with clay (100 g clay per litre water was used as a 'standard' mixture) and synthetic interstitial claywater (representing about 20 vol % of the Boom clay). No corrosion tests have been performed in the interstitial claywater (ICW), since its composition initially was poorly defined ([12]; some results for corrosion of waste glasses in a 'precursor' ICW are given in [13]), and since it was expected that the presence of clay particles was needed to buffer the redox conditions (through their pyrite content). To evaluate the specific role of the clay particles in the solution, glasses SM58 and SAN60 were corroded in three different clay-water media, with clay to water ratios (g/l) of 10 ('diluted' clay-water mixture, DCWM), 100 (CWM) and 500 ('concentrated' CWM, CCWM). The composition of the liquid fraction of both CCWM and DCWM is listed in Table 5.8.

The reference conditions in the present experiments were similar to those used upon corrosion in distilled water (see 5.1.1): 90°C, $SA.V^{-1} = 100$ m^{-1}, total corrosion time up to two years (further experimental details, see [7]). The corrosion tests in CWM and WC were also performed at 40°C (time up to one year) and 120 - 150 - 200°C (up to one month). The samples tested included the five reference glasses SON58, SON64, SAN60, SM58, UK209 and the glass-ceramic C31. At 90 and 200°C the partially crystallized glasses SAN60 and SM58 were tested as well.

As it was found that under oxidizing conditions the presence of clay particles did not provide stable reducing conditions of the medium, other CWM and WC media were prepared using non-oxidized clay, de-aerated water and a nitrogen atmosphere. The E_h of such mixtures ranges between - 120 and - 200 mV, using a Pt electrode and a AgCl/KCl reference electrode [7], and correlates better with the real, reducing conditions of the clay formation at Mol. Screening corrosion tests using thus prepared clay media were performed on glasses SM58 and SAN60.

5.1.6.2 **Corrosion in clay media at 90°C under oxidizing conditions**

The evolution of the mass loss with time in CWM or WC is plotted in

Fig 5.12. In CWM, the slope of the (log ML vs log t) curve for all six waste forms, does not exceed 0.50 during the early stages of corrosion (up to 7 or 42 days), but decreases after this first stage. The leachate pH values drop from initial values of 9.0 - 9.4 to 8.0 - 8.2 after three days of corrosion. Thereafter, the pH increases slowly, and reaches values between 8.3 and 9.4 after one year.

When corroding in wet clay, the slopes for all waste forms range between 0.50 and 0.75.

In a general way, the ML's in both clay media are much larger than the corresponding ML's in distilled water at the same $SA.V^{-1}$ value ($100\ m^{-1}$). The ML's in wet clay are even larger than those in the clay-water mixture, and are a factor 15 to 70 times greater than the corresponding values in distilled water after 8 months at $100\ m^{-1}$.

Partial crystallization of glasses SAN60 and SM58 did not influence (SAN60) or slightly enhanced the corrosion (SM58) in CWM or WC [14].

From the results with various clay to water ratio's for glasses SAN60 - SM58, the following sequence for the mass losses was obtained: $ML_{WC} \geq ML_{CCWM} > M_{CWM} > ML_{DCWM} \geq ML_{DW}$. The increasing mass losses with increasing concentration of clay particles in the leachate, which is also reflected in stronger ML increases with time, can be summarized as follows:

$$ML = \alpha\ (C/W)^{\beta}$$

with C/W the concentration of clay particles, and α and β depending on the corrosion time (see Fig 5.13).

Several reasons may be proposed for the increased corrosivity in both CWM and WC, compared to DW. As far as the oxidizing CWM is concerned, no marked effect from a change of solubility of e.g. Fe, relative to the solubility in distilled water is expected. A major reason for the higher ML's in the clay-water mixture (whose liquid

TABLE 5.8   Composition of the liquid fraction of the diluted and concentrated clay-water mixture (in mg l^{-1})

	DCWM	CCWM
Anions		
SO_4	44	830
NO_3	0.30	< 0.20
Cl	0.50	3.20
F	0.25	1.15
Cations		
Na	21.5	454
K	10.0	54
Sr	0.05	0.50
Ca	1.50	17.8
Mg	3.65	14.5
Ba	0.25	0.16
B	< 0.5	4.25
Al	12.0	< 0.1
Cr	< 0.3	< 0.3
Fe	6.3	< 0.1
Si	28.1	2.2

TABLE 5.9   Elemental distributions in the surface layers formed upon corrosion in clay media

		Enriched with respect to the bulk	Measurable amounts	Barely detectable
SM 58	CWM	Zr,Ti,Nd,Al,Fe,Ca	Si,Ce	Mg,Mo,Na
	WC	Zr,Ti,Nd	Fe,Ca,Ce	Mg,Mo,Na,Al,Si
SON 64	CWM	Zr,Al	Si,Nd,Fe,Gd,Cs	Mo
	WC	Zr,Gd,Al,Fe	Si,Nd,U,Ce,La	Mo,Cs
C 31/3	WC	Zr	Fe,Ca,Si,Ba,Al,Ti	Mg,Mo,Zn

phase already contains some cations also present in the waste forms) relative to distilled water, seems to be the smaller influence of saturation effects (although the ML time evolution also suggests some saturation of the corrosion medium). This is concluded from the tests in distilled water at different $SA.V^{-1}$ values. The ML's in the clay-water mixture at 100 m^{-1} compare better with those in distilled water at 10 m^{-1} than with those at 100 m^{-1} [15]. We suggest that this apparent increase in solubility is mainly caused by the adsorption of certain cations (e.g. Mg^{2+}, Ca^{2+}, Al^{3+}, ...) on the clay particles present in the solution.

The increasing ML's with increasing concentration of clay particles might be caused by the correspondingly increased number of adsorption sites.

It is likely, however, that the leaching mechanisms themselves also change due to the presence of clay particles (see the different slopes in the (log ML vs log t) plots and the different ML's during the initial corrosion sequence, where no saturation effects in the leachate yet inhibit the corrosion of the glass - see 5.1.1.2). Two additional factors may contribute to the enhanced corrosion in clay media: 1) the presence of an increasing amount of anions in the leachant with increasing $C.W^{-1}$ e.g. 45 and 830 mg l^{-1} SO_4^{--} for 10 and 500 g l^{-1}, respectively, and/or 2) the creation of an electric field susceptible to speed up glass dissolution.

The leachate pH time evolution in DCWM is similar to the evolution in CWM. In CCWM, however, the pH continues to decrease with time, and reaches values as low as 4.50 (SM58) and 6.30 (SAN60) after one year [8].

In both CWM and WC sticky surface layers between 15 and 70 μm thick have developed on waste forms SM58, SON64 and C31, representing up to 5 times the measured weight losses (see Figs 5.14 - 15). They are enriched in Al, Zr, Ti, contain Si, the lanthanides (La, Nd, Ce, Gd), U and K (which probably originates from the clay corrosion medium) and are depleted in Cs, Mo, Zn and Mg ([17]; see also Table 5.9). The underlying glass (in case of SM58 and SON64) appears strongly corroded. No such surface layer was observed on glass SAN60. No surface analyses were performed on glasses UK209 and SON58.

5.1.6.3 Influence of the redox conditions

Based on mass loss measurements (see Fig 5.16), no influence of the redox conditions - being either oxidizing or reducing, was found upon corroding both glasses SM58 and SAN60 in wet clay. In the clay-water mixture, however, reducing conditions result in a constant slope of about 0.60 in the (log ML vs log t) plot, while in the oxidizing CWM the curves deflect after a few weeks, resulting in a slope of about 0.25. Therefore, while initially the ML's in both redox conditions are quite the same, the reducing CWM has corroded both glasses about three times more than the oxidizing CWM after about 300d.

The redox conditions in the CWM leachate did not influence the pH evolution upon corrosion, nor the generation of surface layers. As for the oxidizing CWM, a layer of about 65 µm was formed on SM58 after 280d corrosion in reducing CWM. Its composition might be similar as in the oxidizing CWM; a large amount of S was present, too. As for the oxidizing CWM no layer formation occurred in case of SAN60, except for a S-rich zone a few µm thick.

There is no evident explanation for the enhanced attack in the reducing CWM. A similar influence of the redox conditions was observed for ferri-aluminosilicate waste glasses [9]. Enhanced corrosion due to enhanced solubility of redox sensitive elements such as Fe, or due to enhanced sorption of such elements on clay is doubtful, since the quantity of redox sensitive elements in both glasses SM58 and SAN60 is quite small.

5.1.6.4 Influence of temperature

5.1.6.4.1 Corrosion at 40°C

The results in terms of mass losses and leachate pH evolution with progressing corrosion were given in [7,3]. The main indications from these screening tests (the CWM tests were performed in triplicate but the WC tests singly) are:

- the slopes in the (log ML vs log t) plot for all six reference waste forms range between 0.5 and 0.7, for almost any case, in both CWM and WC media. The slopes at 40°C in the CWM medium are somewhat higher than at 90°C (see 5.1.6.2), in WC the slopes are quite similar at both temperatures;

- mass losses in both clay media are larger than in DW - see also 5.1.6.2. Surprisingly, however, ML's in both CWM and WC are almost as large;

- the difference between ML's at 90 and 40°C generally corresponds with a factor of x 10, and is rather independent of time;

- contrary with 90°C, the CWM leachate pH values continuously drop, to reach values as low as 3.50 after six months of corrosion.

5.1.6.4.2 Corrosion at 120 - 200°C

Short term high temperature corrosion tests (120, 150, 200°C) confirmed some trends already deduceable from the long term 40 - 90°C corrosion tests: the relative corrosivity of WC - CWM - DW (see [7] and Table 5.5), the decreasing pH with time of the CWM leachate [7], the relative performance of the various waste forms, the influence of partial devitrification (only 200°C was investigated, and for SAN60 no influence, for SM58 a degradation of the chemical stability was found) and the surface layer structure (see further). In a general way, the corrosion rates (based on mass losses) decrease with time as well. No crystallization was observed on top of the surface layer, which strongly contrasts with the observations on high temperature corrosion in DW - see 5.1.2.

The corrosion data do not allow simple extrapolations from high

temperature to low temperature. Arrhenius plots of the ML's over the
temperature range 40 - 200°C after 7 or 28d of corrosion in both CWM
and WC indicated a linear relationship only in case of SAN60. An
effective activation energy of 35 kJ mol^{-1} was obtained upon
corrosion in CWM during 7d [16]. For the other waste forms this
activation energy is smaller at higher temperatures (interval 90 -
200°C). It should also be pointed out that this picture is leaching-
time dependent. The non-linearity of the Arrhenius plots between 40
and 200°C, and the time dependency of the activation energy may be
due to different cross-over times between corrosion mechanisms, or to
different mechanisms occurring at different temperatures (see also
the large difference in leachate pH - time evolution between 40°C and
90 - 200°C).

Nevertheless comparison of experimental results at different temperatures may prove to be useful. Some similarity was observed indeed
when comparing ML's and surface layer characteristics obtained either
at 200°C after 28d or at 90°C after about 300d (see Table 5.10) in
case of glasses SON64 and SM58 [3]. As a consequence the autoclave
test might be a valuable accelerating test, at least in clay media,
and considering the appropriate temperature interval.

5.1.7 Characterization of Glasses SM58LW11 and SAN602519L$_3$C$_2$

5.1.7.1 Thermal stability

To obtain information about the glass transition range (nucleation
range) and the crystallization temperature(s) of the glasses, thermal
analysis was performed. DTA (differential thermal analysis), DSC
(differential scanning calorimetry) and TMA (thermo-mechanical
analysis) were recorded with a modular Du Pont 990 Thermal Analyser.

As a result, T_g (glass transition temperature), Mg (dilatometric
softening point) and T_c (temperature of maximum crystallization) were
determined (see Table 5.11). Details may be found in [17].

Isothermal annealing treatments were performed, similarly to the

TABLE 5.10   Comparison between specific mass loss and surface layer characteristics

GLASS	SM58				SON64	
Corrosion conditions	WC		CWM		WC	
	200°C 28d	90°C 240d	200°C 28d	90°C 360d	200°C 28d	90°C 240d
ML (gm^{-2})	120	130	19	23	200	120
Surface layer thickness (μm)	30	30	20	50	100	75
Surface layer composition enriched in:	Ti,Zr,Nd, Ce,S	Zr,Ti,Nd, Ce	Zr,Ti,Nd, S	Zr,Ti,Nd, Al,Ca,Si, Ce	Zr,Gd,Al, Ce,Fe,K, Si,S,U	Zr,Gd,Nd, Ce,La,Fe, Al,U,Si
depleted in:	Al,Si,Mg, Na	Al,Mg,Mo, Na,Si	Mg,Si	Mg,Mo		

TABLE 5.11   Characteristic temperatures (in °C)

Glass	T_g	M_g	T_c
SAN602519L$_3$C$_2$	465	495	655
SM58LW11	505	535	-

TABLE 5.12   Crystalline phases

Glass	T (°C)	Phases
SAN602519L$_3$C$_2$	700	(main) 0.7NaAlSi$_3$O$_8$, 0.3CaAl$_2$Si$_2$O$_8$ (plagioclase) (additional) (Ca,Al,Si) rich
SM58LW11	800	(main) SiO$_2$ (tridymite)

TABLE 5.13   Soxhlet corrosion rates (14d period)

SAN602519L$_3$C$_2$	
amorphous	5.68 ± 1.1
annealed (10 d 700°C)	5.22 ± 0.2
SM58LW11	
amorphous	4.17 ± 0.1
annealed (10 d 800°C)	12.6 ± 2.0
UK209	2.6 ± 0.9
UK189	13.0 ± 2.0
SON583020U$_2$	31.0 ± 3.0
VG98/3	19.0 ± 3.0

conditions used in the study of the thermal stability of high level waste glasses performed in the first CEC programme [18]. Glasses SM58 and SAN60 were annealed at either 500, 600, 700 and 800°C during 2 hours, 1, 10 and 100 days. At 500 - 600°C, no crystallization could be measured (by X-ray diffraction analysis) for both SM58 and SAN60. The strongest crystallization occurs after 10 or 100d at 700°C (SAN60) or 800°C (SM58). The identity of the crystals formed is listed in Table 5.12 and Fig 5.17 shows the SEM observations.

5.1.7.2 **Chemical stability**

The chemical stability of glasses SM58 and SAN60, before and after partial devitrification, has been quantified with the standard Soxhlet test. The test was performed in a similar way as in AERE Harwell during the first CEC programme [18].

The corrosion rates based on the mass losses are listed in Table 5.13. Comparison of these results with those from other European reference glasses [18] reveals that glasses SAN60 and SM58 behave quite favourably. The surface layers generated upon corrosion are discussed [7].

It is important to notice that the change in corrosion stability upon partial devitrification is not similar. For SAN60 no measurable change was found, but for SM58 the corrosion rate increases upon partial devitrification. This effect for SM58 was also found by others [19], and can be correlated with the weakening of the glass matrix upon partial removal of SiO_2 from the matrix into the crystals [14].

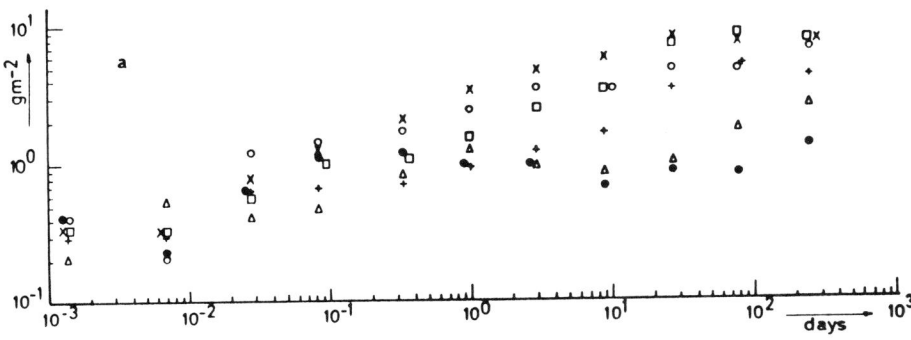

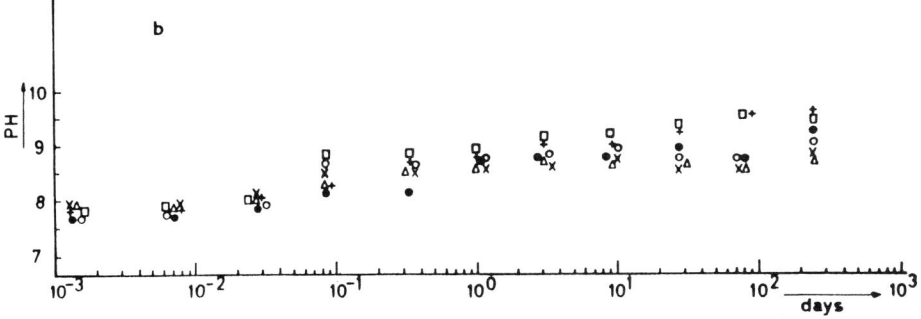

Fig. 5.1 Mass loss (a) and pH (b) as a function of corrosion time
        (DW, 90°C, 100 m^{-1})

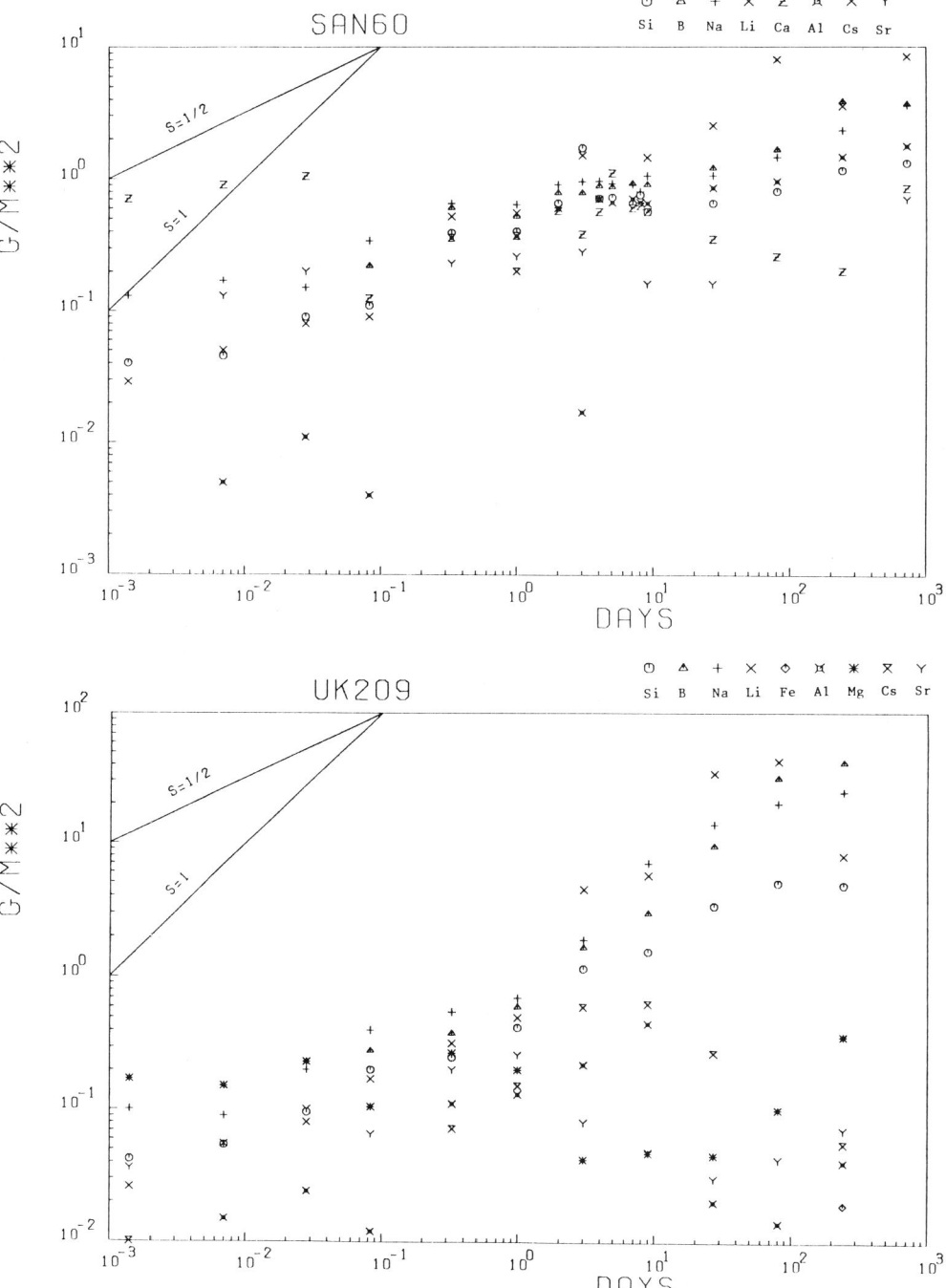

Fig. 5.2 Normalised mass loss vs corrosion time (DW, 90°C, 100 m^{-1})
(a) SAN 60 25 19 L_3C_2, (b) UK209

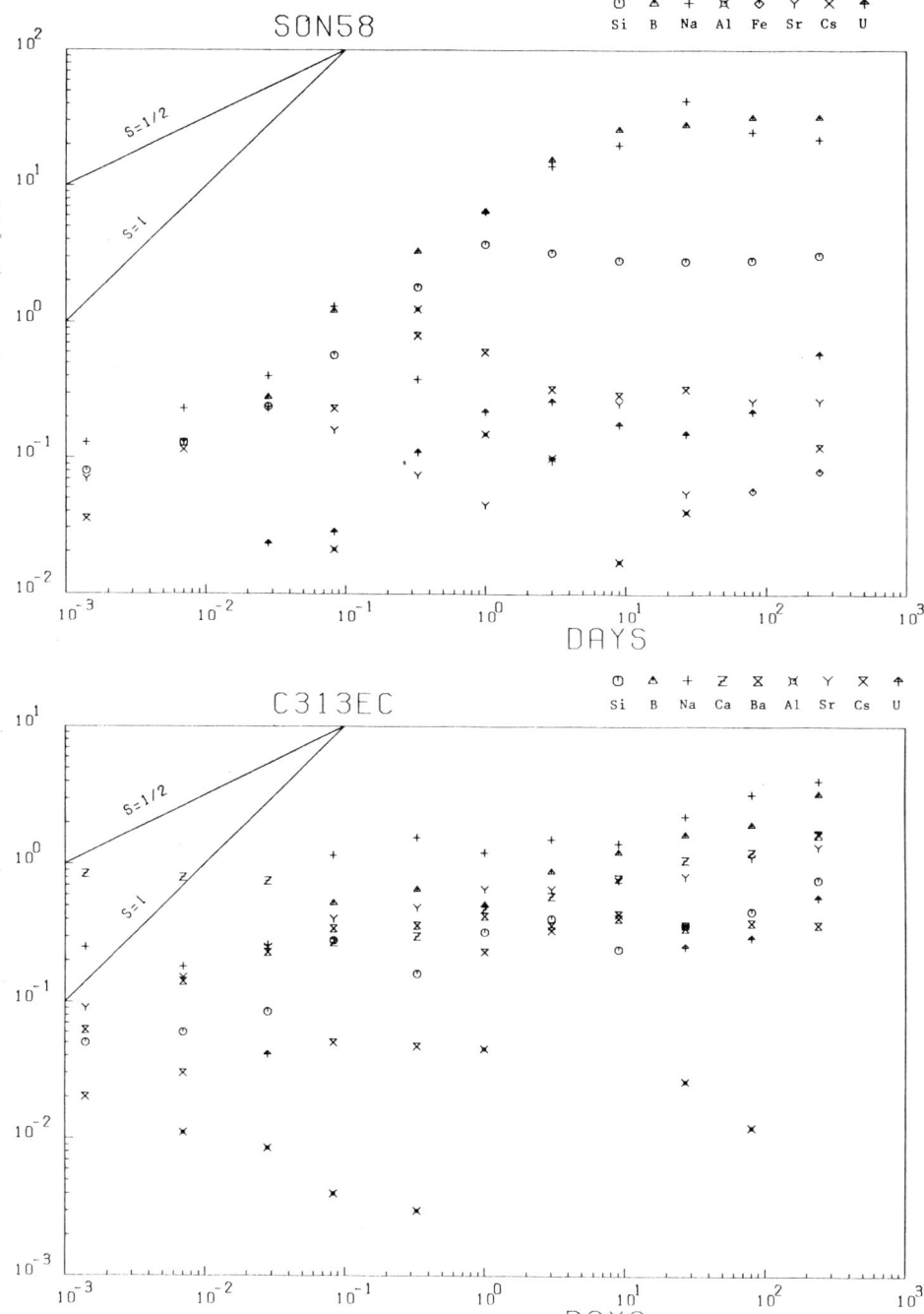

Fig. 5.3 Normalised mass loss vs corrosion time (DW, 90°C, 100 m^{-1})
(a) SON 58 30 20 U2 (b) Glass ceramic C31-3EC

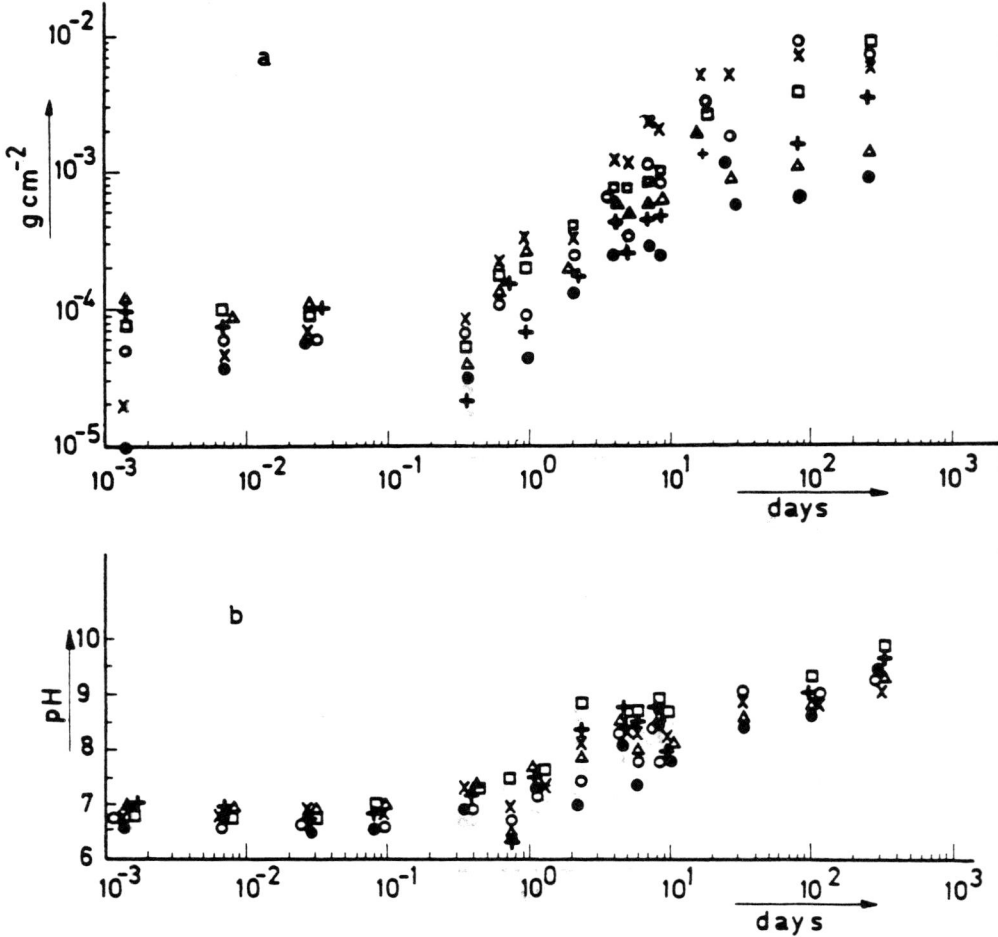

Fig. 5.4 Mass loss (a) and pH (b) as a function of corrosion time (DW, 90°C, 10 m^{-1})
□ SM 58; X SON 58; O SON 65; + UK209; △ SAN 60; ● C31

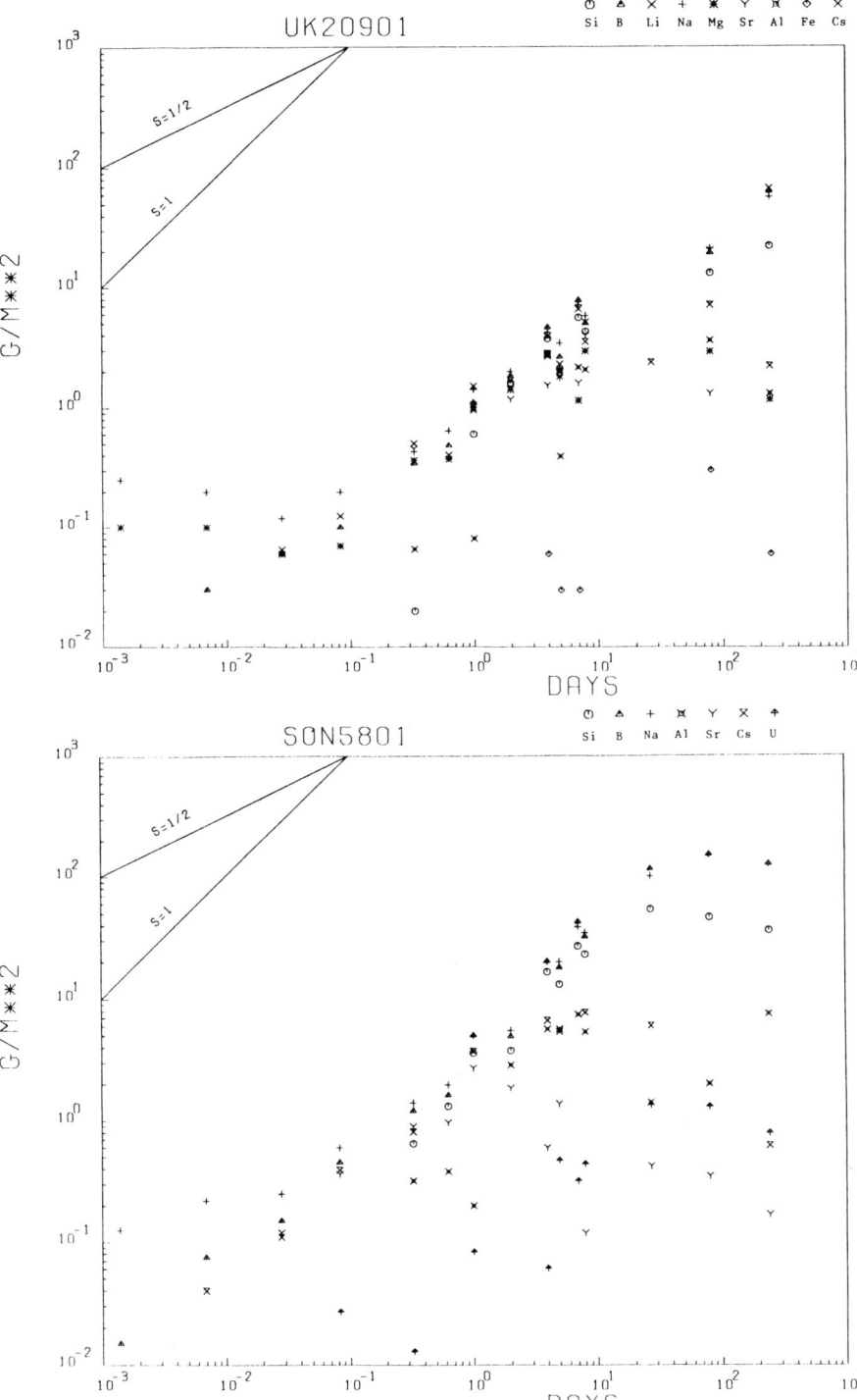

Fig. 5.5 Normalised mass loss vs corrosion time (DW, 90°C, 10 m^{-1}) (a) UK209, (b) SON 58 30 20 U_2

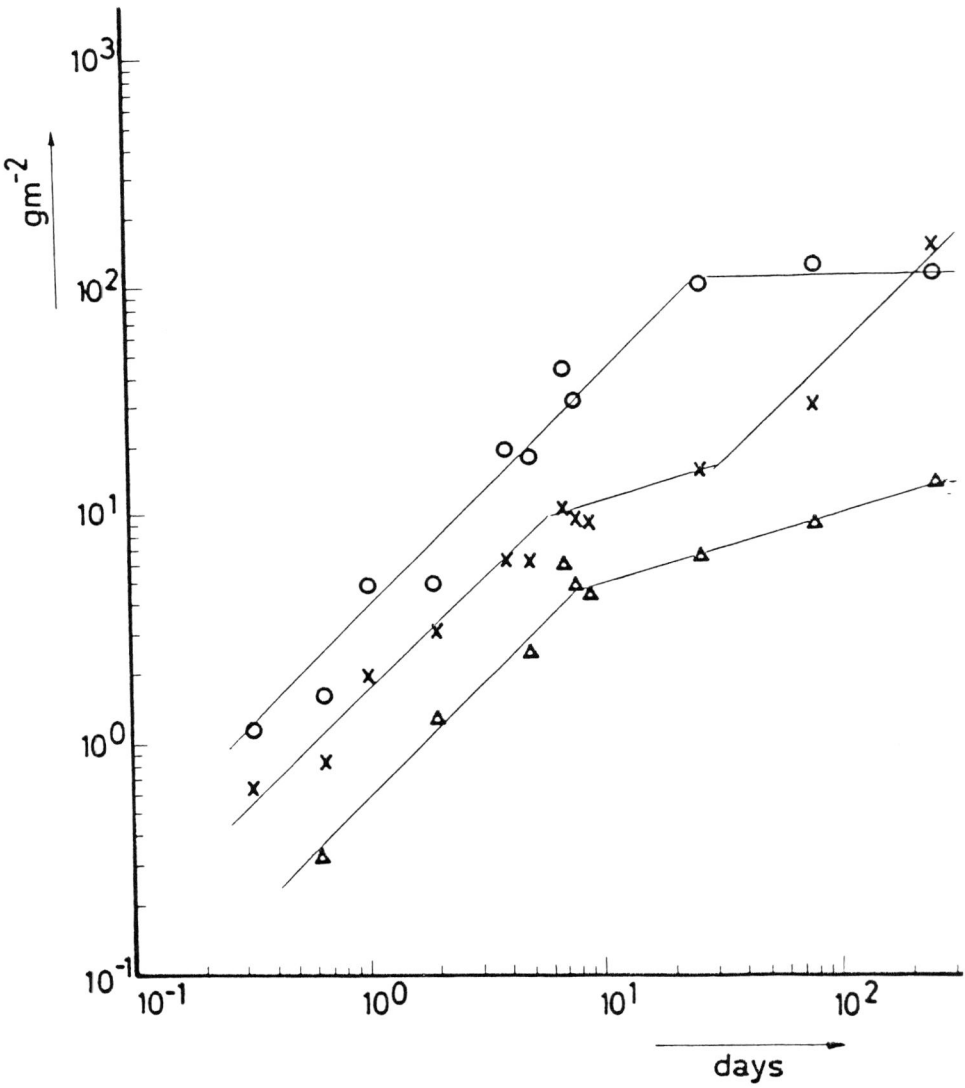

Fig. 5.6 Normalised B mass losses (DW, 90°C, 10 m^{-1})
X SM 58 LW 11; O SON 58 30 20 U$_2$; Δ SAN 60 25 19 L$_3$C$_2$

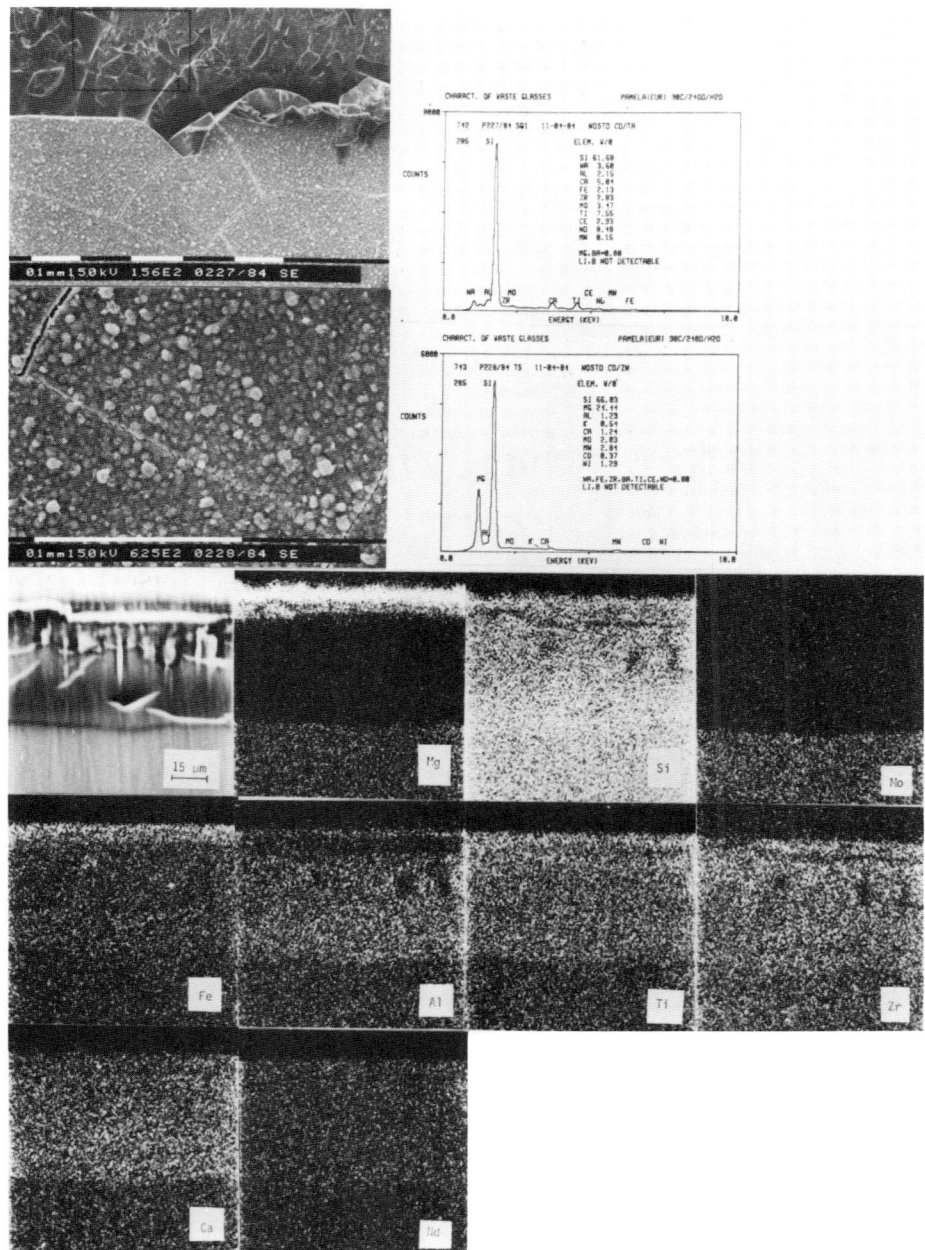

Fig. 5.7 SEM-EDXA (a) and EPMA X-ray cross-section (b) analysis of glass SM 58 LW 11 after 240 days corrosion at 90°C, 10 m^{-1}

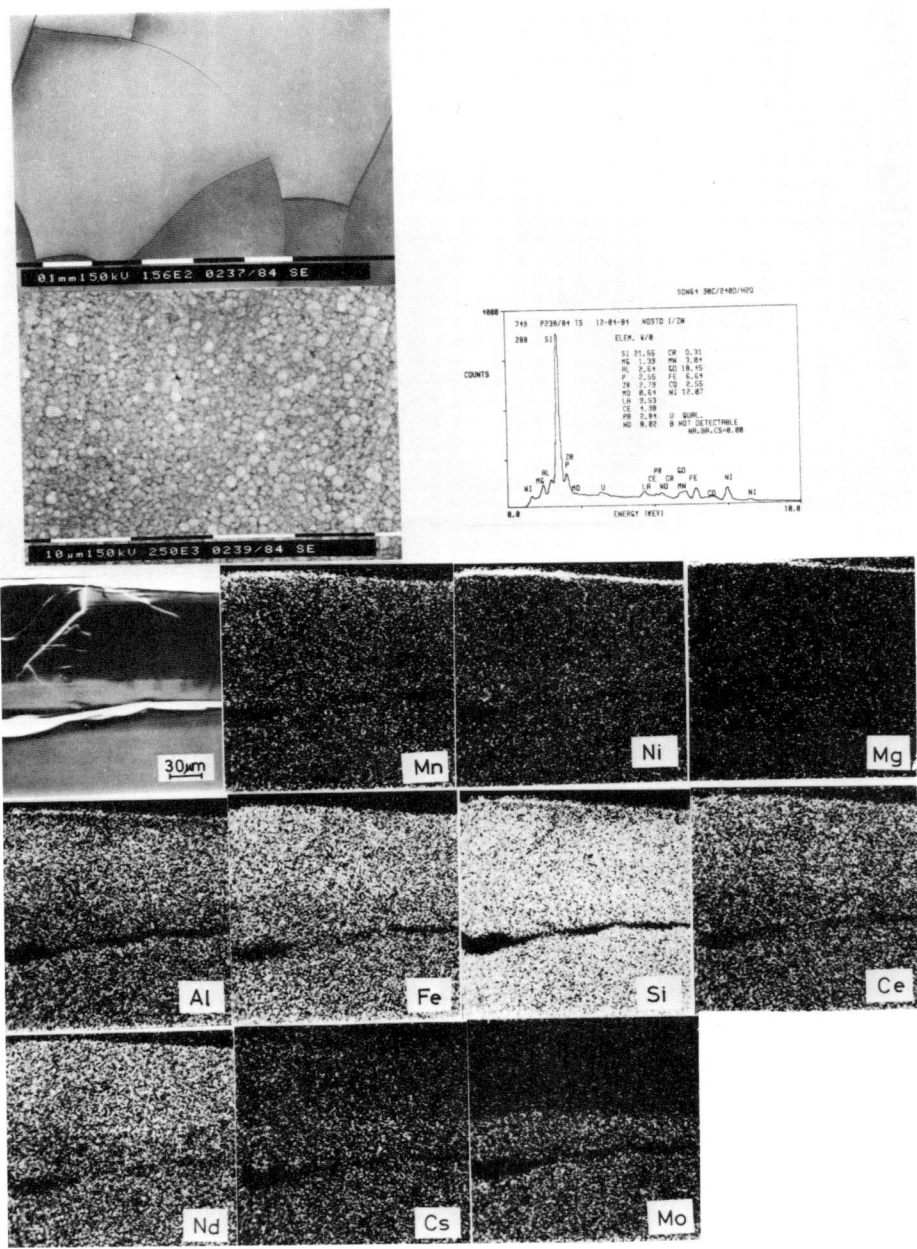

Fig. 5.8 SEM-EDXA (a) and EPMA X-ray cross section (b) analysis of glass SON 64 19 20 F_3 after 240 days corrosion at 90°C, 100 m^{-1}. Both SEM photographs show the outer surface of the corroded glass

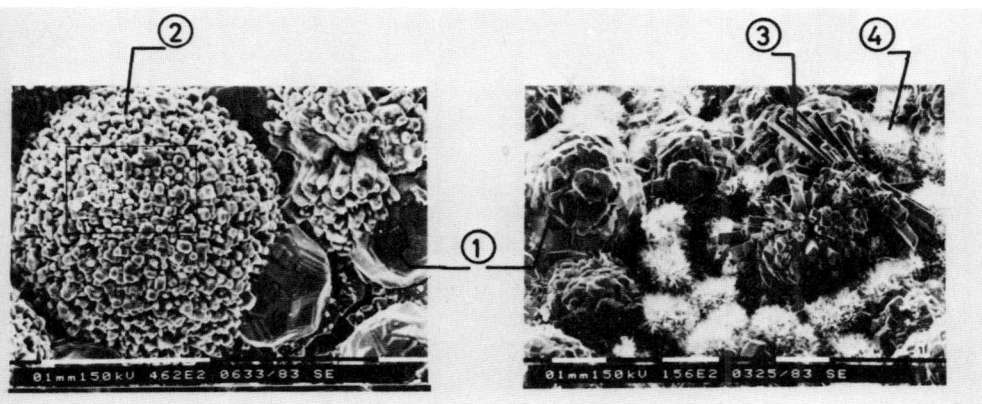

Fig. 5.9 SEM views of glass SAN 60 corroded for 28 days in distilled water. (a) 150°C, (b) 190°C. The phases indicated correspond to (1) Analcime ($NaAlSi_2O_6 \cdot H_2O$); (2) Zeolite ($Na_3Al_3Si_5O_{16} \cdot 6H_2O$); (3) Eucryptite ($LiAlSiO_4 \cdot 2H_2O$); (4) a Ca,Si-based crystal

Fig. 5.10 SEM cross-section view of Glass SAN 60 corroded at 190°C for 28 days in a 50% pre-concentrated leachate (SA/V = 20 m^{-1})

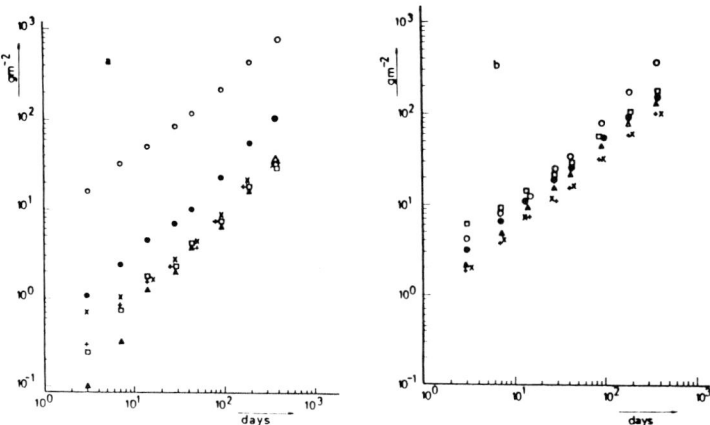

Fig. 5.11 Mass losses in different pH conditions: (a) pH 5.7; (b) pH 9.5 (90°C, 10 m^{-1}). □ SM 58; X SON 58; O SON 64; + UK209; Δ SAN 60; ● C31

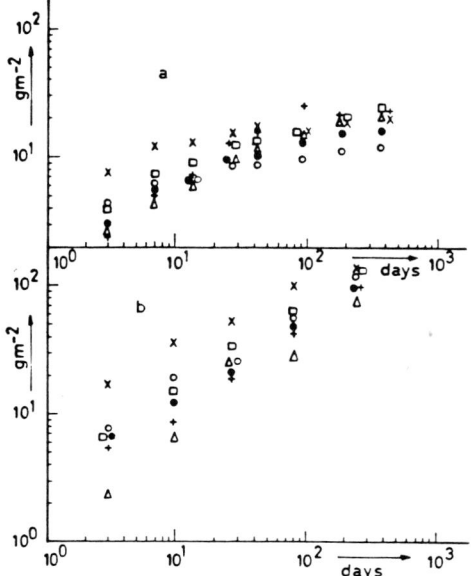

Fig. 5.12 Mass losses vs corrosion time at 90°C, 100 m^{-1} for (a) clay-water mixture and (b) wet clay.
□ SM 58; X SON 58; O SON 64; + UK209; Δ SAN 60; ● C31

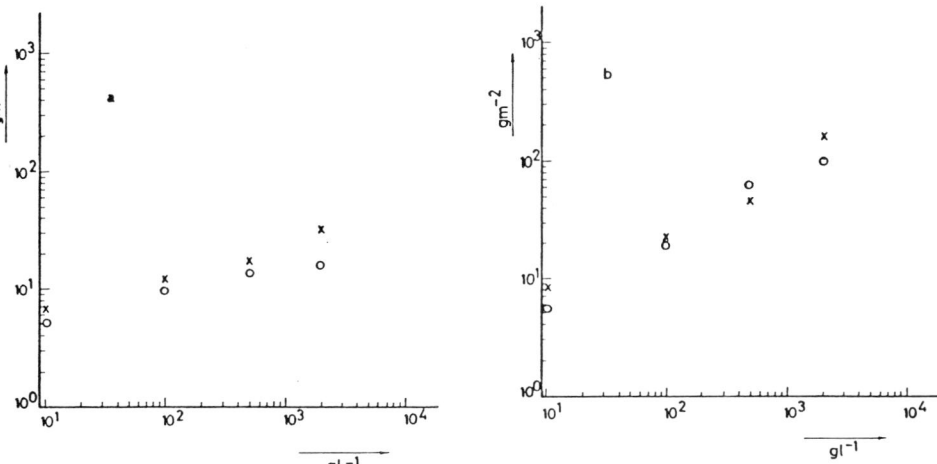

Fig. 5.13 Mass losses as a function of clay to water ratio, (a) 28 days; (b) 364 days. X SM 58; O SAN 60

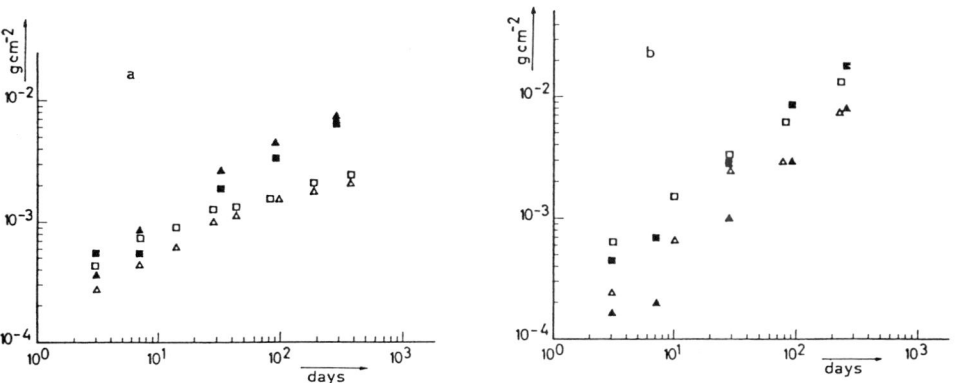

Fig. 5.16 Mass loss vs corrosion time at 90°C, 100 m^{-1} in (a) clay-water mixture and (b) wet clay under different redox conditions for glass SM 58
(□ oxidising, ■ reducing) and for SAN 60 (△ oxidising, ▲ reducing)

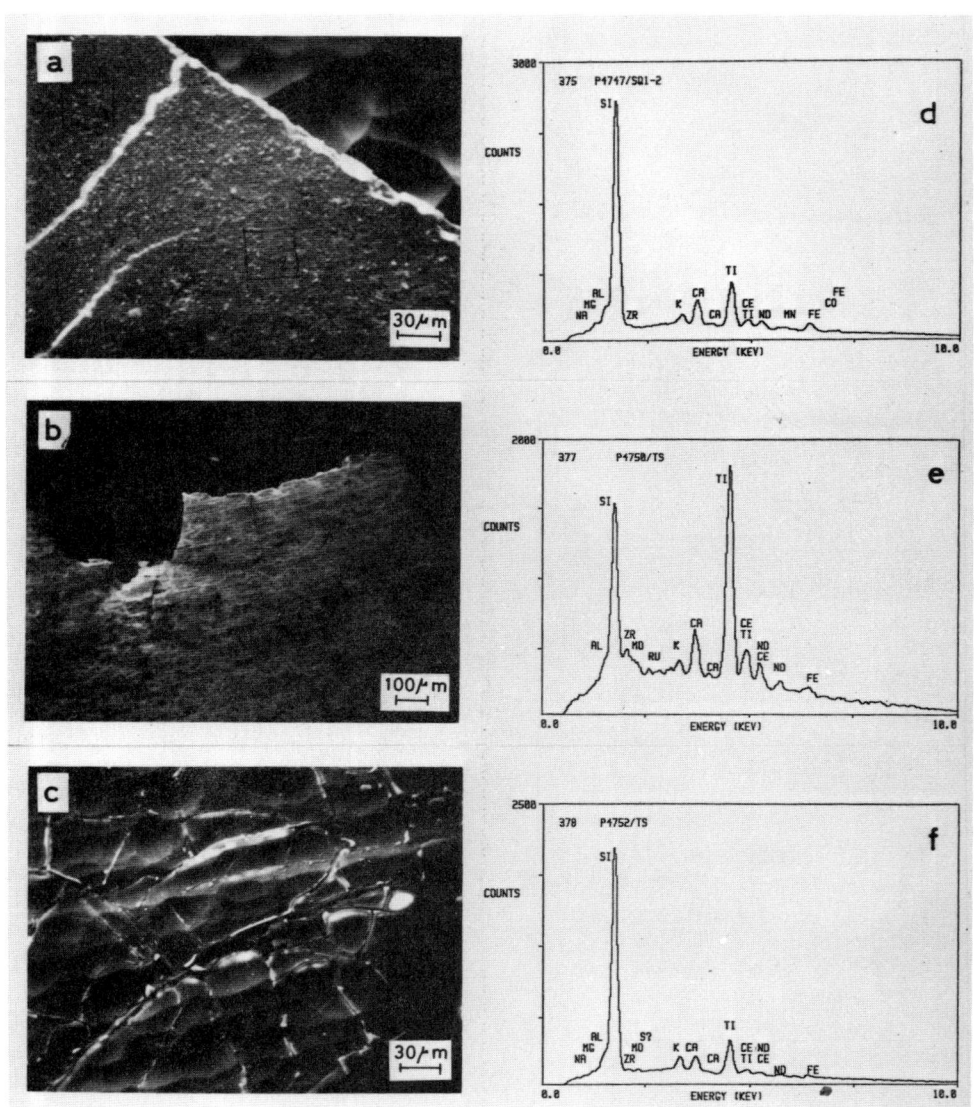

Fig. 5.14 Surface of glass SM 58 LW 11 after leaching at 90°C, 100 m^{-1}, in (a) CWM for 12 months and (b) WC for 8 months. (c) represents the glass beneath the surface layer of (b). The corresponding EDXA analyses are shown in (d), (e) and (f)

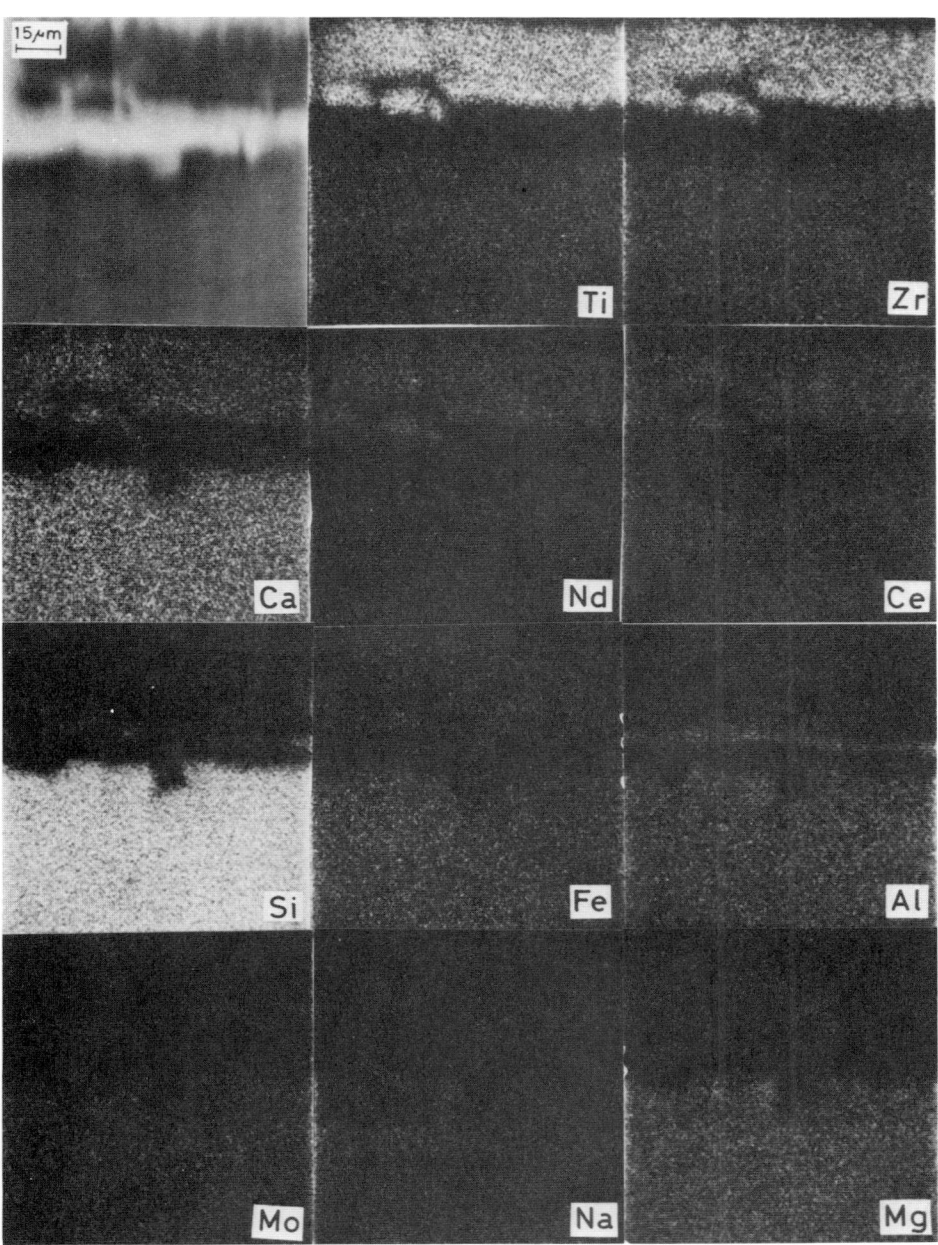

Fig. 5.15 EPMA X-ray cross-section images of glass SM 68 LW 11 after 8 months corrosion in wet clay (90°C, 100 m^{-1})

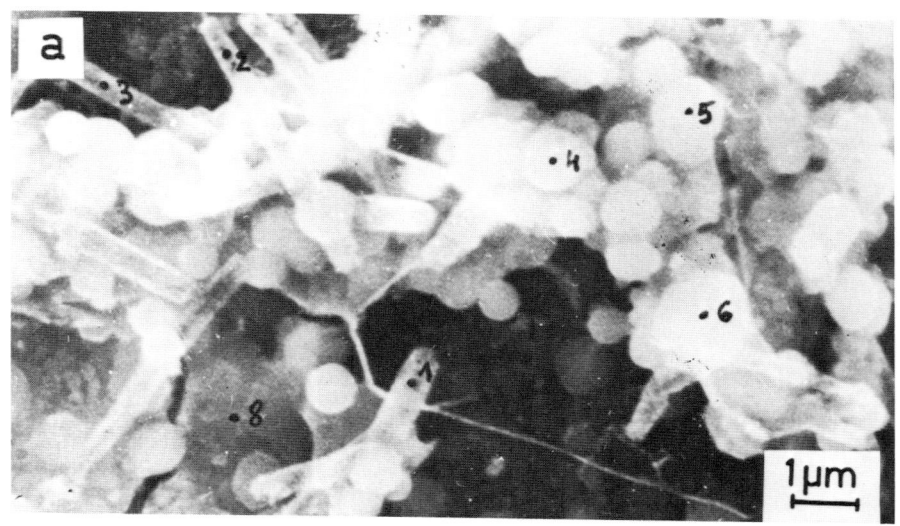

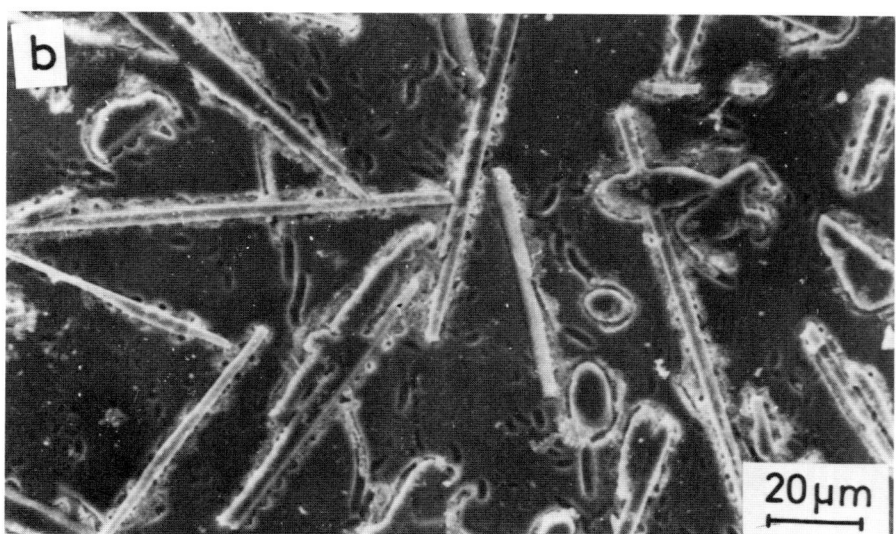

Fig. 5.17 SEM observations of glasses SAN 60 25 19 L_3C_2 (upper) and SEM 58 LW 11 (lower) after partial crystallisation

5.2 LEACHING OF AMERICIUM FROM GLASSES UK209 AND C31-3EC. (HMI BERLIN)

Americium-doped glasses of the type UK209 and C31-3EC were leached in saturated NaCl and Q-brine at 200°C and in a Soxhlet apparatus at 100°C. The composition of the solutions and the details of the samples and their preparation are given in [7].

α-emission spectra of the pristine glasses exhibit significant differences for the two glasses as shown in Fig 5.18. Although the nominal Am-concentrations are identical, the glass UK209 has a much higher yield/channel. Before attributing the yield to a concentration value the energy (channel) scale has to be converted to a corresponding depth scale using the actual value of the stopping power. Even for a constant concentration the spectrum has a final slope caused by the energy dependence of the stopping power. Only the product of the yield and the local stopping power is proportional to the concentration of the α-emitting nuclide.

The parameters of the samples used in the Soxhlet test are given in [7]. After 23 days the average weight loss of the glass C31-3EC was about $4.10^{-5}.cm^{-2}.d^{-1}$ and 2.10^{-4} $g.cm^{-2}d^{-1}$ for the glass UK209. Both values were also measured for inactive samples. The Am-241 activity in the Soxhlet solution was monitored during the run of the experiment. Whereas for the glass UK209 a continuously decreasing solubility was observed the opposite was the case for the glass C31-3 where the concentration of Am increased linearly. However, in both cases the major amount of Am collected in the bottom container was adsorbed on the wall of the container. Even in the siphon system itself a high amount of precipitated Am was found. Final results after 23 days are summarized in Table 5.14. Nevertheless, for both glasses only a few percent remained in solution. Am apparent retention factor of 91% for the glass UK209 and 62% for the glass C31-3 is calculated by comparing the amount of americium removed from the glass to the amount of Am corresponding to the total mass loss, ie even in a Soxhlet test most of the Am stays on the surface of the leached glass as clearly demonstrated in the α-emission spectrum of the glass UK209 after leaching (Fig 5.19).

The amount of Am in the surface peak is almost equivalent to the total loss and is concentrated in a layer of about 3 to 5 μm thickness.

TABLE 5.14   Activity balance of the glasses C31-3 and UK209 after leaching for 23 days in a Soxhlet apparatus

	Am-241 activity/µCi	
	UK 209	C-31-3
Activity/g glass	2.1	1.5
Mass loss equivalent	28.6	3.6
Total in solution	0.043	0.053
In Soxhlet container	0.97	0.68
In Soxhlet siphon	1.5	0.7
Saturation/ml	0.035	not reached
Retained in surface layer / %	91.0	62.0
Ratio (solved/precipitated) / %	2.0	4.0

TABLE 5.15   Data on the leaching of Am spiked glasses

	SAMPLE-ID	SOLUTION	TIME/DAYS	MASS LOSS/MG	SPEC. MASS LOSS/MGCM^{-2}	S/V/CM^{-1}		AM-COUNTS IN SOLUTION	AM-COUNTS IN SURF. LAYER	RETENTION FACTOR/%	
UK209	UK3	NACL	3	4.25	2.63	0.04		- 60	1008	94.4	UK209
	UK1	NAGL	3	6.53	2.79	0.06		20	2026	99.0	
	UK3	NACL	10	20.26	6.51	0.08		0.8	4620	99.9	
	UK2	Q	3	6.76	3.33	0.05		1980	32	1.6	
	UK3	Q	3	5.03	3.12	0.04		1774	21	1.2	
	UK2	Q	10	7.66	3.77	0.06		1228	15	1.2	
C31-3 EC	KA3	NACL	10	27.75	10.08	0.06		24	8786	99.7	C31-3 EC
	KA4	NACL	10	21.13	7.62	0.06		60	7373	99.2	
	KA10	NACL	47	40.30	25.34	0.04		9	4769	99.8	
	KA17	NACL	47	47.80	31.60	0.04		6	8033	99.9	
	KA14B	Q	3	10.07	6.71	0.04	*	3007	176	5.5	
	KA15B	Q	3	10.40	6.89	0.04		3158	57	1.8	
	KA7B	Q	10	13.13	7.77	0.05		1845	179	8.9	
	KA8B	Q	10	13.45	8.15	0.05		3251	90	2.7	
	KA7A	Q	30	12.52	8.00	0.04		2247	30	1.3	
	KA8A	Q	30	11.70	8.23	0.04		2034	36	1.7	
	KA14A	Q	30	11.88	7.29	0.04		3339	26	0.8	
	KA15A	Q	30	13.90	9.50	0.04		3929	50	1.3	

In the autoclave tests the glasses were leached in saturated NaCl-solution and Q-brine from 3 to 47 days. The volume of the leachate was always 40 ml and the sample surface to brine volume ratio ranged from 0.04 to 0.08 cm^{-1}. Total mass losses were determined after removing the surface layers by ultrasonic methods or simple scraping. The results are given in Table 5.15. Comparison with Fig 4.15 shows that, for the glass C31-3, the mass losses follow the same kinetics as measured for inactive samples. In NaCl the initial leach rate is very high, but has already decreased significantly after three days. The slope between 30 and 47 days is almost identical to the slope between 90 and 360 days observed for the inactive samples (Section 4.1.3).

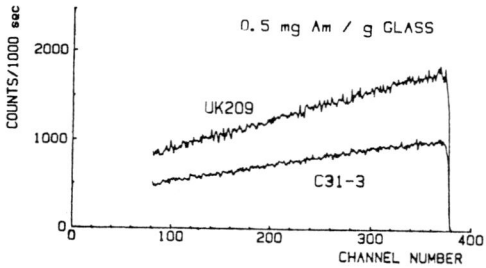

Fig 5.18
α-emission spectra of the pristine glasses C31-3 and UK209 doped with Am-241.

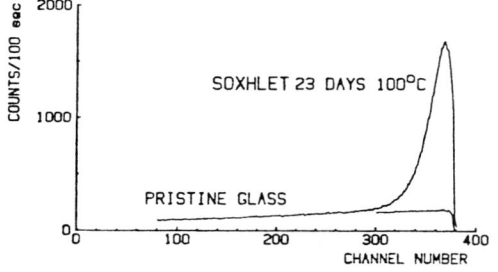

Fig 5.19
α-emission spectrum of Am-241 doped UK209 after a Soxhlet test.

Total mass losses in Q-brine are smaller than in NaCl and the leach rates of the glass UK209 are always smaller than those measured for the glass C31-3. pH-values were determined after the leaching experiments at room temperature by means of a Ross electrode. The values ranged from 8 to 9 in NaCl solutions and 4 to 5 in Q-brine and may be used to explain the solubility of Am in the two solutions. For both glasses more than 99% of the Am corresponding to the total mass loss was found in the surface layers after leaching in NaCl.

After leaching in Q-brine the opposite was the case. Only a few percent of the total amount of Am stayed in the surface layer. More than 90% was in solution for the glass C31-3 and more than 98% for the glass UK209. Data of all experiments are summarized in Table 5.15.

α-emission spectra were recorded before the removal of the surface layers. In layers grown during leaching in Q-brine almost no Am could be detected as already expected due to the high concentration of Am in the leachate.

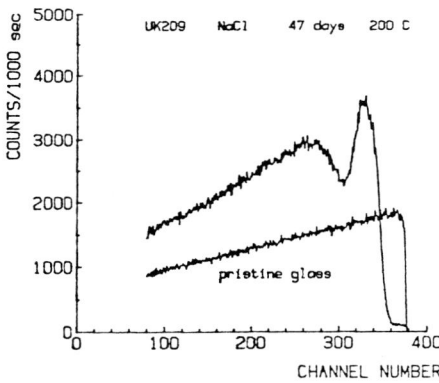

Fig 5.20   α-spectra of glass UK209 before and after leaching 47 days in NaCl at 200°C.

The α-emission spectrum from the glass UK209 after leaching in NaCl for 47 days is shown in Fig 5.20 and compared with the spectrum emitted from the pristine glass. The low yield in the near surface region (channels 360 - 380) is probably caused by analcime crystals covering the surface. The region is followed by a peak indicating a thin layer of possible Am-enrichment (channels 320 - 360). From channels 280 - 320 an apparent Am-depletion is observed whereas the high yield from channels 100 - 360 may be caused by a remarkable change of the stopping power due to the water uptake in the gel layer. Although α-emission spectra reflect in a certain way a layered structure close to the surface of a material, quantitative evaluation is often ambiguous as the local stopping power is not known. Additionally, the integral of a relatively large area can be measured and complementary methods (SEM, EPMA) are needed for correct interpretation of the spectra.

5.3 EXPERIMENTAL STUDY OF THE INFLUENCE OF VARIOUS PARAMETERS ON THE AQUEOUS LEACH RATES OF RADIOACTIVE GLASS (MARCOULE)

5.3.1 Introduction

The glasses used are of the first and second generation 'SON' type (SiO_2, FP Oxide, Na_2O, B_2O_3) suitable for the vitrification of waste from the reprocessing of Light Water Reactor fuel. The 3rd generation LWR glasses are those developed for the R7T7 vitrification plant at La Hague. The compositions used are detailed in Table 5.16.

The active glasses, being made in VULCAIN with real fission product solutions, contain all the aluminium resulting from the process even if its theoretical composition contains none; the Al_2O_3 content, however, never exceeds 2 wt%. Some solutions contain significant quantities of iron oxide (up to 6 wt%).

For the active glass blocks, the tests include the following radioanalyses:

γ-spectrometry for Cs, Ru, Ce, Sb, Co
Total β counting
Radiochemical measurements of Sr
Total α counting

in studying the effect of a given parameter, the same sample is always re-used when the parameter is changed.

TABLE 5.16   Detailed compositions of glasses used

Verre\Oxydes	SON 45 30 14 L2 (A 65)	SON 58 30 20 U2 (A 84)	SON 60 20 18 F3 (A 100)	SON 60 30 14 U2 (A 79)	SON 61 19 20 F3 (A 97)	SON 61 30 14 a (A 53)	SON 62 20 24 F3 (A 98)	SON 64 19 20 F3 (A 96)	SON 60 19 20 F 3G3 (A 87)	SON 68 24 A 2 (A 75)
SiO_2	36,91	43,3	45,11	48,27	44,96	50,82	43,21	47,16	43,8	45,2
Al_2O_3	1,78	1,0	1,69	1,62	1,70	-	1,70	1,69	1,2	4,8
B_2O_3	12,86	18,8	16,50	13,09	18,43	13,39	22,11	18,42	17,15	21,9
Na_2O	19,49	9,3	15,04	8,05	14,74	8,25	12,60	12,53	11,4	8,68
CaO	0,03	-	0,01	0,03	0,01	-	0,01	0,03	0,01	0,03
NiO	-	-	0,26	-	0,24	-	0,24	0,24	-	-
Fe_2O_3	0,66	0,6	5,44	0,73	5,07	-	5,07	5,07	5,86	0,60
Cr_2O_3	0,13	0,2	0,47	0,05	0,43	-	0,43	0,43	0,49	0,10
P_2O_5	-	0,6	0,38	-	0,36	-	0,36	0,36	0,4	-
F^-	0,06	-	0,04	-	0,04	-	0,04	0,04	-	0,06
SO_4	-	-	0,02	-	0,02	-	0,02	0,02	-	-
AmO_2	-	-	-	-	-	2,83	-	-	-	-
MgO	0,24	0,1	-	0,15	-	-	-	-	0,06	0,18
OX PF + ACT	26,4	26,1	15,0	28,0	14,0	24,7	12,7	14,0	19,6	18,7
Cs_2O	2,33	2,0	1,36	2,21	1,27	2,41	1,27	1,27	1,24	1,44
SrO	0,70	0,7	0,44	0,72	0,41	0,72	0,41	0,41	0,40	0,47
BaO	1,21	1,0	0,67	1,10	0,63	1,25	0,63	0,63	0,52	0,71
ZrO_2	2,01	3,1	2,11	3,41	1,96	3,51	1,96	1,96	1,92	2,22
MoO_3	4,35	4,0	2,48	4,30	2,31	4,48	2,31	2,31	2,36	2,80
NiO	1,00	1,1	0,36	1,05	0,33	1,02	0,33	0,33	0,52	0,69
MnO_2	2,05	1,9	0,68	2,02	0,64	2,11	0,64	0,64	0,62	1,31
SnO_2	0,04	Ng	0,03	0,04	0,03	0,05	0,03	0,03	0,02	-
Ag_2O	0,03	0,1	0,03	0,04	0,03	0,03	0,03	0,03	0,02	-
CdO	0,07	0,1	0,04	0,06	0,04	0,07	0,04	0,04	0,03	
Te_3O	-	-	0,33	-	0,31	-	0,31	0,31	-	-
CeO	0,31	0,3	0,11	0,33	0,10	0,32	0,10	0,10	0,10	0,69
Y_2O_3	0,39	0,4	0,25	0,40	0,23	0,40	0,23	0,23	0,23	0,26
La_2O_3	1,89	1,8	1,22	1,98	1,14	1,95	1,14	1,14	1,12	1,29
Pr_2O_3	0,94	0,9	0,60	0,97	0,56	0,97	0,56	0,56	0,55	0,63
Nd_2O_3	3,22	3,0	1,99	3,22	1,85	3,32	1,85	1,85	1,81	2,10
Gd_2O_3	-	-	-	-	-	-	-	-	5,86	-
Ce_2O_3	2,01	2,1	1,39	2,25	1,29	2,07	1,29	1,29	1,27	1,46
ThO_2	-	-	0,46	-	0,43	-	0,43	0,43	-	-
UO_2	3,84	3,6	0,48	3,85	0,45	-	0,45	0,45	0,89	2,63
Activité Ci	40 a	512,3 a	249 a	25,6 a	407,9 b	150 a	318 b	274 a	60 b	95,8 a

For inactive glasses the tests comprise

    Element analysis of the leachate
    Mass loss

In studying the effect of a given parameter a different sample is used for each value of the parameter.

Operating conditions and apparatus used are described in the text for each parametric variation.

The work has been described in two annual reports [20,21] and three half-yearly reports [22,23,24].

5.3.2 Effect of time

The effect of time at ambient temperature has been investigated for two glass compositions:

    SON 62 30 14 actively on glass spiked with ^{241}Am.
    SON 58 30 20 U2 on βγ-active glass and on inactive glass.

The leaching mode is static with daily leachant renewal. The SA/V ratio is 0.6 cm^{-1}, leachant is the filtered industrial water available in the VULCAIN cell.

    Glass SON 61 30 14, α-active (A53)

This glass block was made in VULCAIN in 1975. It has a specific activity of 0.086 Ci.g^{-1}. Prior to this study it had undergone 472 days leaching in various tests.

This material has thus undergone approximately 5×10^{17} α-disintegrations per gramme and having developed a hydrated layer under irradiation.
The apparatus used for this study is the static VULCAIN system shown in Fig 5.21.

During this test the glass block has undergone a total of 739 days leaching as follows:

1. From day 1 to day 161. Static leaching with daily leachant renewal.

2. From day 162 to day 270. Static leaching in a new vessel, without leachant renewal.

3. From day 271 to day 730. Static leaching with daily leachant renewal.

The curve giving the changes in daily leach rate is given in Fig 5.22. The calculated cumulative dose during this period is from 5.3×10^{17} α/g to 7.3×10^{17} α/g.

In the third period (271 - 730 days) the mean leach rate has fallen by a factor of 2.

Comparison of the activities found in the leachates and in the acid used to rinse the apparatus are given in Table 5.17. This gives the order of magnitude of the deviations between the activities measured for the leach rates and those which have actually left the glass in the form of glass grains, complexing products, or debris from the interface layer.

However, the very rapid (several days) dissolution of the fixed activity during acid rinsing suggests that it is a thoroughly degraded reaction product that is being dissolved, ie complexing products or debris from the interfacial layer.

βγ-active glass SON 58 30 20 U2 (A84)

This glass has undergone 500 days static leaching at ambient temperature.

TABLE 5.17   Activities and leach rates. Glass A53

Period (days) Duration (days) Leaching mode	1 - 161 161 Static, with daily leachant renewal	162 - 270 109 Static, no leachant renewal	271 - 730 469 Static, with leachant renewal
α-activity in leachant (μCi)	984	3.5	693
Mean leach rate $\bar{\ell}$	1.5×10^{-7}	7.8×10^{-10}	3.6×10^{-8}
α-activity in rinsing acid (μCi)	1390	58	2630
Total α-activity recovered (μCi)	2374	61.5	3323
Mean leach rate ℓ (g.cm^{-2}.day^{-1})	3.6×10^{-7}	1.3×10^{-8}	1.7×10^{-7}

59, 94 and 79% respectively of the activity extracted during the first, second and third periods remained fixed on the leaching apparatus.

The leaching curves for total β, Cs, Sr, Ru, Ce, Sb and Co are given in Fig 5.23. The leachant was renewed daily, except for a few periods of no renewal recognisable by discontinuities in the curves. The level of the horizontal trace gives the daily leach rate at the end of the period without renewal.

Table 5.18 summarises the values of leach rates after 35, 200, 300, 400 and 500 days. The time intervals over which the mean values are calculated are indicated in parentheses.

The acid rinsings of the leach vessel after 500 days leaching showed that the fixed $\beta\gamma$-activity on the vessel was negligible by comparison with the leached activity.

TABLE 5.18   Leach rates after various times

Period	T.Ambient (estimated) °C	Leach rate $\bar{x}$ (x 10^{-7} g.cm^{-2}.day^{-1})						
		β	Cs	Sr	Ru	Ce	Sb	Co
(30-40)	25	2.2	6.5	1.1	2.0	0.2	30.0	9.0
(195-200)	23	1.3	2.8	0.79	0.94	0.04	7.5	5.5
(295-305)	20	0.86	2.2	0.5	0.85	0.03	8.1	2.4
(395-405)	28	2.9	5.9	2	0.64	0.04	13.0	5.5
(495-505)	25	2.4	5.8	1.5	0.75	0.04	8.2	6.4

It can be seen that the total-β leach rate decreases during the first 300 days, then increases between 300 and 400 days. The ambient temperature in VULCAIN varied considerably during this period. The values of this temperature, estimated from records made during the relevant periods, are given in Table 5.18. It can be seen that the leach rates for total-β, Cs and Sr exactly follow this temperature. However, one cannot be sure that this temperature variation is the only parameter concerned, having regard for the magnitude of the phenomenon (a factor of 3 for Cs, 4 for Sr).

Leach rates for Ru and Ce decrease during the first 200 days and change little thereafter. It has already been observed that they change but little with temperature. The values clearly show a non-congruent dissolution.

The much reduced leach rates during periods of no leachant renewal are particularly conspicuous for Ce and Co, less so for the other elements. This is a surprising result since generally Ce and Ru are more affected than Cs by the frequency of leachant renewal. The peculiar behaviour of this glass is not explained. Note that at the resumption of daily renewal the leach rate is generally higher.

Cumulative leach curves are plotted in log.co-ordinates in Fig 5.24. The slopes of these curves represent values of the exponent in the expression

$$\Sigma \ell = Kt^n$$

Some values of the slopes are given in Table 5.19.

TABLE 5.19 Slopes of curves taken from Fig 5.24

Element Time (days)	Values of n at various times						
	β	Cs	Sr	Ru	Ce	Sb	Co
2	0.4	0.5	0.3	0.3	0.4	0.3	0.3
35	0.5	0.5	0.5	0.5	0.5	0.2	0.1
200	0.6	0.6	0.7	0.4	0.3	0.1	0.3
300	0.6	0.6	0.5	0.5	0.4	0.2	0.2

Table 5.19 shows that even after 300 days leaching congruent dissolution is not obtained.

Values of the slopes beyond 300 days have not been taken into consideration, having regard for the large variations in ambient temperature.

Inactive glass SON 58 30 20 U2 (A84)

This glass has undergone 365 days leaching.

The apparatus, constructed from Teflon, is shown in Fig 5.25.

With the procedure used (daily leachant renewal, ambient temperature), no measure of leach rate could be made; the coupon was weighed before and after 365 days leaching.

The weights were:

Initial weight (after heating for 1 hour at 120°C)   14.6190 g
Final weight (after drying, before heating)   14.6436 g
Final weight (after heating for 1 hour at 120°C)   14.6424 g

This result can only be explained by formation in the interface layer of compounds, more or less hydrated, either amorphous or crystalline.

It is also possible that there has been an ion exchange between ions carried by the industrial water and those released by the glass.

The value of pH attained after 24 hours leaching did not vary throughout the test.

$t = 0$, pH = 7.8
$t = 24$ hours, pH = 8.4

Optical examination of the glass surface (Fig 5.26) revealed a beige granular layer (or deposit?) of irregular texture and thickness, of maximum thickness 100 to 120 microns.

Conclusions

Within the system studied (ambient temperature, static leaching with daily leachant renewal, industrial water, SA/V = 0.6 cm^{-1}) the leaching of the two active glasses ($\beta\gamma$ and α) of different chemical compositions show that congruent corrosion is not attained and that elements whose hydroxides are insoluble in alkaline media have the lowest leach rates (actinides, rare earths). The persistence of this deviation between the rates of dissolution of different oxides, even after 500 hours in the $\beta\gamma$ glass, show that the thickness of the interfacial layer continues to increase and that the profile concentrations of the elements of which it is composed are still changing. However, the solubility limit for an element is relative,

for it depends on its chemical form (simple or complex ion) and of the anion with which it can combine (carbonate, hydroxide, sulphate, etc).

The gain in weight observed with the inactive sample is somewhat surprising. Unfortunately, it is not possible to determine whether this effect also occurs with an active sample because of the difficulty of measuring a weight loss in the region of 10^{-5} g, in the VULCAIN cell.

The most likely explanation for this phenomenon is that the hydrolysed layer resulting from the attack of the glass has adsorptive properties for certain of the species present in industrial water (notably calcium).

Finally, the presence of the α-emitter in the americium glass shows no phenomenon indicating any radiolytic effect on time-dependence.

5.3.3 Effect of Temperature at Atmospheric Pressure

Two glass compositions were used (Table 5.16).

    SON 64 19 20 F3
    SON 60 20 18 F3

The tests were made on both active and inactive glasses. The leachate (industrial water ex VULCAIN, Table 5.32) is renewed weekly. Figure 5.27 shows the assembly used for leaching the active samples. This was set up in the 'storage' cell of the VULCAIN facility. The block is only removed from the vessel at the completion of the experiments, typically after 300 days leaching at various temperatures.

The inactive tests are made in 'Mecabar' reaction vessels (Figure 5.28) comprising an 18/8 Mo stainless steel body enclosing a Teflon container. A different specimen was used for each temperature.

(I)   Active glass SON 64 19 20 F3 (A96)

Leach rate values are given for individual elements in Figure 5.29. Table 5.20 gives the mean leach rates obtained at various temperatures. Table 5.21 gives the multiplication factors found in relation to the ambient temperature value for a cycle in which the temperature is increased without returning to ambient between steps.

TABLE 5.20 Effect of Temperature on the Mean Leach Rates of the Active Glasses SON 64 19 20 (A96) and SON 60 20 28 Fe (A100) ($\mu g.cm^{-2}day^{-1}$)

Tempéra-ture °C	β		Cs		Sr		Ru		Ce	
	A 96	A 100	A 96	A 100	A 96	A 100	A 96	A 100	A 96	A 100
25	0,63	7,15	1,1	8,9	0,6	6,0	1,1	2,15	0,021	0,30
50	5,8	22,0	17,0	28,0	8,1	24,0	< 1,1	21,9	≤ 0,045	1,19
70	27,0	30,0	63,0	58,0	25,0	16,0	5,9	9,75	≤ 0,35	0,66
100	22,0	47,0	58,0	79,0	13,0	30,0	< 8,7	9,4	≤ 0,36	2,2
120	31,0	/	64,0	/	23,0	/	< 9,6	/	0,54	/
150	30,0	/	59,0	/	25,0	/	< 4,9	/	0,74	/
180	94,0	/	210,0	/	65,0	/	26,0	/	5,2	/
25	23,0	/	23,0	/	27,0	/	< 3,2	/	0,35	/
70	23,0	/	29,0	/	24,0	/	< 3,8	/	0,17	/
150	26,0	/	45,0	/	24,0	/	≤ 3,5	/	0,25	/
25	37,0	/	26,0	/	43,0	/	≤ 3,1	/	0,28	/

TABLE 5.21 Factors of increase with Temperature of the Leach Rates Compared to 25°C

Tempéra-ture °C	β		Cs		Sr		Ru		Ce	
	A 96	A 100	A 96	A 100	A 96	A 100	A 96	A 100	A 96	A 100
25	1	1	1	1	1	1	1	1	1	1
50	9	3	15	3	14	4	~ 1	10	≤ 2	4
70	43	4	57	7	42	3	5	5	~ 17	2
100	35	7	53	9	22	5	< 8	5	~ 17	7
120	49	/	58	/	38	/	< 9	/	~ 26	/
150	48	/	54	/	42	/	< 4	/	35	/
180	150	/	191	/	108	/	24	/	248	/

Behaviour of Cs and Sr

A large increase is seen between 25° and 50°C (factor of 15), less between 50° and 70°C and no increase at all between 70° and 150°C. A single experiment at 180°C gave a much higher leach rate. A second experiment made from 70 to 150°C confirmed the zero increase found between these two temperatures under static conditions with weekly leachant renewal. The greatly increased leach rate after return to ambient temperature should be noted. One possible explanation of this behaviour could be the re-precipitation of certain elements extracted at high temperatures, on or within the surface layer, which would limit the leach rates above 70°C. After return to ambient temperature the dissolution of these elements, governed by solubility limits, would maintain the leach rates at the same order of magnitude as those obtained at higher temperature.

Behaviour of Ruthenium

The leach rate of this element is increased by a factor of approximately 5 at 70°C: there is no further increase between 70°C and 150°C. The multiplication factor is 24 at 180°C.

Behaviour of Cerium

Up to 150°C a behaviour intermediate between that of ruthenium and strontium is seen. A single point at 180°C appears to indicate a very much larger increase at this temperature.

Effect of time

Comparison of the first experiment at 70°C to 150°C with the second experiment after 175 days leaching shows a decrease in leach rate for all elements except strontium.

(II) Active glass SON 60 20 18 F3 (A100)

This glass has been leached successively from ambient to 150°C.

TABLE 5.22   Effect of Temperature – Glass SON 60 20 18 F3 (A100)

Time (weeks)	T°C	\multicolumn{7}{c}{Leach Rate (g.cm^{-2}.d^{-1} × 10^{-6})}						
		B	Cs	Sr	Ru	Ce	Sb	Co
1	25	9,0	13	7,0	3,3	0,47	2,6	2,0
2	25	5,3	4,8	5,0	1,0	0,12	1,2	-
3	25	10,0	10	8,4	< 2,3	< 0,17	5,1	2,6
4	25	5,4	9,5	7,7	3,5	0,20	2,2	2,8
5	25	7,5	5,5	6,4	< 1,4	< 0,10	3,1	2,5
6	50	20,0	27	8,0	3,8	< 0,17	9,0	3,3
7	50	24,0	29	20	20	2,2	13	3,2
8	50	23	25	22	3,2	0,27	12	2,9
9	50	17	19	15	< 1,8	0,52	13	5,2
10	50	18	19	17	< 2,4	0,22	16	0,56
11	70	25	53	15	< 8,5	< 0,66	60	3,5
12	70	35	63	17	< 11	< 0,66	64	1,9
13	70	35	56	22	< 14	< 1,1	64	1,7
14	70	42	63	26	< 12	< 0,75	69	2,9
15	70	37	.58	28	< 12	< 0,75	47	2,9
16	70	40	59	32	< 12	0,8	35	4,1
17	100	ND	ND	ND	ND	ND	ND	ND
18	100	47	79	30	< 9,4	2,2	66	2,8
19	100	57	85	39	23	4,0	47	< 2,6
20	100	70	110	54	< 30	2,5	42	< 10
21	100	51	85	48	< 24	< 1,6	37	< 10
22	120	\multicolumn{7}{l}{No sample – (Leakage at 120°C)}						
23	25	61	95	48	/	/	/	/
24	120	\multicolumn{7}{l}{No sample – (Leakage at 120°C)}						
25	25	65	159	/	/	/	/	/
25/27/28	120	\multicolumn{7}{l}{No sample – (Leakage at 120°C)}						
29	25	62	.140	30	/	/		

Glass dry for 12 days – Vessel changed

31	120	20	33	12	21	< 1,2	140	14
32	120	24	23	22	< 13	1,6	63	19
33	150	37	45	31	46	4,0	74	17
34	150	70	74	94	34	2,9	26	46
35	150	45	48	31	< 20	2,4	41	15
36	150	30	62	11	< 23	2,3	35	11
37	150	26	41	17	< 14	2,5	89	13
38	Ambient	49	26	48	< 13	< 1,1	5	19
39	"	35	16	24	< 9	< 1	< 2	< 4
40	"	29	13	17	< 8	< 1	2	6
41	"	27	14	20	< 6	< 1,1	4	< 1
42	"	21	11	14	< 6	< 0,5	2	3
43	"	20	9	13	< 5	< 0,5	1	< 1

Leaks in the vessel appeared at 120°C [21]. After 2 months of unsuccessful experiment, we were obliged to change the equipment. The 'Pressure Vessel' (Figure 5.34) was equipped for temperature control, the leachant being in equilibrium with its own saturated vapour. Using the same glass block the temperature was raised immediately to 120°C for 2 weeks, then 150°C for 5 weeks and finally returned to ambient temperature for 6 weeks. Results are given in Table 5.22. As for the preceding glass, there appears to be a 'ceiling' in the leach rates beyond 100°C. Furthermore it is evident that the change of vessel is accompanied by a change in the results. This change is unexplained. A plausible hypothesis would be the presence of debris at the bottom of the vessel, accumulated in the course of the first 30 days leaching.

The mean leach rate values calculated over the first two weeks, together with the corresponding multiplication factors, are given in Tables 5.20 and 5.21.

(III) Results for inactive glasses

Values of leach rates by mass loss and by element for the two glasses are summarised in Table 5.23. The results obtained confirm that the temperature dependence varies with the element.

(a)   The leach rate for silicon increases only slightly and is approximately double the mass leach rate.

(b)   Leach rates for boron, sodium and caesium are greater than the mass leach rate, caesium being less mobile than sodium.

(c)   Leach rates for aluminium and above all cerium are always much lower than the mass leach rate.

Examination of the leached specimens shows the presence of small crystals at the surface. At higher temperatures their numbers diminish but their size increases.

TABLE 5.23 Effect of Temperature on the Leach Rates of Two Inactive Glasses

T°C	Nombre de jours	Lm	L Si	L Al	L B	L Na	L Cs	L Ce	pH
25°C	0-7 7-14	≤ 4,0 ≤ 4,0	/ /	/ /	/ /	/ /	/ /	/ /	8 7,45
50°C	0-7 7-14	19,5 30,9	5,87 33,6	< 0,6 < 0,6	34,6 48,6	32,4 51,6	37,1 64,9	< 0,3 < 0,3	8,38 8,35
70°C	0-7 7-14	64,5 40	87,8 87,8	< 0,6 < 0,6	518 462	124 90	121 83,5	< 0,3 < 0,3	8,75 8,72
100°C	0-7 7-14	75,1 63,5	119 111	22 18	379 273	316 266	145 156	< 0,3 < 0,3	8,89 8,83
120°C	0-7 7-14	155 71,7	186 109	25 8	531 155	458 153	130 195	< 0,3 < 0,3	8,96 8,44
150°C	0-7 7-14	236 81,4	271 126	19 < 0,6	677 91,5	596 127	197 265	< 0,3 < 0,3	9 8,22

Leach rates by mass loss and by elements. Inactive glass SON 20 18 F3 (A 100) ($\mu g \cdot cm^{-2} \cdot day^{-1}$)

T°C	Nombre de jours	Lm	L Si	L Al	L B	L Na	L Cs	L Ce	pH
25°C	0-7 7-14	≤ 4,0 ≤ 4,0	/ /	/ /	/ /	/ /	/ /	/ /	7,95 7,7
50°C	0-7 7-14	3,1 6,2	25,7 5,18	< 0,6 < 0,6	< 18,5 < 18,5	7,68 5,38	< 9,94 < 9,94	< 0,3 < 0,3	9,41 8,07
70°C	0-7 7-14	41,9 25,9	71 65,6	< 0,6 < 0,6	106 64,3	97,8 97,8	94,4 64,6	< 0,3 < 0,3	8,20 8,13
100°C	0-7 7-14	32,6 56,5	82,2 93	11,3 13,3	144 174	159 194	115 119	< 0,3 < 0,3	6,36 8,41
120°C	0-7 7-14	136 68,2	171 104	20 10,6	405 114	391 119	119 195	< 0,3 < 0,3	8,64 8,01
150°C	0-7 7-14	193 95,7	225 134	9,3 0,6	486 119	460 142	133 244	< 0,3 < 0,3	8,77 7,91

Leach rates by mass loss and by element. Inactive glass SON 64 19 20 F3 (A96) ($\mu g \cdot cm^{-2} \cdot day^{-1}$)

Comparison of results and conclusions

The mean leach rates for all samples are summarised in Table 5.24. Factors of increase with temperature relative to 50°C are given in Table 5.25 for both active and inactive glasses.

TABLE 5.24   Effect of Temperature on the Leach Rates of Two Glasses
SON 64 19 20 F3 and SON 60 20 18 F3 ($\times 10^{-6} \mathrm{g \cdot cm^{-2} d^{-1}}$)

Glass	θ °C	m	Si	Al	B	Na	Cs	Ce	β	Cs	Sr	Ru	Ce
SON 64 A 96	25	<4	-	-	-	-	-	-	0,6	1,1	0,6	1,1	0,02
	50	4,7	15,4	0,6	18,5	6,53	9,94	0,3	5,8	17	8,1	1,1	0,05
	70	33,9	68,3	0,6	85,1	97,8	79,5	0,3	27	63	25	5,9	0,35
	100	44,5	87,6	12,3	159,0	176,5	117,0	0,3	22	58	13	8,7	0,36
	120	102,1	137,5	15,3	259,5	255,0	157,0	0,3	31	64	23	9,6	0,54
	150	144,4	179,5	4,9	302,5	301,0	188,5	0,3	30	59	25	4,9	0,74
SON 60 A 100	25	<4	-	-	-	-	-	-	7,2	8,9	6,0	2,15	0,30
	50	25,2	19,7	0,6	41,6	42,0	51,0	0,3	22	28	24	21,9	1,19
	70	52,3	87,8	0,6	490,0	107,0	102,3	0,3	30	58	16	9,8	0,66
	100	69,3	115,0	20,0	326,0	291,0	150,5	0,3	47	79	30	9,4	2,2
	120	113,4	147,5	16,5	343,0	305,5	162,5	0,3	-	-	-	-	-
	150	158,7	198,5	9,8	384,3	361,5	231,0	0,3	-	-	-	-	-
		inactive glass							active glass				

The main observations which can be made are as follows:

(a)   The leach rates for glass A96 are approximately 5 times more temperature-dependent than those of glass A100, for mass loss, boron, sodium, caesium and strontium.

TABLE 5.25   Factors of Increase with Temperature of the Leach Rates Compared to 50°C SON 19 20 F3 (A96) and SON 60 20 18 F3 (A100)

θ °C	m A96	m A100	Si A96	Si A100	B A96	B A100	Al A96	Al A100	Na A96	Na A100	Cs A96	Cs A100	Sr A96	Sr A100	Ru A96	Ru A100	Ce A96	Ce A100
50	1,0	1,0	1,0	1,0	1,0	1,0	1,0	1,0	1,0	1,0	1,0 ≠1,0	1,0 ≠1,0	≠1,0	≠1,0	≠1,0	≠1,0	? ≠1,0	? ≠1,0
70	7,4	2,1	4,4	4,5	)4,6	11,8	?	?	17,1	2,5	7,9 ≠3,8	2,0 *2,5	≠3,0	≠1,4	≠5,0	≠1,9	? ≠)8	? ≠1,2
100	9,7	2,8	5,7	5,8	)8,6	7,2	)21	)34	27,2	6,9	11,7 ≠3,5	2,9 *3,8	≠1,6	≠2,6	≠(8	≠3,5	? ≠)8	? ≠3,8
120	22,4	4,5	8,9	7,5	)14	8,2	)25	)29	39,2	7,3	15,7 ≠3,9	3,1 *1,1	≠2,7	≠1,0	≠(9	≠2,7	? ≠)13	? ≠2,1
150	31,4	6,3	11,7	10,0	16,3	9,2	)8	17	46,3	8,6	18,8 ≠3,6	4,5 ≠2,3	≠3,0	≠2,2	≠(4	≠4,4	? ≠)17	? ≠4,3

*for active samples

(b)   Silicon loss is very similar for the two glasses and appears to be governed solely by temperature for this type of composition.

(c)   Ruthenium and cerium are much less temperature-dependent above 50°C.

(d)   A break in slope in the $\ell = f/T$ curves between 70° and 100°C for both glasses and for all elements is interpreted as a change in mechanism (Figures 5.30 - 5.33). Table 5.26, giving the activation energies calculated from these curves, indicates this phenomenon clearly: from 25° to 70° approximately the mean apparent energy is 80 ± 40 kJ mole^{-1} according to the element and glass composition. Above 70°C this energy does not exceed 20 kJ mole^{-1}, it is even negative in the case of aluminium.

TABLE 5.26   Apparent Mean Activation Energies of Leaching
($kJ\ mole^{-1}$)

glass	θ °C	m	Si	Al	B	Na	Cs	Ce	β	Cs	Sr	Ru	Ce
SON 54 (A 96)	25 to 70°C	75,6	66,6	-	74,1	134,2	96,3	-	66,3	69,6	61,3	44,8	56,5
	70/100 to 150°C	21,8	14,2	24,9	20,1	16,6	14,4	-	3,7	1,2	3,7	1,9	12,8
SON 60 (A 100)	25 to 50/70°C	118,0	70,5	-	113,1	46,2	34,1	-	35,7	35,9	44,3	70,5	44,3
	50/70 to 150°C	19,3	12,4	18,0	4,0	5,3	11,8	-	15,2	18,3	4,3	17,8	12,7
		Inactive							Active				

(e)   For both glasses the increase in leach rate for caesium is much
smaller for the active than for the inactive glasses. This is
probably linked to the procedure used: in the inactive case a fresh
sample is used for each temperature whereas in the active case the
same sample is successively leached and thus its surface is
progressively depleted in mobile elements such as caesium.

The essential conclusion from these experiments is that leach rate is
not a simple function of temperature. This behaviour is probably
linked to the appearance of retarding mechanisms, such as:

   (a)   Increase in solution concentration

TABLE 5.27   Effect of Pressure.
$\beta\gamma$-active SON 61 19 20 F3 (A97)

Week No	P (Bars)	Leach Rates ($\times 10^{-7}$ g.cm^{-2}d^{-1})					Leach Rates (Arbitrary Units)	
		B	Cs	Sr	Ru	Ce	Sb	Co
1	1	22	23	19	< 8,1	2,5	31	12
2 + 3 + 4	1	8	9,2	6,9	< 4,0	1,6	25	2,8
5	1	11	11	11	< 8,6	0,43	24	2,9
6	1	9,8	8,4	9,5	4,0	0,36	8,5	1,3
7	1	11	8,0	11	3,6	0,61	9,3	1,7
8	20	14	10	13	< 1,7	0,30	13	1,9
9	20	11	8,7	11	< 1,2	0,67	10	1,7
10	20	12	9,5	12	< 1,6	0,38	15	1,7
11	20	16	9,5	14	< 1,8	0,25	10	1,5
12	20	16	12	15	< 1,8	0,35	15	1,8
13	50	14	13	12	2,8	0,22	12	2,0
14	50	16	10	8,8	< 1,2	< 0,10	12	1,5
15	50	10	10	8,8	< 2,0	< 0,19	15	1,8
16	50	7,1	8,7	7,2	< 1,4	' ـ	10	1,7
17	100	8,7	8,7	7,2	< 2,3	0,17	11	1,0
18	100	11	11	8,8	2,6	0,22	10	1,5
19	100	8,9	8,7	8,2	1,3	0,20	9,4	1,4
20	100	10	9,5	8,2	< 2,0	0,80	9,5	1,4
21	200	10	10	8,8	1,5	0,45	13	1,6
22	200	12	12	9,4	< 2,3	< 0,24	13	1,4
23	200	11	12	8,8	< 2,6	/	11	1,5
24	200	12	14	10	< 9,4	< 0,80	24	3,1

(b)   Re-precipitation of species in solution.

Probably the effects of these secondary mechanisms would be even more important when the SA/V ratio is large and leachant renewal infrequent. It is therefore clear that the factors of increase given depend on the experimental conditions and should be used with caution. These phenomena have already been verified in a previous study [26].

5.3.4 Effect of Pressure at Ambient Temperatures

Glass SON 61 19 20 F3 (active)

The apparatus used for these experiments is shown in Figure 5.34. Leaching is static and leachant renewal is weekly. The results are summarised in Table 5.27 and Figure 5.35. No increase of leach rates with pressure is found.

Glass SON 61 19 20 F3 (inactive)

The apparatus used is similar to VULCAIN. However, a fresh sample is used for each pressure.

The results are presented in Table 5.28. Up to a pressure of 100 bars no significant variation is found, except for sodium which is reduced. At 200 bars a mass loss is measured but this does not appear significant. In fact the leach rates of the elements have not increased and the sensitivity of the balance used (approx. 0.01 g) is poor compared to that of ion analysis. The continued reduction of the sodium leach rate as the pressure increases must be confirmed.

Conclusions

Under the test conditions at ambient temperature and in the absence of backfill materials increase in pressure to 200 bars appears to have no effect on the leach rate of any element. The increase in mass loss of the inactive glass at 200 bars remains unexplained: this still remains to be confirmed.

TABLE 5.28 Effect of Pressure-inactive SON 61 19 20 F3 ($\times 10^{-6}$ g.cm^{-2} d^{-1})

Pressure (bars)	Element	Mass Loss	Si	Al	B	Na
1	1st week	< 5	1,25	< 0,6	1,25	13,5
	2 week	< 5	0,45	< 0,6	1,05	17,0
	Average	< 5	0,85	< 0,6	1,15	15,2
20	1st week	< 4	1,25	< 0,6	4,8	9,6
	2nd week	< 4	0,45	< 0,6	3,5	9,6
	Average	< 4	0,85	< 0,6	4,15	9,6
50	1st week	< 3	3,4	< 0,6	6,9	8,5
	2nd week	< 3	1,9	< 0,6	2,1	7,2
	Average	< 3	2,65	< 0,6	4,5	7,8
100	1st week	< 3	3,2	< 0,6	2,5	1,5
	2nd week	< 3	1,5	< 0,6	1,5	8,9
	Average	< 3	2,35	< 0,6	2,0	5,2
200	1st week	51	2,3	< 0,6	4,6	0,1
	2nd week	19	1,1	< 0,6	1,2	1,5
	Average	35	1,7	< 0,6	2,9	3,8

5.3.5 Effect of pH at Ambient Temperature

Leaching of βγ-active specimens

In view of the high activities of leachates at pH 1 and pH 14 the leach tests were made on two specimens of approximately 30 g reserved from the casting of glass A96. Sample No 1 was used successively for all acid tests and No 2 for the basic tests.
The leach test is dynamic (water recirculated with a period of 1.5 minutes [21]) with daily renewal.

The leachate is 'VULCAIN' industrial water, adjusted to the required pH by addition of HNO_3 or NaOH. Mixing is achieved directly in the 40 litre stainless steel supply tank.

Results obtained at various pH are given in Tables 5.29 and 5.30. The pH of the solution after leaching is also given. The acidic pH values change very little, whereas the neutral or basic pH values do change and tend towards pH 8.5, which is the pH of the water equilibrated with atmospheric CO_2.

Acid leaching

In general a considerable increase in leach rates is found between pH 7 and pH 4. Between pH 4 and pH 2 the rates increase only slightly, they then again increase greatly between pH 2 and pH 1. However, the effect of pH differs greatly for the various elements.

(a) Caesium and ruthenium: Leach rates for these elements increase by a factor of 10 between pH 7 and pH 4, and again by a factor of 10 between pH 4 and pH 1. Both elements increase by the same factors; the leach rate of ruthenium remains lower than that of caesium, whatever the pH.

(b) Strontium and cerium: Decrease of pH leads to a large increase in leach rate for these elements. At pH4, the rate for strontium is

TABLE 5.29   Leach Rates at Various Acid pH
Sample 1 (A96) ($\times 10^{-6}$ g.cm^{-2}.d^{-1})

Initial pH	Final pH	β	Cs	Sr	Ru	Ce
7	8.6	0.5	2.0	0.2	ca0.4	0.015
5	5	40 (x80)	7.0 (x3.5)	5.0 (x25)	ca1.5 (x3)	0.5 (x33)
5	4	20 (x40)	25 (x12)	18 (x60)	5.0 (x10)	20 (x1300)
4	3.5	20 (x40)	25 (x12)	18 (x60)	5.0 (x10)	20 (x1300)
3	3	30 (x60)	35 (x17)	25 (x125)	8 (x16)	25 (x1600)
2	2	30 (x60)	35 (x17)	25 (x125)	8 (x16)	25 (x1600)
1	1	200 (x400)	220 (x110)	190 (x950)	50 (x100)	150 (x10^4)
8	8.3	1.0 (x2)	1.4 (x0.7)	1.0 (x5)	5.0 (x10)	0.075 (x5)
7	8.5	0.5 (x1)	0.7 (x0.35)	0.45 (x2)	5.0 (x10)	0.15 (x10)

found to increase by a factor of 60, and that of cerium by 1300. From this value of pH and below, the leach rates of caesium, strontium and cerium reach the same order of magnitude. Only the very low leach rate of ruthenium, at whatever pH, prevents this from being a congruent corrosion regime.

Basic leaching

In general a slow regular increase in leach rate is found as the pH is increased. The increase reaches a maximum at pH 12.5 but is much smaller in extent that that found in acid pH. The curve with changing pH shows no discontinuities.

A second experiment at pH 9 after 360 days leaching gave the same result as at the start. By contrast, a second experiment at pH 12 gave a much smaller value than previously found.

The decrease in leach rate between pH 12 and pH 14 is quite surprising and contrary to the results generally found for industrial glasses. It was demonstrated that this variation was not due to the sodium concentration (possible limitation of the $Na^+ \rightleftharpoons H^+$ exchange) by repeating the experiment at the same pH using ammonia; there was no significant difference between the two results. Probably this phenomenon originates from the presence of fission products having very low solubility in alkaline media which protect the silica from leaching.

(a) Caesium: The leach rate increases up to pH 13 but the increase is small. It decreases from pH 13 to pH 14.

(b) Strontium: The leach rate increases more sharply with pH and is a maximum near pH 11.

(c) Ruthenium: Although this element could not be measured with precision the increase in its leach rate is small and of the same order of magnitude as that found for caesium.

(d) Cerium: this is the element which showed the greatest increase in leach rate (a factor of 100 at pH 12). However, congruent corrosion is not found and the leach rate for cerium remains lower than that of caesium or strontium.

Leaching of inactive samples

A fresh sample was used for each pH value. Leaching was static with daily leachant renewal. The results are collected in Table 5.31. The sensitivity to pH varies greatly from one element to another.

TABLE 5.30 Leach Rates at Various Basic pH Sample II (A96) (x 10^{-6} g.cm^{-2}.d^{-1})

Initial pH	Final pH	β	Cs	Sr	Ru	Ce
Ind. water	8.6	0.03	0.08	<0.05	<0.15	0.01
9	9.1	0.4 (x0.8)	0.7 (x0.35)	0.4 (ca x2)	≤0.3 (ca x1)	0.02 (x 1)
10	(9.0 (9.6 (8.9	0.6 1.0 0.6	0.9 1.6 1.0	0.5 0.9 0.7	<0.4 <0.6 <0.4	0.026 0.09 0.035
11	9.5	1.0 (x2)	1.6 (x0.8)	1.0 (x5)	<0.5 (<x1.5)	0.12 (x7)
12	10	3.0 (x6)	4.5 (x2)	3.0 (x15)	<1.0 (<x2.5)	1.0 (x67)
13	10.5	5.0 (x10)	6.0 (x3)	4.5 (x22)	<1.5 (<x4)	2.0 (x133)
13	12.5	7.0 (x14)	14.0 (x7)	3.8 (x19)	<3.0 (<x22)	2.0 (x133)
14	13.7	2.0 (x4)	6.0 (x3)	2.5 (x12)	<1.5 (<x4)	1.0 (x67)
9	8.5	0.4	0.6	0.42	<0.5	≤0.025
(NaOH) 12	9.8	0.6	1.6	0.41	<0.8	≤0.04
(NH$_4$OH) 12	9.6	–	1.1	0.86	<0.8	0.065

(a) Silicon: From pH 1 to pH 10 the leach rate is practically constant, at pH 12 it increases by a factor of 10.

(b) Aluminium: This is relatively soluble for all pH values tested. This probably arises because of the amphoteric nature of its hydroxide, which dissociates into soluble Al^{3+} in acid and into aluminate AlO_2^- in basic pH.

(c) Cerium: This is bound in a chemical form which is relatively soluble in acid of pH < 3 but practically insoluble in basic media.

TABLE 5.31 Leach Rates at Various pH (Ambient Temperature) for Inactive SON 64 19 20 F3 ($\times 10^{-6}$ g.cm^{-2}.d^{-1})

pH	Day No.	Mass Loss	Si	Al	B	Na	Cs	Ce	Initial pH	After 24 h
pH = 1 (0.1N Nitric)	1		8.3	130.4	87.4	136.1	111.3	75.7	1.1	1.1
	2		8.3	121.1	133.9	115.3	68.6	90.8		
	8		15.1	65.2	71.4	143.4	61.2	-		
	9		13.6	65.3	61.2	73.5	55.7	49.9		
	11	28.0	-	-	-	-	-	-	1.1	1.12
pH = 3 (10^{-4} Nitric)	1		<7.6	88.5	29.1	30.5	<27.8	15.1	3.1	3.15
	2		<7.6	41.9	11.1	19.7	<27.8	7.6		
	7		<7.6	32.6	17.5	20.6	<27.8	13.6		
	9		<7.6	20.5	3.4	17.9	<27.8	3.8		
	12	9.1	-	-	-	-	-	-	3.1	3.13
Vulcain water	12	5.2	-	-	-	-	-	-	7.3	8.25
pH = 10 (10^{-4} NaOH)	1		<7.6	32.6	6.4	-	<27.8	<1.5	9.8	8.5
	2		<7.6	46.6	5.2	-	<27.8	<1.5		
	7		<7.6	195.7	3.5	-	<27.8	<1.5		
	9		<7.6	-	5.2	-	<27.8	<1.5		
	12	7.4							9.8	8.59
pH = 14 (1N NaOH)	1		59.5	73.2	235.0	-	42.4	<15		
	2		89.2	163.3	272.2	-	47.9	<15		
	7		193.4	137.5	286.2	-	41.1	<15	12.1	12.1
	9		223.1	113.5	183.2	-	<27.4	<15		
	12	17.4	-	-	-	-	-	-	12.1	12.05

(d) Boron: Boron appears to be in a very soluble form at pH > 10. Examination of stability diagrams for B_2O_3 in aqueous solution at 25°C confirms this observation, the metaborate (BO_3^{3-}) or tetraborate ($B_4O_7^{2-}$) forms being the most soluble at pH 14.

Conclusions

The following conclusions may be drawn from comparison of the active and inactive results.

(a)   Cerium leach rates are much more sensitive to the action of acid media than to that of basic media.

(b)   In general the active species are more than 10 times more sensitive to the action of an acid medium than to that of a basic medium. This behaviour is the inverse of that generally found with classical industrial glasses, which resist acid leaching much better than basic. This difference in behaviour can only be explained by the compositional differences between these two types of glass. The usual soda-lime-silica glasses, of high silica content, resist acid media well. On the other hand the silicon matrix is destroyed above pH 9-10. Fission product glasses have low silica content (generally < 50%) and a large number of network modifiers. In acid media the H^+ modifier exchanges are accelerated and the glass is rapidly attacked. However, the presence in nuclear glasses of numerous elements whose hydroxides have very low solubility in basic media (Zr, transition elements, rare earths and actinides) results in very good resistance to alkaline media.

It is very interesting to note that, in closed systems, borosilicate FP glasses will buffer the medium towards pH 9-10. This behaviour, peculiar to nuclear waste glasses, will particularly affect the actinides.

5.3.6 Effect of Type of Water

Leach tests were made in the 'VULCAIN' cell at ambient temperature using the usual 'dynamic leaching' equipment with daily leachant renewal.

The glass was the $\beta\gamma$-active SON 62 20 24 F3. This was leached successively with waters of various types, a period of leaching with industrial water being interpolated at every change.

The following leachants were used:

(a) 'Industrial water' is the water distributed on the Marcoule site, filtered. 200 litres have been stocked in a plastic reservoir to ensure a uniform supply.

(b) 'Siliceous water' is simulated by Mont Dore water (mineral water of the Massif Central, not commercially used).

(c) 'Granite water' is simulated by 'Charrier' water (Massif central).

(d) Sea water: from Toulon.

(e) 'Clay water' is a synthetic water whose composition was given by the CEC.

Analysis of these waters is given in Table 5.32.

Figure 5.36 shows the total β leach rates for the various tests.

Variations in leach rate during individual tests, probably caused by temperature fluctuation, were larger than those caused by changing the type of water. However, in an effort to isolate the effect of water type, we have compared the mean leach rate obtained with each water to the average of those obtained before and after with the reference industrial water. The first 10 days leaching after each water change is eliminated from the calculation.

In view of the small variations observed by comparison with the fluctuations with time, these values should only be taken as an indication. However, the following observations may be made:

(a) The concentration of dissolved mineral salt has no effect; the least concentrated (granite water) and the most concentrated (sea water) are the leachants having the most pronounced increases.

TABLE 5.32 Analysis of Waters used in Leach Tests

	VULCAIN industrial water	Siliceous water (Mont Dore)	Granitic water (Charrier)	Sea water (Toulon)	Clay water (synthetic)
Resistivity at 20°C (Ω/cm)	2 440	680	20 600	26	60
pH	8,29	8,58	6,46	8,04	7,31
Total Solids at 110°C (mg.l^{-1})	285	1 150	34	31 850	18 500
SiO_2 (mg.l^{-1})	3,3	182,0	3,6	0	5,1
Ca^{++} (mg.l^{-1})	77	55	5	480	380
Mg^{++} (mg.l^{-1})	7,8	34	1	2 020	1 878
Na^+ (mg.l^{-1})	10,2	349	3,4	8 255	4 335
K^+ (mg.l^{-1})	1,6	48	0,35	432	532
Fe^{++} (mg.l^{-1})	0,016	0,114	0,008	0,007	0,09
Al^{+++} (mg.l^{-1})	0,410	0,190	0,01	4,74	0,28
CO_3H^- (mg.l^{-1})	201,3	915	12,2	159	213,5
Cl^-	30	197	5,2	17 200	130
$SO_4^=$	46,5	87,5	3,9	3 700	15 900
NO_3^-	3,7	0	3,7	0	8,9
Total Boron	0,22	0,11	0,07	5,3	-

Table 5.33 shows the ratios for the various types of water calculated in this way.

TABLE 5.33

Element	Siliceous Water (b) ÷ (a)	Granite Water (c) ÷ (a)	Sea Water (d) ÷ (a)	Clay Water (e) ÷ (a)
β	0.5	1.6	2	0.5
Cs	0.5	1.3	2	0.4
Sr	0.5	1.7	3	0.6
Ru	-	-	-	-
Ce	<1	2.2	1.8	-
Sb	<0.25	1.5	2	0.35

(b) Very high concentrations of alkalis and alkaline earths have no particular effect, since although the sum of the concentrations (Na + K + Ca + Mg) is little different for sea water and clay water, the leach rates for both Cs and Sr differ by a factor of 5. There may be an effect due to the chlorides present in sea water following βγ radiolysis.

(c) The effect of sea water is not due to its pH since both reference water and siliceous water have slightly higher pH.

(d) The effect of sulphates does not appear decisive.

In conclusion, for the experimental conditions used, no pronounced effects due to the various types of water have been demonstrated.

3.7 Effect of crystallization

The following glass compositions were used in this study:

A65:   SON 45 30 14 U2
A75:   SON 65 23 24 A201
A79:   SON 60 30 14 U2

The compositions are given in Table 5.16.

The exact composition of Glass A79 is in doubt, due to an error in the records.

Glasses of considerably different compositions have been chosen, to represent extreme conditions. A65 is very low in Si and high in Na; A75 is low in Na but high in B_2O_3 and Al_2O_3; A79 is fairly high in Si and low in Na. In all three glasses the F.P. oxide content is higher than that actually used for LWR waste vitrification. However, as these elements favour devitrification, it could be thought that this can only enhance the effect of crystallization on the leach rates.

The glasses were given the following treatment:

(a)   A leach test before heat treatment.

(b)   Nucleation for 6 hrs at 600°C and growth for 24 hrs at 800°C, followed by a leach test.

(c)   Second heat treatment, 24 hrs at 800°C, followed by a leach test.

(d)   Third heat treatment, 48 hrs at 800°C, followed by a leach test.

After the 800°C heat-treatments the glass is furnace-cooled to 600°C, then from 600°C to ambient at 13°C/hour.

The leach tests are recirculated flowing water tests at ambient temperature using VULCAIN water with daily leachant renewal.

I.   Leaching of $\beta\gamma$-active block SON 45 30 14 U2 (A65).

The following leach tests were made

(a)   First test (45 days)
(b)   Test before heat treatment (60 days)
(c)   Test after first heat treatment (24 days)
(d)   Test after second heat treatment 21 days)
(e)   Test after third heat treatment (38 days)

Results are given in Table 5.34. These should be taken as indicative only, since in most cases equilibrium was not reached.

The leach rates do not increase in proportion to the heat treatment, but increase with the first treatment and then decrease as a result of the later treatments. The increase is large for caesium and strontium, small or zero for ruthenium and cerium.

II.   Leaching of $\beta\gamma$-active block SON 65 23 24 A201 (A75)

The following leach tests were made

TABLE 5.34 Mean Leach Rates for Glass A65 ($g.cm^{-2}.day^{-1}$)

Test Element	(a) $\times 10^{-6}$	(b) $\times 10^{-6}$	(c) $\times 10^{-6}$	(d) $\times 10^{-6}$	(e) $\times 10^{-6}$
β	0.24	0.3	4.4	3.7	2.0
Cs	0.24	0.3	4.8	3.3	1.9
Sr	0.28	0.36	6.3	3.7	2.5
Ru	-	2.6	3.5	7.4	3.2
Ce	0.07	0.65	2.0	1.7	0.55

(a) First test, (43 days)
(b) Test before heat treatment (22 days)
(c) Test after first heat treatment (36 days)
(d) Test after second heat treatment (29 days)
(e) Test after third heat treatment (22 days)

Results are given in Table 5.35

TABLE 5.35 Mean Leach Rates for Glass A75 ($g.cm^{-2}.day^{-1}$)

Test Element	(a) $\times 10^{-6}$	(b) $\times 10^{-6}$	(c) $\times 10^{-6}$	(d) $\times 10^{-6}$	(e) $\times 10^{-6}$
β	0.16	0.18	0.36	0.39	0.57
Cs	0.43	0.38	1.1	1.2	2.2
Sr	0.15	0.10	0.23	0.22	0.41
Ru	0.05	0.90	0.20	0.14	0.37
Ce	0.048	0.30	0.077	0.025	0.062

The effect of crystalisation increases in proportion to the heat treatment, as revealed by an increase in leach rates for caesium and strontium and a decrease for ruthenium and cerium.

III. Leaching of βγ-active block SON 60 30 14 U2 (A79)

The following leach tests were made:

(a) First test (54 days)
(b) Test before heat treatment (98 days)
(c) Test after first heat treatment (24 days)
(d) Repeat under the same conditions after 4 months dry (41 days)
(e) Test after second heat treatment (44 days)
(f) Test after third heat treatment (56 days)

Results are given in Table 5.36

TABLE 5.36   Mean Leach Rates for Glass A79 ($g.cm^{-2}.day^{-1}$)

Test Element	(a) $\times 10^{-6}$	(b) $\times 10^{-6}$	(c) $\times 10^{-6}$	(d) $\times 10^{-6}$	(e) 10^{-6}	(f) $\times 10^{-6}$
β	6.0	3.0	2.5	2.0	2.0	1.0
Cs	40.0	10.0	9.0	7.0	5.0	2.0
Sr	3.0	2.0	0.7	0.6	0.8	0.7
Ru	0.2	0.6	0.2	0.5	0.7	1.5
Ce	0.6	0.7	0.12	0.2	0.13	0.17

The leach rates of all elements except ruthenium decrease after crystallization for this glass. For caesium this decrease is proportional to the heat treatment. For strontium and cerium there is no further change after the first treatment.

Conclusions

Table 5.37 gives the factors of variation between test b (before heat treatment) and test e or f (after the third heat treatment). A(+) indicates an increase, a(−) shows a decrease.

TABLE 5.37 Factors of Variation due to Crystallisation

Glass	Element β	Cs	Sr	Ru	Ce
A65	(+)6	(+)6	(+)7	(+)1.2	(−)1.2
A75	(+)3	(+)6	(+)4	(−)2.4	(−)5
A79	(−)3	(−5)	(−3)	(+)2.5	(−)4

In no case does the change in leach rate after heat treatment at 800°C exceed one order of magnitude. The leach rate of cerium decreases for all glasses. The formation of mixed crystals $(Ce,U)O_2$ has been previously mentioned; it is not impossible that crystallization immobilises this element in stable phases.

It will be noted that after the third heat treatment the caesium leach rate is 2×10^{-6} for all three glasses. This is perhaps a coincidence; one could, however, envisage that this value corresponds to the alterability of a crystalline form of this element. Crystallization would increase the caesium leach rate of glasses where it is lower than 2×10^{-6} (A65 and A75) but would decrease it for those glasses for which it is greater than 2×10^{-6} (A79).

The effect of crystallization on ruthenium, sometimes negative, sometimes positive, is less pronounced.

This study shows that the effect of crystallization can be positive or negative, depending on glass composition and on the element studied.

5.3.8 Effect of the Type of Leach Test and Frequency of Leachant Renewal

The $\beta\gamma$-active glass block (A87) SON 64 19 20 F3 G3 was leached at ambient temperature according to two different methods in the apparatus shown in Figure 5.21.

(a) The dynamic mode where the water periodically streams over and bathes the glass every 1.5 minutes. The leachate is replaced by fresh industrial water every 24 hours.

(b) The static mode where the water bathes the glass without agitation; the leachate was renewed after 1, 3, 7 and 30 days.

The mean leach rates are given in Table 5.38.

TABLE 5.38 Mean Leach Rates for Glass A87 Following Two Test Methods

Method	Renewal frequency (days)	Leach rates $\times 10^{-7}$ g. cm^{-2}. day^{-1}			
		Cs	Sr	Ru	Ce
Dynamic	1	4.1	1.3	2.0	0.50
Static	1	4.1	1.3	8.7	0.25
	3	3.1	1.1	1.8	0.05
	7	5.2	1.3	2.6	0.10
	30	4.5	1.3	0.4	0.05

The sensitivity of the leach rate to the test method varies greatly according to the element. These fall into two groups, i.e.

(a) Rate unaffected by the test method and frequency of leachant renewal (at least until the silica solubility limit is reached): elements considered as mobile and found in a soluble chemical form. Caesium and strontium are of this type.

(b) Rate significantly decreased when changing from dynamic to static mode; here the decrease is even greater when the renewal frequency is lowered; relatively immobile elements bound in chemical forms having low solubility for the prevailing pH. Cerium and ruthenium are of this type, also americium (see section 5.3.2).

5.3.9 Conclusions

The objective of this work was to estimate the influence of the main parameters capable of affecting the leach rates of FP species incorporated in the glasses.

We have sought to vary each parameter separately so as to isolate its effect. It is, however, very difficult to obtain a 'pure' effect for at least two reasons. Firstly, the studies have been made with glasses of different composition. Secondly, modifying one parameter often causes modification of another, for example concentration of dissolved species in the case of temperature variation. Nevertheless, for all ambient temperature experiments the methods used (dynamic mode with daily leachant renewal, SA/V of 0.6 cm^{-1}) lead to low concentrations of dissolved species and secondary effects were found to be limited. This is proved by the pH of the leachates remaining at 8.5 approximately, essentially fixed by the water/CO_2 equilibrium.

It was not possible to provide sufficient samples to use one per experiment in the active cell. Thus the same sample was tested successively at different temperatures or pH values etc. As a result a time effect is superimposed on that of other parameters for active tests. Also in general the conclusions should be used with caution, remembering that the leach rate is only the expression of the quantity of the element passed into the liquid phase.

Despite these limitations, one may extract qualitative information from this work on the sense and magnitude of leach rate variations with certain parameters even if, in strict accuracy, the data are only valid for the glass-water systems studied.

Effect of Time

Beyond an equilibriation period, which varies with the element (30 days for Cs and Sr, 200 days for Ru, Ce, Sb, Co, Am), no significant change in daily leach rates is found, apart from that arising from temperature variation. This confirms the poor protection characteristics of the hydrolysed layer which has been demonstrated by others [25].

The generally incongruent corrosion behaviour is confirmed. There is a factor of 100 between the rates for Cs and Ce, even after 500 days.

At ambient temperature with frequent water renewal and after an initial transitory period, the leach rates vary very little with time.

Effect of temperature at atmospheric pressure

The leach rates appear to follow an Arrhenius relationship with an apparent activation energy between 40 and 120 kJ.mole^{-1}. However, the larger decrease in mean activation energy above 70°C ± 20°C (15 ± 10 kJ.mole^{-1} depending on the element and the glass composition) is explained by the appearance of secondary mechanisms having an inhibitory characteristic when the concentration of dissolved species increases (for these experiments water renewal was weekly).

Effect of pressure at ambient temperature

Between 1 and 200 bars no significant effect of pressure is seen. This was predictable, since the chemical reaction constants do not change significantly until 1000 bars. However, possible mechanical effects on backfill material, or compaction of hydrolysed layers cannot be ignored.

Effect of pH

These experiments confirm the peculiar behaviour of nuclear glasses which, unlike classical glasses, have poor resistance to acids but very good resistance to alkaline solutions. In acidic media corrosion rapidly becomes congruent whereas in a basic medium elements with insoluble hydroxides, including actinides, remain substantially confined in the surface layer.

Effect of type of water

By comparison with the reference industrial water, leaching is retarded in siliceous water and clay water, but increased in 'Charrier' (granite) water and sea water. These changes are small with daily water renewal. Different results would probably be obtained in completely static media where solubility limits could be reached for many elements.

Effect of crystallisation

These results show that the effect of crystallization can be positive, negative or zero, depending on glass composition and element, without ever exceeding a factor of 10.

Effect of leach test method and frequency of water renewal

At ambient temperature and SA/V of 0.6 cm^{-1} the silica solubility limit is far from being reached, even with monthly water renewal and generally no change in corrosion rates of mobile species such as caesium is seen. However, the leach test method does have a considerable effect on elements of low solubility (Ce, Ru, Am) whose leach rates are decreased even further with less frequent leachant renewal.

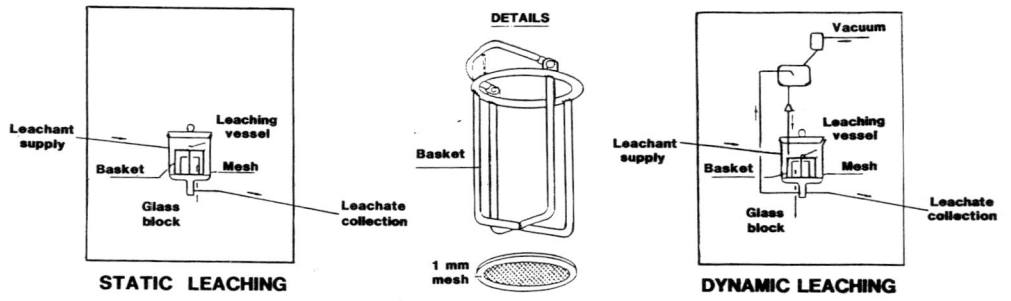

Fig. 5.21   'VULCAN' cell ambient temperature leaching vessel

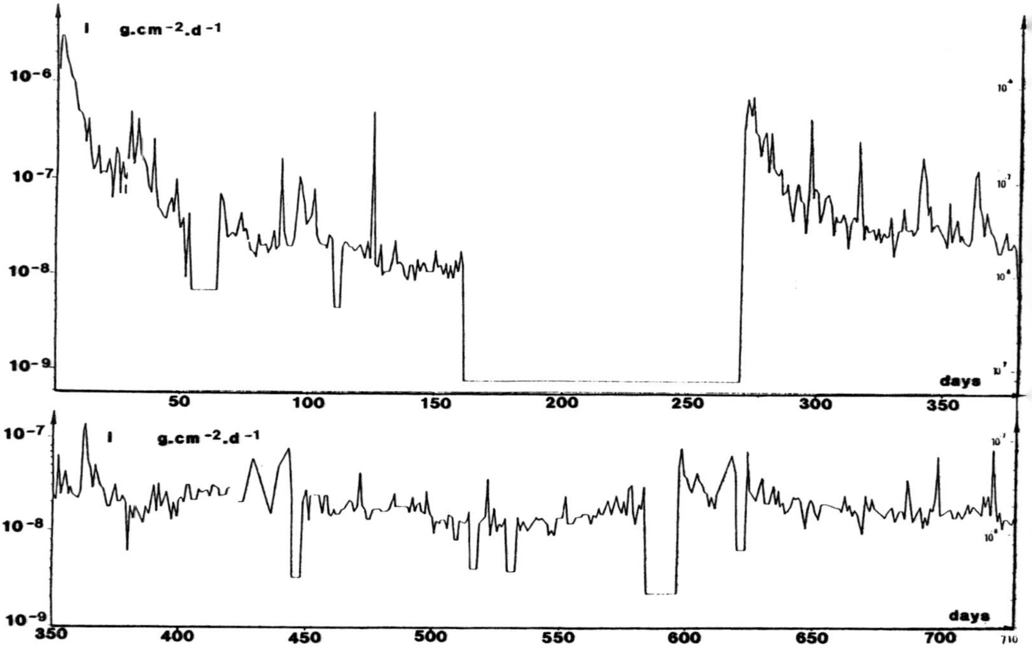

Fig. 5.22   Long-term leaching of α-doped glass A53

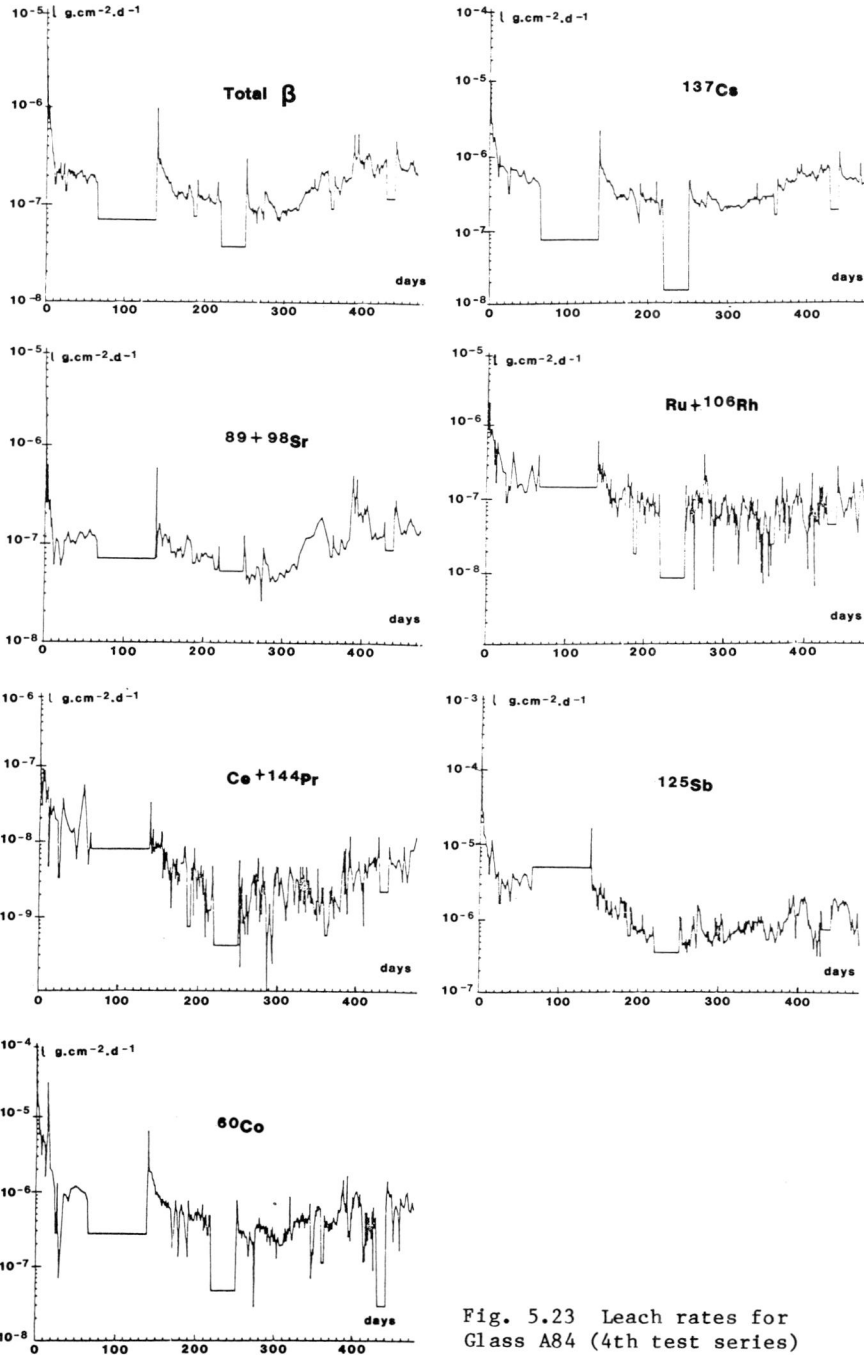

Fig. 5.23 Leach rates for Glass A84 (4th test series)

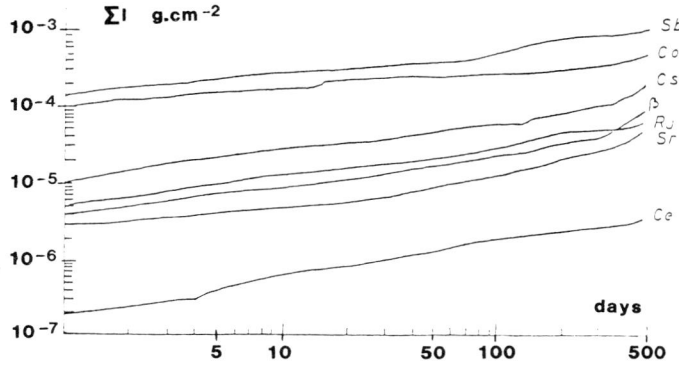

Fig. 5.24 Cumulative quantities leached, glass A84 (4th test series)

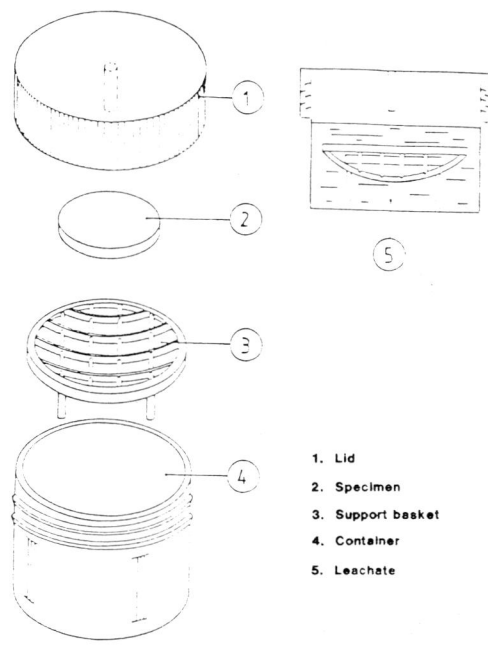

Fig. 5.25 Apparatus for the determination of the effects of time (inactive)

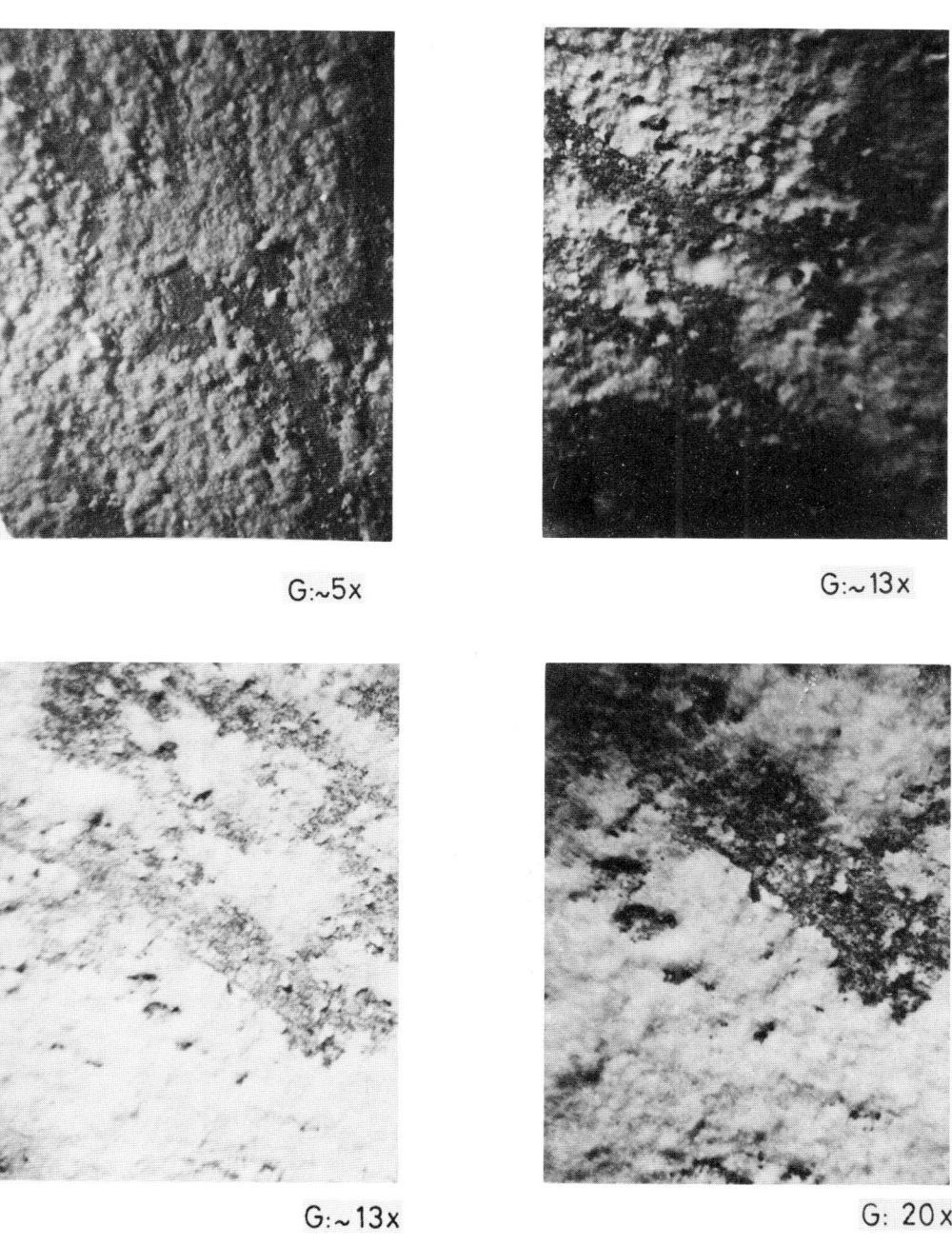

Fig. 5.26 Surface of glass SON 58 30 20 U2 after 1 year's leaching at ambient temperature

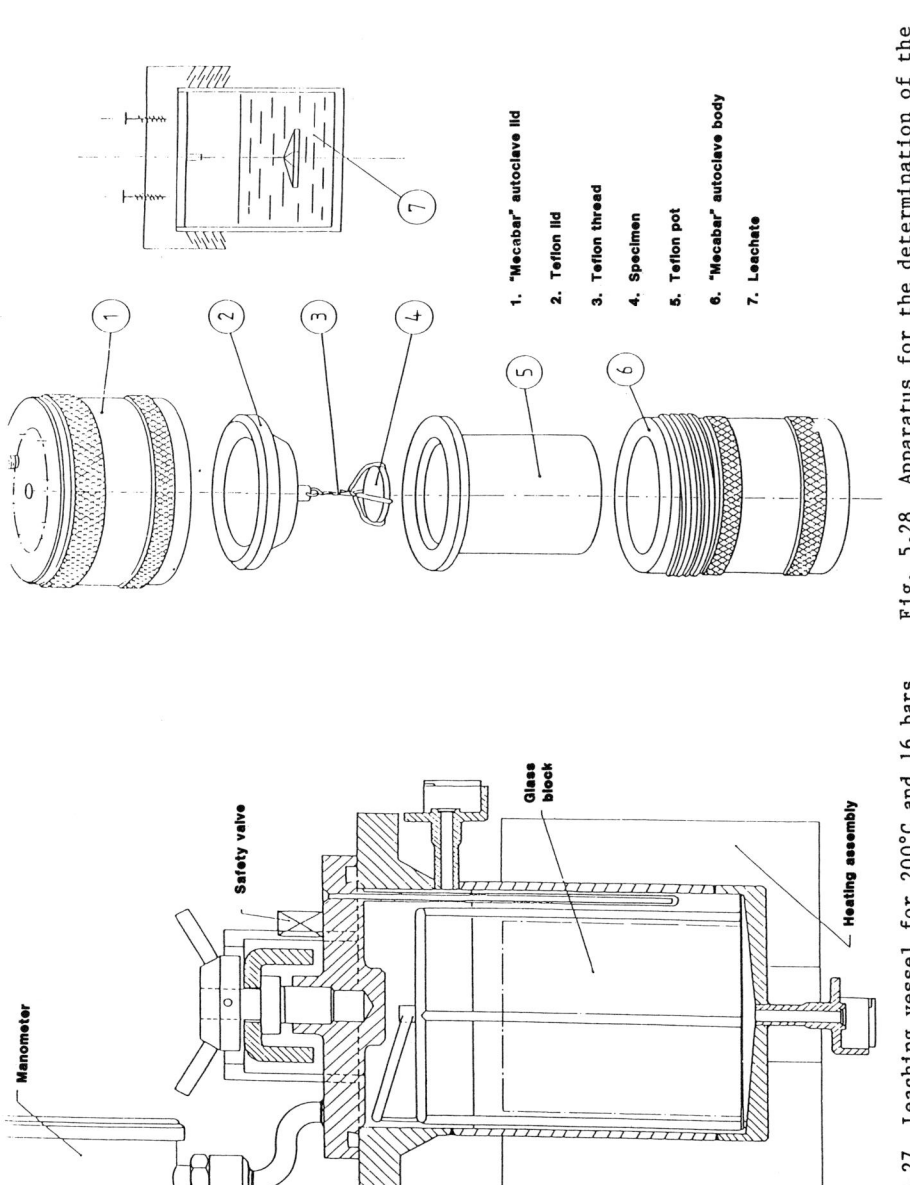

Fig. 5.28 Apparatus for the determination of the effects of temperature (inactive)

1. "Mecabar" autoclave lid
2. Teflon lid
3. Teflon thread
4. Specimen
5. Teflon pot
6. "Mecabar" autoclave body
7. Leachate

Fig. 5.27 Leaching vessel for 200°C and 16 bars

Total β

Caesium

Strontium

Ruthenium

Cerium

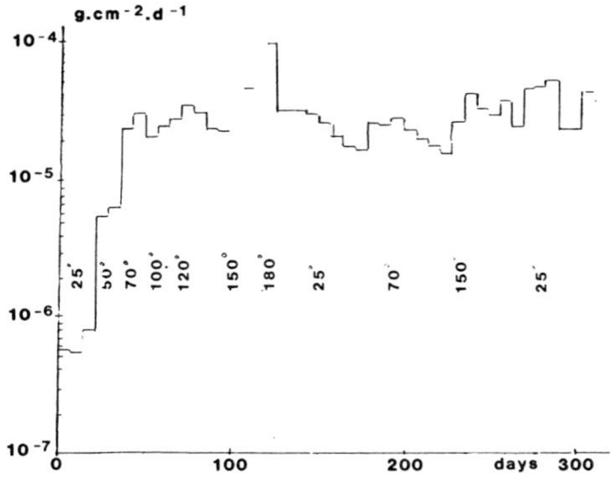

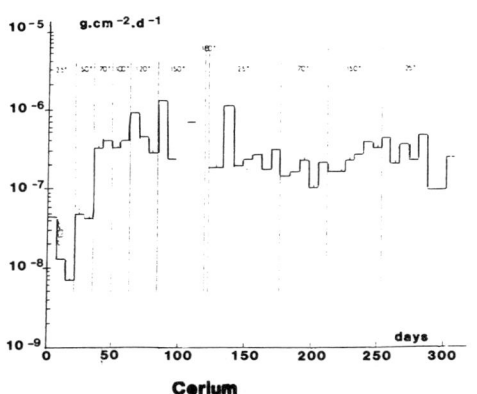

Fig. 5.29 Effect of temperature on leach rate for glass SON 64 19 20 F3 (A96)

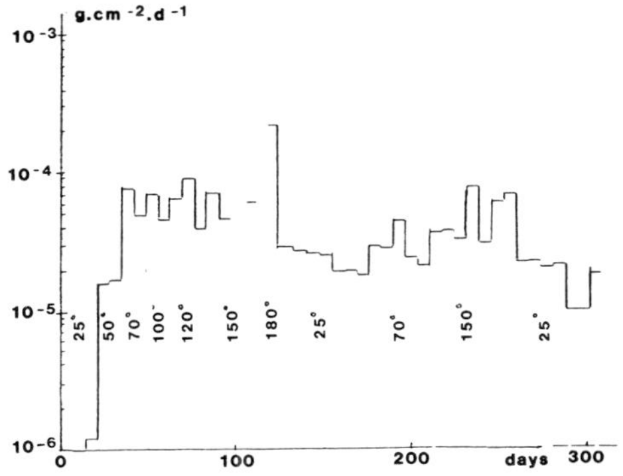

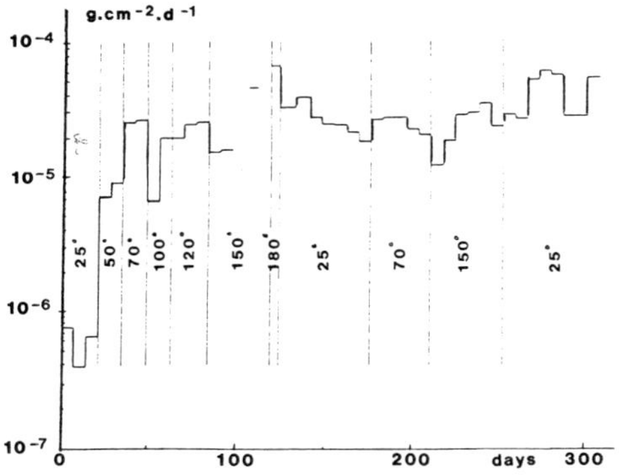

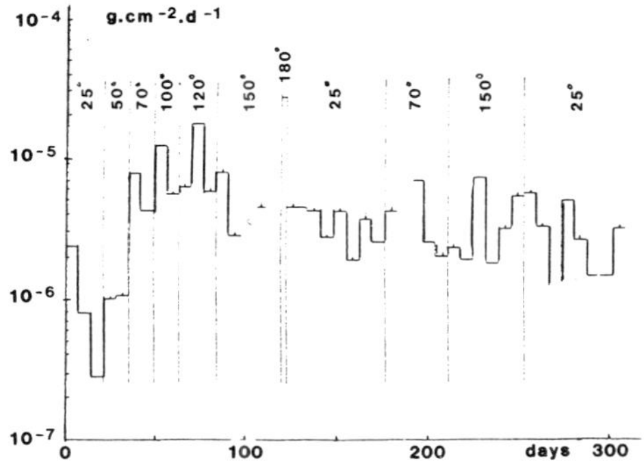

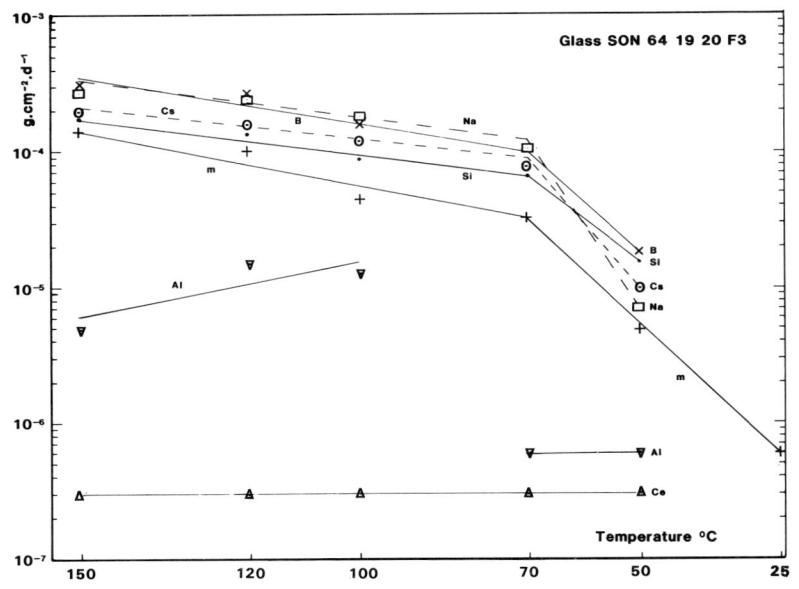

Fig. 5.30 Arrhenius plot of leach rates for glass SON 64 19 20 F3

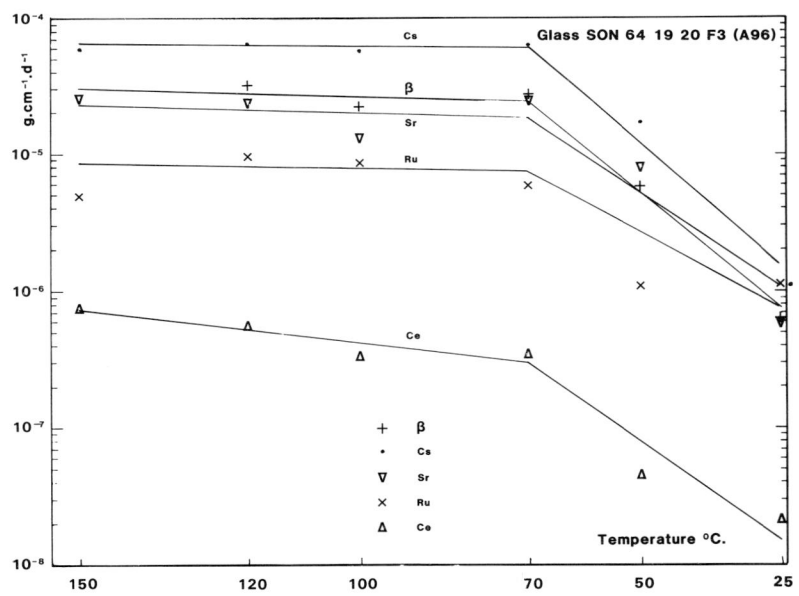

Fig. 5.31 Arrhenius plot of leach rates for glass SON 64 19 20 F3 (A96)

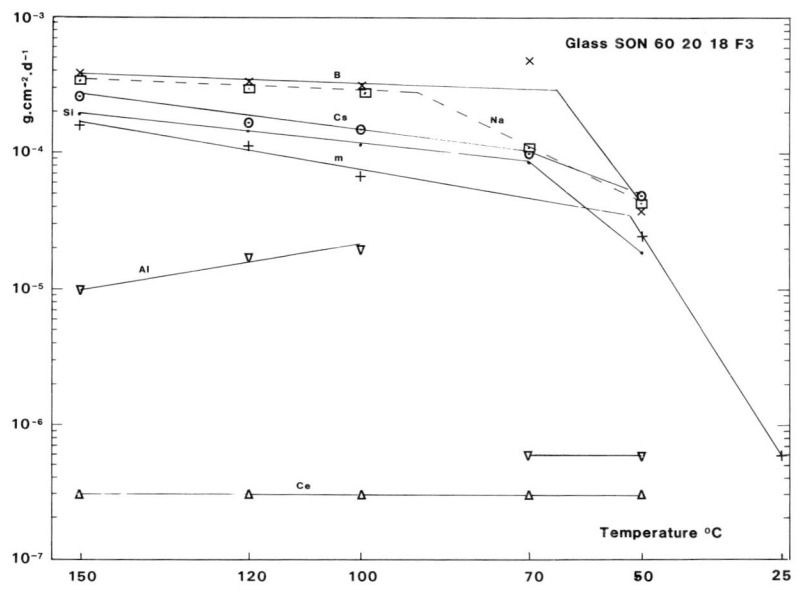

Fig. 5.32 Arrhenius plot of leach rates for glass SON 60 20 18 F3

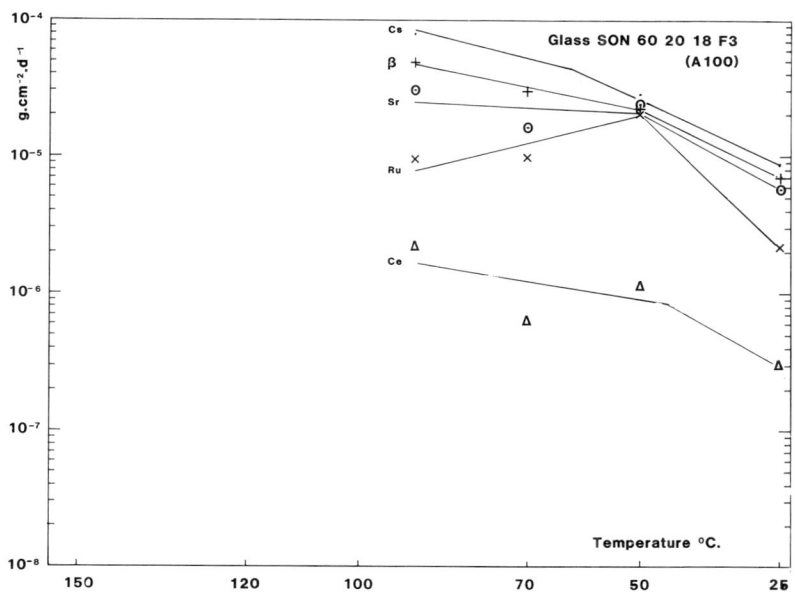

Fig. 5.33 Arrhenius plot of leach rates for glass SON 60 20 18 F3 (A100)

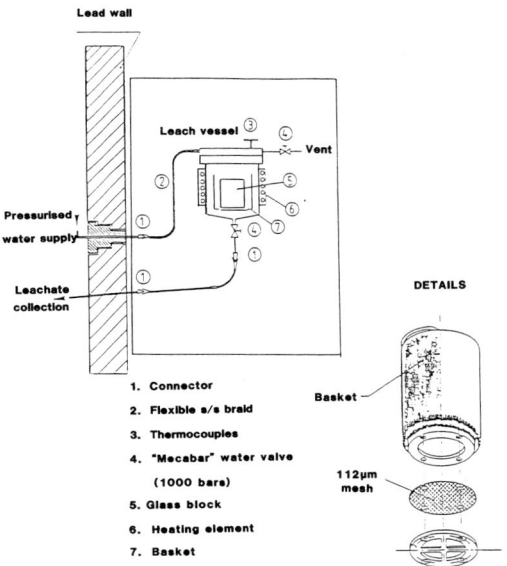

Fig. 5.34 'VULCAN' cell high-pressure leaching vessel

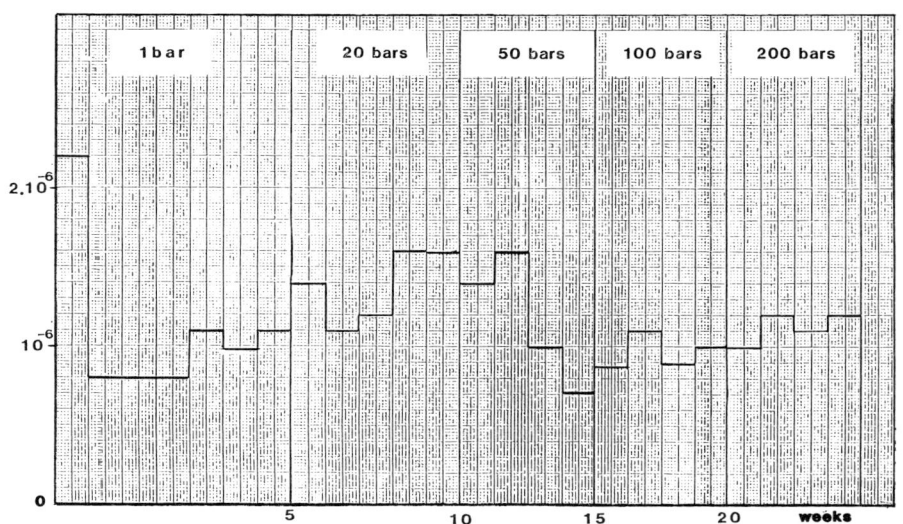

Fig. 5.35 Effect of pressure on the total-β leach rate for glass SON 61 19 20 F3 (A97)

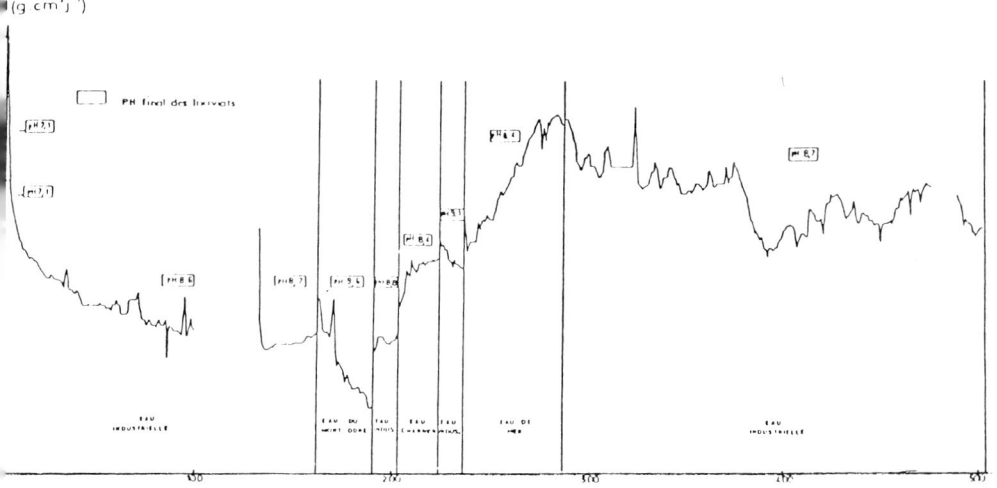

Fig. 5.36 Effect of type of water on the total-β leach rate for glass SON 62 20 24 Fe

REFERENCES - CHAPTER 5

[1] DOE/TIC 11400 - Nuclear Waste Materials Handbook - Test Methods, PNL-Richland (Washington) (1981).

[2] D.E. Clark et al., "Corrosion of Glass", Books for Industry (New York) (1979).

[3] G. Malow et al., EUR 10038 EN (1985).

[4] P. Van Iseghem et al., Riv. della Staz. Sper. Vetro nr 5, 1984, 163-170.

[5] A. Barkatt et al., PNL 4382, 20-40 (1982).

[6] P. Van Iseghem et al., Scientific Basis for Nuclear Waste Management VII, Ed. G. McVay, North-Holland (1984) 527-534.

[7] Ch. Engelmann et al., EUR 9268 EN (1984).

[8] P. Van Iseghem et al., Progress Report to Action No. 5, Jan-June 1984.

[9] P. Van Iseghem et al., EUR 8979, 27-42 (1984).

[10] Data submitted for Nuclear Waste Materials Handbook, J.E. Mendel, Ed., PNL 3990 (1983).

[11] L.L. Hench et al., Nucl. & Chem. Waste Manag. $\underline{5}$, (1984), 149-173.

[12] R. De Batist et al., Progress Report to Action No. 5, Jan-June 1981

[13] P. Vejmelka and R.A.J. Sambell, EUR 9423 EN (1984).

[14] W. Lutze et al., Radioactive Waste Management and Disposal, ed. R. Simon, Cambridge University Press (1985), 323-250.

[15] P. Van Iseghem et al., Scientific Basis for Nuclear Waste Manag. V, ed. W. Lutze, Elsevier Science (1982), 219-227.

[16] R. De Batist et al., presented at the ANS Intern. Topical Meeting on High-Level Nuclear Waste Disposal - Technology and Engineering, September 1985, Seattle, in press.

[17] R. De Batist et al., EUR 8424 EN (1983).

[18] J.A.C. Marples et al., European Appl. Res. Rept. - Nucl. Sci. Technology, 3 (1981), 395-484.

[19] R. Conradt et al., Final Report to Action No. 5 (1985).

[20] Experimental study of the influence of various parameters on the aqueous leach rate of fission product glasses. Contract CCE WAS 123 80 55F, Annual Report 1981 (DGR No. 248).

[21] Experimental study of the influence of various parameters on the aqueous leach rate of fission product glasses. Contract CCE WAS 123 80 55F. Annual Report 1982 (Technical Report SDHA/SEMC 83.01).

[22] 1st semestrial report (period ending 30th Sept. 1981). Contract CCE WAS 123 80 55F.

[23] 2nd semestrial report (1st Oct. 1981 to 31st June 1982). Contract CCE WAS 123 80 55F.

[24] 3rd semestrial report (1st July 1981 to 31st Dec. 1982). Contract CCE WAS 123 80 55F.

[25] J.L. Nogues, "Les mechanismes de corrosion des verres de confinement des products de fission". Thèse Docteur-Ingenieur - Montpellier France 1984.

[26] Development of highly active glasses in the VULCAIN and PIVER cells - Leaching at three temperatures of certain glasses in the VULCAIN cell.
Contracts CCE 040-77-11 WAS F
CCE 70-78-9 WAS F
Final Report

6. INFLUENCE OF REPOSITORY CONDITIONS ON THE LEACHING OF HIGH LEVEL WASTE FORMS

6.1 LEACHING OF RADIOISOTOPE-SPIKED GLASSES UNDER SIMULATED REPOSITORY CONDITIONS (AERE)

6.1.1 Introduction

After disposal, the radiological hazard from vitrified high-level waste will be dominated by the isotopes Cs-137 and Sr-90 for the first few hundred years, then by Am-241 and -243 and Pu-239 and -2 for several millenia and finally by Tc-99 and Np-237. The half-li of these isotopes are, in years:

Sr-90	Tc-99	Cs-137	Np-237	Pu-239	Pu-240	Am-241	Am-24
28.5	2×10^5	30.1	2×10^6	2.4×10^4	6.6×10^3	458	8×10^3

To investigate the specific isotope leach rates, specimens of glass were doped separately with these isotopes and then leach tested un conditions relevant to those likely to occur after disposal in har rock. The base glasses were made with a full spectrum of inactive isotopes of the fission products and were subsequently doped with active isotopes mentioned above. For Sr and Cs, the levels were chosen arbitrarily, for Tc, Pu and Am the levels were approximatel those likely to occur in the real waste whilst for Np the level wa ten times that, even assuming that all the Np produced in the fuel concentrated in the high-level waste stream. This was done so tha the leachates from the Np-containing glass were active enough to 'count' The levels were thus:

	Sr-90	Tc-99	Cs-137	Np-237	Pu	Am-241
mCi/g	0.05	0.05	0.05	0.003	0.05	2
Bq/g	2×10^6	2×10^6	2×10^6	1×10^5	2×10^6	7×10^7

The doped glass was cast into rods 12 mm in diameter which were th cut into discs 1 - 2 mm thick with a total surface area of about 3 cm^2. These discs were used for all the experiments described below. For each experiment, six similar apparatuses were used, on for each of the six isotopes.

6.1.2 Flowing leachate experiments

The apparatus is shown diagrammatically in Fig. 6.1. Water from the reservoir is pumped by a peristaltic pump into the specimen chamber which is held in a temperature controlled water bath and then out to a collecting vessel. The specimen chamber contains about 2 ml of leachate and the disc specimen is mounted upright so that almost all the surface is freely exposed to the leachate. The leaching vessel and all the pipework are made of PTFE (Teflon). Initially for the UK189 specimens, the flow rate was about 9 ml per day but this was reduced to 1 ml per day and then to 1 ml per week, the latter reduction being achieved by running the pump on a time switch for only about 2 hours per day. The other glasses were only studied at a flow rate of 1 ml per week.

At intervals the collecting vessel was replaced with a new one, the contents were acidified and a 1 ml aliquot evaporated to dryness on a counting tray. This was then counted in an α- or β-counter as appropriate; these were calibrated from time to time using standard sources. The count rates obtained from the leachates were converted to leach rates by dividing by the area of the specimen, the time over which the sample had been collected and the specific activity of the glass. This latter was obtained by dissolving a small weighed quantity of each doped glass in 5 M HCl to which a few drops of HF had been added, diluting the solution and counting as before.

An example of the results obtained is shown in Fig. 6.2 for glass 209.

The procedure of changing the collecting vessel at each sampling meant that any material adsorbed on the walls was brought into solution and accounted for. However, for the first 150 days of the experiment with UK209 this was not done, the vessel being emptied with a syringe. At the first sampling after the procedure was altered, slightly higher values were found for the leach rate because the amounts of radioactive isotopes that had been adsorbed on the vessel walls over several months were included in the total for that

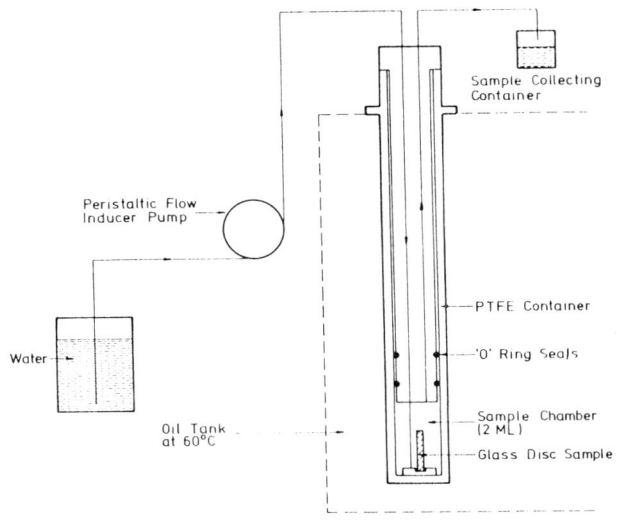

Fig. 6.1 Flowing leachant experiment

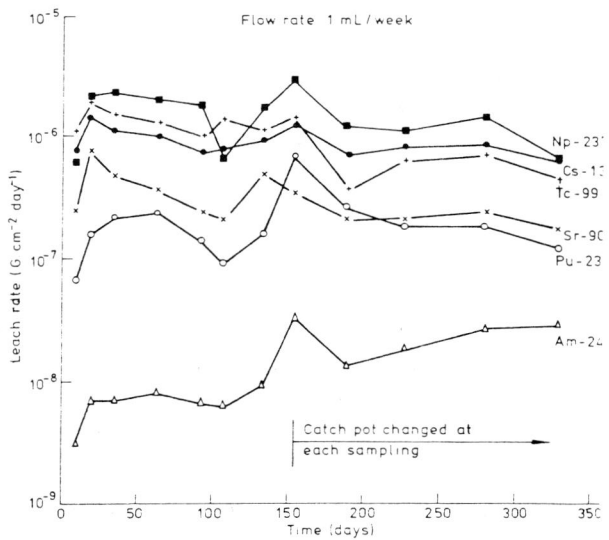

Fig. 6.2 Leaching of UK209 at 60°C

month. This can be seen in Fig. 6.2.

The leach rates obtained for glass 189 at the various flow-rates are given in Table 6.1.

TABLE 6.1 Leach rates for glass 189 at various flow rates ($60°C$: values in $g.cm^{-2}\ day^{-1}$)

Flow rate	9ml per day	1ml per day	1ml per week
Tc-99	7×10^{-5}	7×10^{-5}	6×10^{-5}
Sr-90	5×10^{-5}	2×10^{-5}	5×10^{-5}
Cs-137	6×10^{-5}	4×10^{-5}	2×10^{-5}
Np-237	1.8×10^{-5}	1×10^{-5}	3×10^{-6}
Pu	(5×10^{-7})	1.3×10^{-7}	(1.5×10^{-8})
Am-241	4×10^{-9}	6×10^{-10}	6×10^{-10}

Values in brackets () are uncertain.

The leach rates obtained in rapidly flowing water at 60°C with pH 7 was $6.5 \times 10^{-5}\ g.cm^{-2}.day^{-1}$, closely similar to the values obtained for Sr and Cs at a flow rate of 9 ml per day and to that for Tc at all flow rates.

The leach rate obtained by counting for Tc did not change with flow rate, suggesting that, in this range of flow-rates, the release of Tc is limited by its rate of escape from the glass rather than by its solubility in the leachant. The leach rates for Cs, Sr and Np decreased by x 3, x 10 and x 6 for a 60-fold decrease in flow-rate suggesting that the solubility of the species in the leachant may have some effect.

The leach rates obtained at 60°C at a flow rate of 1 ml per week for all the glasses tested are given in Table 6.2 the values being taken once apparent equilibrium had been reached. For B1/3, both the parent glass and the glass ceramic, after about a year at 60°C, the temperature was increased to 90°C and leaching continued for a

TABLE 6.2 Leach rates in water flowing at 1 ml per week (g.cm^{-2} day^{-1})

Glass		At 60°C					At 90°C	
		189	209	SON 68/18/17	Parent Glass B1/3	Glass Ceramic B1/3	Parent Glass B1/3	Glass Ceramic B1/3
Apparent leach rates (amounts found in collecting vessel only)	Sr	5x10^{-6}	4x10^{-7}	4x10^{-7}	1x10^{-7}	3x10^{-7}	8x10^{-7}	1.7x10^{-6}
	Tc	6x10^{-5}	1.3x10^{-6}	6x10^{-7}	4x10^{-7}	3x10^{-6}	8x10^{-7}	8x10^{-6}
	Cs	2x10^{-5}	9x10^{-7}	1x10^{-6}	5x10^{-7}	5x10^{-6}	4x10^{-6}	1x10^{-5}
	Np	3x10^{-6}	2x10^{-6}	7x10^{-6}	3x10^{-8}	3x10^{-8}	2x10^{-9}	4x10^{-8}
	Pu	1.5x10^{-8}	1.5x10^{-7}	2x10^{-7}	6x10^{-9}	6x10^{-9}	2x10^{-9}	1x10^{-9}
	Am	8x10^{-10}	8x10^{-9}	3x10^{-8}	2x10^{-9}	6x10^{-9}	8x10^{-10}	1x10^{-9}
True leach rates (including amounts adsorbed on apparatus)	Sr	5x10^{-6}	4x10^{-7}	7x10^{-7}	1.2x10^{-7}	4x10^{-7}	1x10^{-6}	2x10^{-6}
	Tc	6x10^{-5}	1.4x10^{-6}	8x10^{-7}	4x10^{-7}	3x10^{-6}	9x10^{-7}	8x10^{-6}
	Cs	2x10^{-5}	1.3x10^{-6}	1.3x10^{-6}	5x10^{-7}	5x10^{-6}	4x10^{-6}	1x10^{-5}
	Np	3x10^{-6}	2.5x10^{-6}	–	6x10^{-8}	5x10^{-7}	4x10^{-7}	7x10^{-7}
	Pu	1x10^{-7}	9x10^{-7}	3x10^{-7}	1.7x10^{-8}	6x10^{-8}	5x10^{-9}	1x10^{-8}
	Am	2x10^{-8}	3x10^{-8}	3x10^{-7}	5x10^{-9}	1.2x10^{-8}	2x10^{-9}	2x10^{-9}
Fraction retained in gel layer	Sr	0.62		0.02	Values on the right are after 1 year at 60°C and 8 months at 90°C		0.31	0.52
	Tc	0.15		0.04			0.66	0.41
	Cs	0.31		0.004			0.47	0.58
	Np	0.91		–			0.91	0.89
	Pu	0.99		0.29			0.98	0.99
	Am	0.998		0.17			0.98	0.99
Wt loss Not including gel layer		2.1x10^{-5}	8x10^{-7}	5x10^{-7}			1.5x10^{-6}	3.2x10^{-6}
Including gel layer		4.6x10^{-5}	1.1x10^{-6}	8x10^{-7}			2.0x10^{-6}	7x10^{-6}

– 280 –

further 8 months. For the fission product isotopes and for Np-237 the increase in temperature produced increases in leach-rate, the increase being larger for the parent glass than for the glass ceramic. For Pu and Am, however, the leach rates continued to decrease slowly with time despite the increase in temperature. Presumably, the total elapsed time of 20 months was not sufficient for equilibrium to be reached at such low leach rates.

The following conclusions may be drawn from the data in Table 6.2
(a) There is a large difference between the release rates of the different isotopes. Tc, Cs and sometimes Np are more readily leached than Sr and particularly Pu and Am.

(b) The glass-ceramic does not retain the isotopes studied as well as its parent glass - by a factor of ten in some cases. This suggests that the good properties of the parent glass may deteriorate when it is, of necessity, slow cooled after manufacture.

(c) In this test, the parent-glass B1/3 is better at retaining almost all the other isotopes than the other glasses, although for some isotopes there is not a large difference between B1/3, SON 68 18 17 and 209.

At the end of the experiments, the specimens were removed from the containers, the apparatus was rinsed with acid and this was then evaporated and counted in the same way as the leachate to determine the amount of the various isotopes held on the walls of the apparatus; the amounts in the specimen chamber, the outflow pipe and the collecting container were measured separately. The values obtained were divided by the concentration of that isotope in the glass to give a mass balance although the various components of the glass would in fact be adsorbed to different extents. To complete the mass balance, the gel layer was removed, dissolved in acid and counted and the total amounts that had been found in the leachates were summed. Finally the specimens were weighed to determine the total amounts lost. These data are given for all six specimens of glass 189 in Table 6.3 as an example. In general, in this experiment,

TABLE 6.3 Apparent mass balance for samples of glass 189 (grams)

Isotope in sample	Tc-99	Sr-90	Cs-137	Np-237	Pu	Am-241
In sample container	2.2×10^{-4}	5.8×10^{-4}	3.0×10^{-4}	3.0×10^{-4}	2.0×10^{-4}	5.8×10^{-5}
In outflow pipe	3.8×10^{-5}	1.8×10^{-5}	4.9×10^{-5}	1.7×10^{-5}	3.7×10^{-5}	5.1×10^{-6}
In catch pot	9.3×10^{-5}	1.3×10^{-4}	1.5×10^{-4}	3.9×10^{-5}	8.4×10^{-5}	4.6×10^{-6}
Total in leachate	5.5×10^{-2}	1.4×10^{-2}	$2.6_5 \times 10^{-2}$	6.8×10^{-3}	5.7×10^{-5}	1.9×10^{-6}
In gel layer	9.5×10^{-3}	2.3×10^{-2}	1.2×10^{-2}	7.5×10^{-2}	6.8×10^{-2}	3.2×10^{-2}
Sum	6.5×10^{-2}	3.7×10^{-2}	3.9×10^{-2}	8.2×10^{-2}	6.8×10^{-2}	3.2×10^{-2}
Actual weight loss	$5.0_5 \times 10^{-2}$	$3.5_5 \times 10^{-2}$	3.5×10^{-2}	4.8×10^{-2}	4.7×10^{-2}	2.4×10^{-2}
% retained in gel layer	15%	62%	31%	91%	99.4%	99.8%

The quantity of each isotope for each location, as determined by counting techniques, was divided by the specific activity of the glass to obtain a notional equivalent weight of glass.

the quantities of glass deduced from counting agreed fairly well with
those obtained by direct weighing. The percentages of each isotope
retained in the gel-layer are given in Table 6.2: these data were not
available for 209. 189, the glass-ceramic and its parent glass
retained a large fraction of the activity in the gel-layer,
particularly for the actinides. For SON 68 18 17 this amount was
much smaller, probably because the gel-layer was very thin, typically
7×10^{-5} g.cm^{-2} compared to 720×10^{-5} for 189, 38×10^{-5} for B1/3 parent
glass and 770×10^{-5} for the glass-ceramic.

The quantities of Tc, Sr, Cs and Np adsorbed on the apparatus are
negligible compared to the totals found in the leachate but more Pu
and Am are retained adsorbed on the teflon walls of the apparatus
than was found in the leachate. Thus, to obtain the amounts actually
released from the glass, the values in Fig. 6.2 and in the upper part
of Table 6.2 must be increased. This has been done in the lower half
of the table and makes a substantial difference for the actinides.

6.1.3 Leaching in a repository simulation

In this experiment the glass discs were positioned in a small chamber
(volume 1.5 ml) at the top of a column of crushed granite saturated
with water, shown schematically in Fig. 6.3. There was no water flow
through the chamber but at about monthly intervals 1 ml samples of
water were withdrawn from the chamber, evaporated to dryness and
counted to determine their radioactivity content. The flow past the
specimen could thus be said to be 1 ml per month: before reaching the
specimen chamber the water to replace the sample had been in contact
with granite for many months. In this apparatus, there was nothing
to prevent the activity leached from the glass diffusing back down
the column and indeed, as wil be shown below, this did occur.

The leach rate was calculated by dividing the activity in each
monthly leachate sample by the specific activity of the specimen, the
specimen surface area and the time since the last sample was taken,
to give a leach rate expressed in grams (of glass) per cm^2 per day.

Four glass compositions have been tested by this technique 189, 209 and B1/3 both as the parent glass and as a glass ceramic. As an example, the leach rates for the B1/3 parent glass are shown in Fig. 6.4 for a period of a year. The 'leach rates' obtained in this way when the system had reached equilibrium are given in Table 6.4. These values are reduced from the 'true' ones because of adsorption on the granite and on the rest of the apparatus but a similar phenomenon will occur in a full scale repository.

TABLE 6.4 Apparent leach rates at 60°C in repository simulation (in $g.cm^{-2}.day^{-1}$)

Isotope	189	209	B1/3 glass	B1/3 glass ceramic
Sr-90	3×10^{-8}	1.3×10^{-8}	5×10^{-9}	
Tc-99	2.5×10^{-6}	1.7×10^{-7}	4×10^{-8}	
Cs-137	1.5×10^{-8}	2×10^{-8}	7×10^{-9}	2×10^{-7}
Np-237	8×10^{-9}	2.5×10^{-8}	4×10^{-9}	8×10^{-9}
Pu	6×10^{-8}	2×10^{-8}	2×10^{-9}	2×10^{-9}
Am-241	6×10^{-8}	2×10^{-10}	8×10^{-10}	2×10^{-9}
Weight loss	6×10^{-6}∓	*	$1.6 \pm 0.6 \times 10^{-6}$	$6.0 \pm 2.0 \times 10^{-6}$

∓ 2 specimens gained weight
* all specimens gained weight

After about a year the apparatus was dismantled and the granite columns cut into thick slices. The apparatus was rinsed with acid (HNO_3/HF) to take the adsorbed radionuclides into solution and the granite powder in each slice was similarly treated. Aliquots from each solution were then dried and 'counted' and divided by the specific activity to give a notional weight of glass adsorbed in each location, this method being used for easier intercomparison between the isotopes. The glass samples were dried and weighed and the overall leach rate calculated both before and after the gel-layer had been removed.

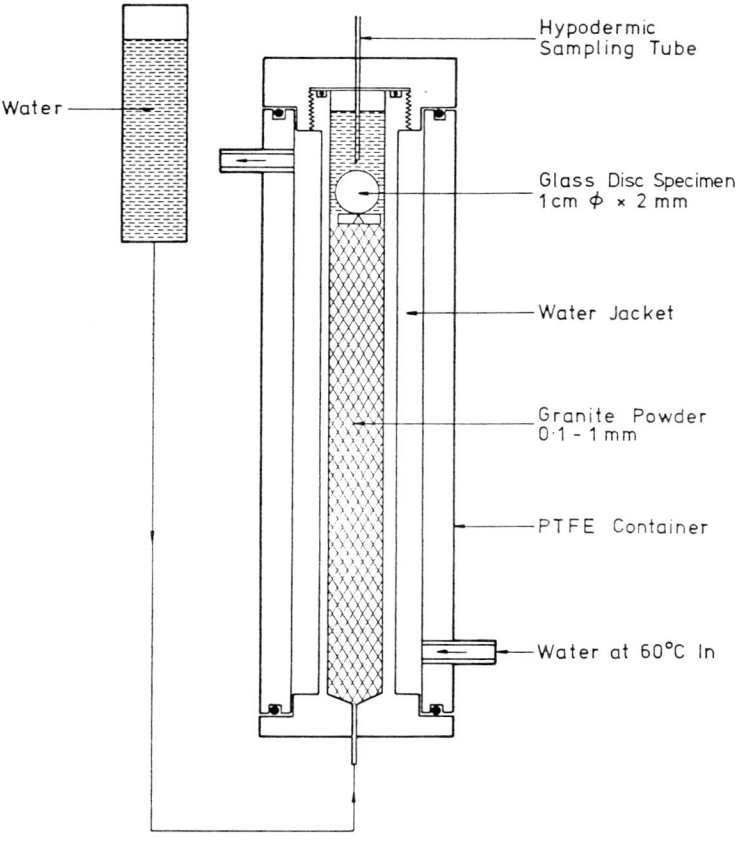

Fig. 6.3 Granite repository simulation

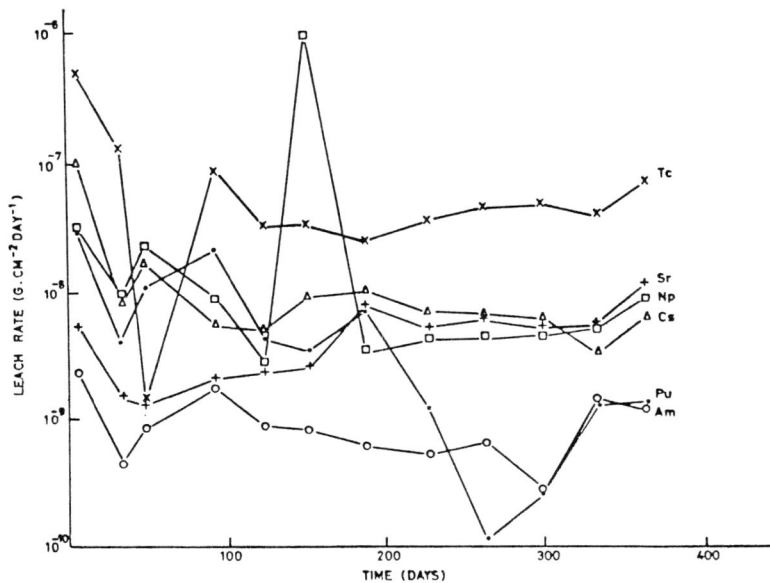

Fig. 6.4 Leach rates of B1/3 in repository simulation at 60°C

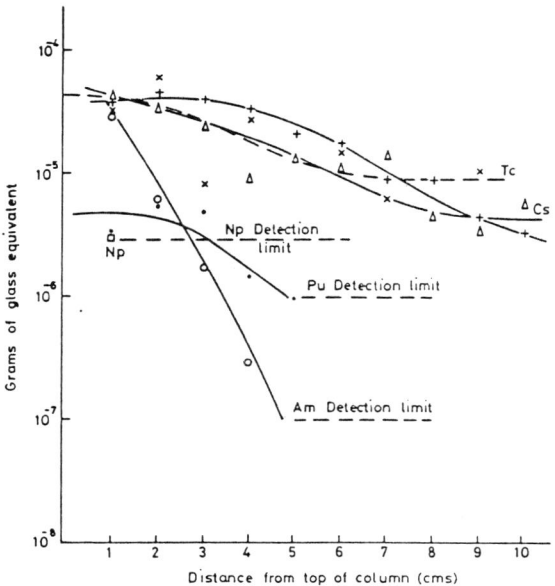

Fig. 6.5 Glass B1/3 in simulated granite repository at 60°C: Distribution of isotopes in the column

The results are given in Table 6.5 and the distribution of activity down the columns for each isotope is shown in Fig. 6.5. The recovery of the activity from the system is not good: the 'weight loss' calculated from the activity measurements is smaller than that obtained by direct weighing by about a factor of 4 to 6. The leach rates calculated from weight losses should have been the same for each specimen because the amount of dopant added was not sufficient to affect the leach rate and the differences are in fact not very great.

From the weight loss measurements and the amount of each isotope retained in the gel layer on the surface, leach rates can be calculated using the amount of each isotope released from the specimen, i.e. eliminating the amounts adsorbed on the granite etc.

$$\text{leach rate} = \frac{\text{total weight loss}}{\text{area} \times \text{time}} (1 - \text{fraction retained in layer})$$

These values are given in Table 6.7 and are those which should be used in safety assessments where allowance is made for adsorption on the backfill and other components of the near field. The fraction of activity retained on the gel-layer is given in Table 6.6.

The true leach-rates for the three glasses given in Table 6.7 are remarkably similar to each other and there is only a spread of about ten times between the various isotopes. The results for the glass-ceramic are significantly higher than for its parent glass except for Np. Except for 189, the values are in general somewhat higher than those in slowly flowing leachant described in Section 6.1.2.

Although the apparatus was sealed this was probably not sufficient to maintain a reducing atmosphere. The pH of the water in contact with the granite and the specimens was 7.8 ± 0.5.

6.1.4 The effect of backfill

In a repository, the encapsulated cylinders of glass will eventually be surrounded by some form of backfill whose purpose is both to reduce the access of water to the glass and also to adsorb some of

TABLE 6.5 Leach rates and mass balance for German glass B1/3 after a year at 60°C in a simulated granite repository

Dopant Isotope	Sr-90	Tc-99	Cs-137	Np-237	Pu	Am-241
Weight loss (mg)	1.84	1.52	1.25	3.62	1.35	2.26
Layer weight (mg)	0.25	0.35	0.65	0.45	0.50	0.60
Total loss (mg)	2.09	1.87	1.90	4.07	1.85	2.86
Leach rates ($g \cdot cm^{-2} day^{-1}$):						
From weight loss	1.5×10^{-6}	1.2×10^{-6}	1.0×10^{-6}	2.9×10^{-6}	1.1×10^{-6}	1.8×10^{-6}
				AVERAGE $1.6 \pm 0.6 \times 10^{-6}$		
From weight loss including layer	1.6×10^{-6}	1.5×10^{-6}	1.6×10^{-6}	3.3×10^{-6}	1.5×10^{-6}	2.3×10^{-6}
				AVERAGE $2.0 \pm 0.6 \times 10^{-6}$		
From counting	5×10^{-9}	4×10^{-8}	7×10^{-9}	4×10^{-9}	2×10^{-9}	8×10^{-10}
Weight loss deduced from activity measurements (mg)	0.3	0.5	0.5	0.6	0.4	0.8
Activity found:						
% in leachate	2	15	2	18	2	0.1
on container	10	30	17	2	8	2
in water	5	5	8	46	2	5
on granite	76	36	33	1	5	5
in layer	7	13	40	33	83	88

TABLE 6.6 Fraction of isotopes retained in gel-layer at 60°C in repository simulation

	189	209	B1/3 glass	B1/3 glass-ceramic
Sr-90	0.14	0.35	0.07	--
Tc-99	0.02	0.63	0.13	--
Cs-137	0.61	0.82	0.40	0.06
Np-237	0.66	0.73	0.33	0.84
Pu-239	0.96	0.73	0.83	0.63
Am-241	0.98	0.94	0.88	0.28

TABLE 6.7 True leach-rates at 60°C in repository simulation (in $g.cm^{-2}.day^{-1}$)

Isotope	189	209	B1/3 glass	B1/3 glass ceramic
Sr-90	1.8×10^{-6}	1.2×10^{-6}	1.5×10^{-6}	
Tc-99	1.2×10^{-5}	1.0×10^{-6}	1.3×10^{-6}	
Cs-137	5.5×10^{-7}	3.2×10^{-7}	9×10^{-7}	4.4×10^{-6}
Np-237	3.2×10^{-6}	4.6×10^{-7}	2.2×10^{-6}	8.5×10^{-7}
Pu	6×10^{-7}	4.6×10^{-7}	2.5×10^{-7}	3.4×10^{-6}
Am-241	4×10^{-7}	1.1×10^{-7}	2.8×10^{-7}	6.0×10^{-6}

the active isotopes leached from the glass. The canister and overpack that encase the glass will protect it from attack by water for many hundreds of years and will subsequently act as a buffer, giving a reducing environment as they oxidise. Two backfills that have been suggested are bentonite and concrete and we have been investigating the effects of these on the leaching of the glass.

(a) Bentonite

Bentonite would almost certainly be emplaced dry, so that when water seeped into the repository, it would swell and help to seal the glass in place. The apparatus used is illustrated schematically in Fig. 6.6.

Briefly, the doped glass specimens were each placed horizontally in the centre of a column of bentonite 75 mm long and 30 mm diameter. Water that had been equilibrated with crushed granite was then allowed to seep into the bottom of the column, taking 10 to 20 days to saturate it. The apparatus was purged with a mixture of 95% argon 5% hydrogen to keep the conditions slightly reducing. At intervals, the water from on top of the columns (10 ml) was removed and replaced with fresh water. Normally the pH of the water after about one month above the bentonite filled tube was in the range 8.0 to 8.5. An aliquot was taken from this sample, evaporated to dryness and counted. A 'leach rate' was calculated from the value obtained by dividing by the initial concentration of the isotope in the specimen and by the time since the last replenishment. This neglected any net adsorption on the apparatus or any activity diffusing down to the water below the column of bentonite. The system may be said to represent a repository where there is a flow of fresh water (10 ml per month) past the surface of the bentonite backfill but only diffusion through the bentonite and past the specimen.

The leach rates obtained in this way are plotted in Fig. 6.7 for glass 209 as an example. After 350 days, the columns of bentonite were removed from their containers and sliced up. The quantities of the radioisotopes in each slice were determined and these are plotted in Fig. 6.8. The glass specimens were dried and weighed, the gel layer was removed and they were then reweighed. The weights were used to calculate leach rates, both with and without including the gel-layer. The distribution of the isotopes between the leachate, the bentonite, the apparatus and the gel-layer was determined in each case. All the data are tabulated in Table 6.8.

The leach rates at 60°C calculated from weight losses for glass UK209 are as follows:

Technique	Not including gel layer	Including gel layer
Soxhlet	1.8×10^{-5}	2×10^{-5}
Flowing water (1 ml/week)	6×10^{-7}	1.1×10^{-6}
Granite	*	1.9×10^{-6}
Bentonite	6×10^{-6}	9.5×10^{-6}

*The samples above the granite column gained weight, presumably due to precipitation of constituents from the granite.

Thus, although the flow past the sample was very small, the leach rate in bentonite is only a factor of 2-3 smaller than observed by the Soxhlet technique, presumably due to the adsorption of silica onto the bentonite reducing the concentration near the specimen.

The agreement between the measured weight losses (line 6 in Table 6.8 and those calculated from the counting results (line 15) is good except for Pu and Am, probably due to their strong adsorption properties. The last five lines of this table give the location of the radioisotopes found on dismantling. In most cases, much the largest fraction was adsorbed on the bentonite: except for Pu, there was more on this than there was retained in the gel-layer. Except for Tc, only a small fraction was released to the leachate, accounting for the very small apparent leach rates found in this experiment and plotted in Fig. 6.7.

The distribution of the radioisotopes along the column is shown in Fg. 27. This is very similar to that found for glass UK189 [1]. Tc, Cs, Sr and Np were fairly evenly distributed along the column whilst Pu and Am were strongly localised in the vicinity of the sample.

TABLE 6.8 Leach rates and mass balance for glass UK209 in bentonite at 60°C after 348 days

Dopant Isotope	Tc	Cs	Sr	Np	Pu	Am
Surface area (cm²)	3.24	3.15	3.05	3.34	3.26	3.14
Original wt. (g)	.60831	.57286	.44096	.68805	.67995	.57052
Final wt. (g)	.60185	.56755	.43550	.66040	.67405	.56225
Loss (g)	.00646	.00531	.00546	.00765	.00590	.00827
Layer wt. (g)	.00400	.00385	.00305	.00610	.00360	.00360
Total loss (g)	.01046	.00916	.00851	.01375	.00950	.01187
Leach rates (g·cm⁻²day⁻¹) from weight loss	5.73×10^{-6}	4.84×10^{-6}	5.14×10^{-6}	6.58×10^{-6}	5.20×10^{-6}	7.57×10^{-6}
				AVERAGE $5.8 \pm 1.0 \times 10^{-6}$		
From weight loss including layer	9.28×10^{-6}	8.36×10^{-6}	8.02×10^{-6}	1.18×10^{-5}	8.37×10^{-6}	1.09×10^{-5}
				AVERAGE $9.5 \pm 1.5 \times 10^{-6}$		
From counting	4×10^{-6}	3×10^{-8}	6×10^{-9}	3×10^{-7}	$<1.0 \times 10^{-9}$	$<2.0 \times 10^{-11}$
Mass balance (g)						
Wt. In leachate	5.79×10^{-3}	6.04×10^{-5}	9.97×10^{-6}	3.19×10^{-4}	3.0×10^{-6}	7.56×10^{-6}
Wt. on glassware	4.03×10^{-4}	4.49×10^{-4}	1.31×10^{-4}	8.04×10^{-5}	0	5.8×10^{-7}
Wt. on bentonite	3.57×10^{-3}	1.07×10^{-2}	7.24×10^{-3}	1.38×10^{-2}	8.24×10^{-4}	4.4×10^{-3}
Wt. In bottom water	3.47×10^{-4}	2.21×10^{-5}	1.43×10^{-5}	0	0	0
Wt. In layer	1.48×10^{-3}	3.30×10^{-4}	2.13×10^{-4}	4.08×10^{-4}	2.23×10^{-3}	1.83×10^{-3}
Total	.01159	.01156	.00761	.01461	.00306	.00623
From activity found						
% In leachate	50.0	0.5	0.1	2.2	0.1	.001
container	3.4	3.9	1.7	0.5	0	.01
bentonite	30.8	92.6	95.1	94.4	27.0	70.8
water	3.0	0.2	0.2	0	0	0
layer	12.8	2.9	2.8	2.8	73.1	29.2

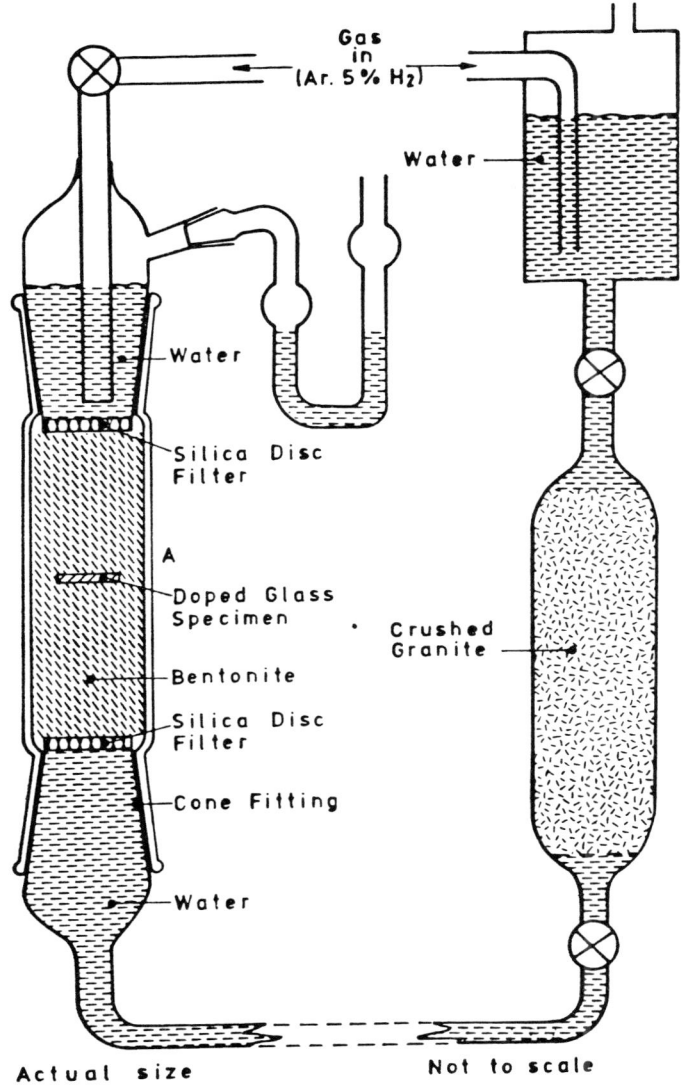

Fig. 6.6 Leach test in water-saturated bentonite

- 293 -

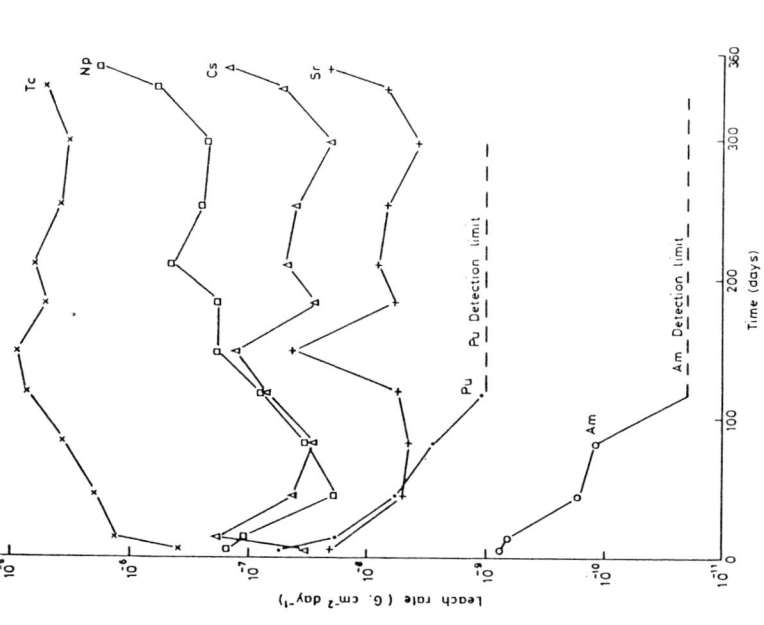

Fig. 6.8 Glass UK209 in wet bentonite column at 60°C: Distribution of isotopes in the column

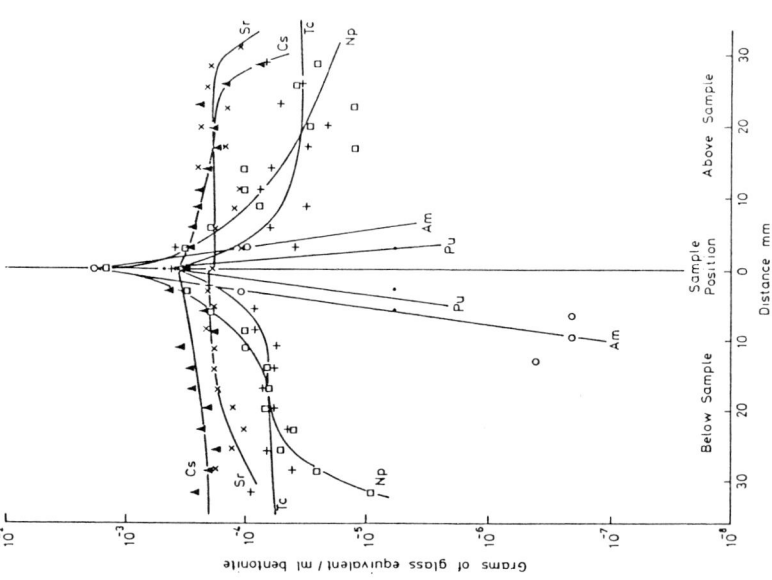

Fig. 6.7 Leach rate of UK209 at 60°C in wet bentonite

Source term leach rates have been obtained by multiplying the 'weight loss' leach rates by the fraction of each isotope released from the glass, i.e. neglecting that retained in the gel layer. These are given in Table 6.9.

TABLE 6.9 Leach rates governed by diffusion through backfill at 60°C ($g.cm^{-2}day^{-1}$)

Isotope	189 (From data in [1])			209		
	Cement Apparent	Bentonite Apparent	True	Cement Apparent	Bentonite Apparent	True
Sr 90	3×10^{-9}	1×10^{-6}		10^{-7}	6×10^{-9}	7×10^{-6}
Tc-99	1×10^{-8}}	3×10^{-6}		$<10^{-9}$	6×10^{-6}	9×10^{-6}
Cs-137	5×10^{-9}	3×10^{-6}	7×10^{-5}	10^{-7}	5×10^{-8}	1×10^{-5}
Np-237	$<5 \times 10^{-9}$	1×10^{-7}	3×10^{-5}	$<5 \times 10^{-9}$	3×10^{-7}	1×10^{-5}
Pu	$<10^{-9}$	6×10^{-9}	3×10^{-6}	$<10^{-9}$	$<10^{-9}$	7×10^{-7}
Am-241	$<2 \times 10^{-11}$	10^{-10}	2×10^{-6}	$<2 \times 10^{-11}$	$<2 \times 10^{-11}$	4×10^{-6}

These values of course do not include any adsorption on the bentonite (or on oxidised canister material etc.) — nor would they be applicable for other backfills.

Analysis of the data was dominated by the activity swept away from the specimen by the initial inflow of water.

(b)   Concrete

The system used to investigate the effect of concrete was much simpler because it was thought that the flow or diffusion of water wold be much slower and therefore that the preconditioning of the water by granite would have only a minimal effect. If concrete is used in the real case then it will probably be emplaced wet in the usual way, using granite chips as the aggregate portion. Because of size problems we decided to omit these and merely used an ordinary Portland cement/sand mixture, in the proportions:

	Cement	Sand	Water
Weight	1	2.5	0.5
Volume (approx.)	1	2.3	0.7

A 2.5 cm layer of cement was poured into a closed ended tube and vibrated to remove air bubbles. The normal disc of doped glass was placed in position and a further 2.5 cm of cement added and vibrated. The cement was then cured for 48 hours at 60°C. The volume of cement was about 40 ml and the weight about 90 g.

10 ml of water was then poured into the tube on top of the cement: about half of this soaked into the cement. The tube was sealed and held in an oven at 60°C. At about monthly intervals the liquid on top of the cement was withdrawn and counted and replaced with 10 ml of fresh water. No further liquid soaked into the cement. At the end of each month the pH of the 'leachate' was 11.9.

The results are shown in Table 6.9 where they are compared with those obtained for the bentonite backfill. In most cases they are extremely low and below the detection limit. Unfortunately, when the apparatus was dismantled it was found that the cement had stuck firmly to the glass and no measurements of weight losses were possible. Thus, it is not possible to discover whether the low apparent results are due to a beneficial effect of the cement on the leach-rate, perhaps by impeding water access or whether they are due to a strong adsorption of the isotopes on the cement. In either case, however, cement seems to be a better backfill than bentonite from this point of view, although of course the latter does not tend to crack.

6.2 INTERACTION BETWEEN REPOSITORY CLAY AND SIMULATED HIGHLY ACTIVE WASTE FORMS (SCK/CEN)

6.2.1 In-situ experiments

6.2.1.1   Surface clay quarry of Terhagen

Introduction

Since 1979 S.C.K./C.E.N. has carried out tests in a surface clay quarry, where the clay ('Boom' clay) has the same composition as the clay considered for the geologic disposal of conditioned radioactive waste. Besides experiments relating to clay characterization, the in-situ corrosion behaviour of the candidate container materials was investigated in contact with a clay derived atmosphere and in direct contact with clay [2]. Using the experimental set-up developed by the team of Dr. Casteels (container materials), some screening tests on the in-situ interaction between glassy waste forms and a clay derived atmosphere have been performed. No experiments on glass in direct contact with clay were performed.

Description of the experiments

The experiments were performed using a loop, similar to those designed for the experiments discussed in 6.2.1.2, horizontally fixed in the clay. Nine waste form samples were fixed in a barge (see Fig. 6.9), freely removable from the loop, and chosen as follows:

- Two HLW glasses, referring to the AVM reference glass SON 64 19 20 F3, but containing 14 and 9% fission products, respectively. The 5.9% Gd_2O_3 was replaced by SiO_2 (glasses BWG 1 and 2).

- A glass with a composition similar to the HLW AVB reference glass SAN 60 25 19 L_3C_2 (glass BWG3).

- Ceramic compacts, obtained by hot pressing at 800°C and 1 GPa of the granular slags resulting from inactive test runs with the FLK incinerator (conditioning of TRUW [3]).

- Five FLK base glasses (WG76-80). They essentially have an SiO_2 content between 65 and 75 mol.%, an alkali (Li, Na, K) oxide content between 6 and 10 mol.%, and an alkaline earth (Mg, Ca, Ba) oxide content of about 10 mol.%. The Al_2O_3 content of glass WG76 was 2 mol.%, and for the other glasses between 6 and 9 mol.%. The glasses were melted during 3 h at 1550°C.

The composition of the clay derived atmosphere (moisture), obtained at different temperatures, was measured at room temperature, after condensation at $-80°C$. The moisture contains 50 to 20000 mg l^{-1} CO_2-CO, < 800 mg l^{-1} $NO-NO_2$, < 100 mg l^{-1} SO_2 and < 100 mg l^{-1} HCl. After about 1.5 year of heating no more sulphur or chloride bearing corrosive products were detected.

Two screening tests were performed. In the first, the samples were exposed to the moisture during 517 d at ambient temperature, removed and weighed, and further exposed during 303 d at 50°C. In another test the samples were corroded during 232 d at 150°C in the clay derived atmosphere, removed and weighed, and further corroded during 597 d at 150°C. The test runs started in March 1980. The last test was terminated in June 1983.

Results

The results of the weight loss measurements, expressed per unit geometrical surface area, are listed in Table 6.10. Some typical results from Infrared reflection spectroscopy surface analysis are given in [1]. From the limited amount of data collected from these screening tests, the following rough tendencies can be deduced:

- The weight losses at 150°C are smaller than those measured after exposure at 50 or 13°C.

- At < 50°C, BWG 1, 2 and the FLK compacts yield the highest weight losses.

- The IRRS surface analyses confirm the ML data; BWG 1 is more heavily corroded at 50 than at 150°C and the formation of a Na depleted surface layer is suggested [4]. The IRRS spectra for BWG 3 and WG 76 before and after exposure to the moisture are almost identical. This suggests a very weak attack and a congruent release of Si and Na. Also for the FLK compacts the IRRS peak wavenumber does not shift upon exposure, but the decreased specimen reflectance indicates stronger attack (see also the ML values).

- In some cases, mainly at 13°C, weight increases were measured.

TABLE 6.10 **Specific weight losses (in $g.m^{-2}$) after in-situ exposure to a clay derived atmosphere (clay quarry of Terhagen)**

Glass	13°C (517 d)	50°C (303 d)	150°C (232 d)	150°C (829 d)
BWG1	4.14	5.93	0.29	0.21
BWG2	3.05	4.43	-0.48	0.81
BWG3	-1.58	0.85	0.16	0.16
WG76	-1.05	0.97	0.29	-0.51
WG77	-0.45	0.37	0.15	0.38
WG78	0.85	0.0	0.0	2.82
WG79	-0.32	0.32	0.34	0.51
WG80	-1.51	1.16	0.15	0.08
FLKHP79	-10.57	18.4	/	/
AFLKHP78	/	/	1.77	20.86

6.2.1.2   Underground laboratory in clay at 220 m depth

Introduction

The construction of the underground laboratory of 30 m long was finished in 1983 [5]. The period 1983-1984 has further been devoted to certain controls (e.g. creep of clay), repairs due to small leakages and the construction of a second, small shaft at the end of the gallery. Final licensing of the underground laboratory was done in the Autumn of 1985.

A description of the experiments with the waste forms is given below (see also [6]). The specimens will be mounted on the same devices as the candidate container materials. The following simulated waste forms will be investigated:

- The reference HLW forms SM 58, UK 209, SAN 60, C31, SON 58 and SON 64; glasses SM 58 and SAN 60 will be tested after crystallization, too.

- The new French reference glass SON 68 18 17 $L_1C_2A_2Z_1$, and the actual Pamela reference glass SM 513 LW 11.

- An inactive ceramic compact as obtained from high temperature FLK incineration of TRUW [3].

- The TRUW FLK base and reference glasses WG 119, 122, 123 and 124, both before and after crystallization [3].

It is also envisaged to include some 'integral' tests, exposing simultaneously waste form and container samples directly to clay, and some waste form samples tracered with Pu, Cs and Sr.

Preparation of the experiments in direct contact with clay.

Fig. 6.10b shows the design of the tube which will be used for the experiments in direct contact with clay. Tests at 170, 90 and ambient temperature ($\sim$ 13°C), during one or two different corrosion times are planned. Four tubes, manufactured by the Technology Department of S.C.K./C.E.N. are ready now. The temperature and the electric power required by the experiment will be regulated within the experimental gallery. Some blank tests have been performed at ground level.

About 64 waste form samples with sizes 40 x 15 x 5 mm, together with metal container samples will be fixed on each tube; the samples will be in contact with the non-frozen clay (a 3 m thick zone around the gallery has been frozen to enable its construction). Each tube will normally be used only once.

It is foreseen to retrieve the tubes, together with a clay layer of about 10 cm thickness after finishing the experiment. To demonstrate

the feasibility of this, an overcoring test of a blank tube is
planned beginning 1985.

Preparation of the experiments in contact with a clay derived
atmosphere

Fig. 6.10a shows the design of the loop to be used in the experiments
in contact with the clay derived atmosphere. The clay derived
atmosphere, entering the loop through the porous outer wall (AISI 304
stainless steel), will be carried over the waste form and container
samples by an Argon flow. The clay derived atmosphere (moisture)
will be sampled and pre-conditioned before analysis in the gallery.

Tests at 170, 90, 50 and ambient temperature during 4 corrosion times
are scheduled. About 60 samples will be fixed on each loop. Each
loop will be used for the distinct corrosion times at a single
temperature. One loop has been finished thus far.

6.2.2 Laboratory Experiments

6.2.2.1 Elemental leaching behaviour in clay media

Introduction

To obtain data on the elemental leaching behaviour from high level
waste forms upon corrosion in clay media, it was decided to
incorporate tracer amounts of the elements of interest into the waste
forms. These elements are the actinides, some β, γ emitting fission
products and some of the main glass matrix components. First, the
combination and concentration of tracers to be added, in view of
relevant radiochemical analyses of the leachates were determined. As
a result, each waste form would be prepared in four-fold; each
preparation corresponding with different tracers:

*In addition to the work presented in this section, a considerable
 amount of work performed by SCK/CEN on corrosion in clay-related
 media has been included in Section 5.1.6, where it can be compared
 more readily with the results obtained by the same authors under the
 same conditions in deionised water.

(1) Pu-239, Cs-134, Sr-90.
(2) Am-241, Eu-154, Ca-45.
(3) Np-237, Tc-99, Fe-55.
(4) U-233, Na-22.

It was attempted to prepare the reference glasses SAN60, SM58 and UK209 in this way.

Experimental Results

The 1983-1984 program has been devoted to the preparation of two of the glasses (SM58, SON68), tracered with Pu-Sr-Cs, and to a screening corrosion test on glass SM58 in the 100 gl^{-1} clay-water mixture (SA/V = 100 m^{-1}; T = 90°C).

First, a suitable glass preparation technique was selected [7]. The technique included the impregnation with the tracers - about 1 MBq per tracer (solutions of $Pu(NO_3)_4$, $SrCl_2$ or CsCl) per g of glass - of the premelted, crushed inactive glass, followed by several consecutive meltings. Homogeneous tracer distributions were observed in the thus prepared glass.

A screening corrosion test has been performed on glass SM58 in the clay-water mixture. The E_h of the clay solution was - 200 mV. Five test durations were carried out, with a maximum of 80 d. After each test duration the leachates were ultracentrifuged at 15000 rpm, thus distinguishing between the leached elements able or unable to migrate through the clay host (the separation at 15000 rpm corresponds with about 10^6 molecular weight units, and is a conservative upper limit to migration [8]). As a result, it was found that the overall mass losses and the total Cs and Sr releases as a function of time follow the diffusion law (releases proportional with $t^{\frac{1}{2}}$). Secondly, the amounts of Cs and Sr sorbed on the clay (> 10^6 MWU) or in solution (< 10^6 MWU) did not reach constant, steady state values after 80 d.

Further experiments are scheduled.

6.2.2.2 Influence of an external γ radiation field on corrosion

This investigation is aimed at studying the influence of a γ-radiation field, which is likely to be present in the nearest environment of the container-waste form system upon geological disposal. The elaboration of the project started beginning 1983 in the Technology Department. Both the metallic container materials and the vitreous or glass-ceramic waste forms are included in the project.

Technical elaboration

It was planned to carry out the experiments in the existing RITA irradiation installation of the BR2 test reactor of Mol. This installation, which is situated in the hydraulic channel in the machine hall of the BR2 consists of four symmetrically disposed Co60 sources at 7 m below the water level. A fairly homogeneous radiation field of 10^5 rad.h^{-1} is produced. A vertical transport system allows periodical removal of the sample container from the radiation field, to perform any required manipulations within the container. Regulation and data recording panels are located on a platform just beside the hydraulic channel.

Fig. 6.11 shows some components of the installation, before its transportation to the test facility. It consists of:

- An air-tight stainless steel container, in which a small overpressure will be generated. Connections for the sampling and controlling systems coming out of the sample basket are seen on the inner wall of the container.

- The sample basket, consisting of aluminium plates, which can be removed freely from the container. Electrical resistance wires are fixed between the Al plates. The basket can contain 37 samples (sample volume 200 cm^3), either metallic container samples in pyrex flasks, or waste form samples in Pt cups. Some of the cylindrical holes, however, will be used to monitor the temperature, or to continuously measure radiolysis gases (in the absence of a sample).

- The connections between the container, containing the sample basket, and the instrumentation panel; the electrical and pneumatic conduits pass over an instrumentation box, situated about 1 m above the container (outside of the radiation field), and which serves as a transition between radiation resistant and non-radiation resistant conduits. The regulation panels consist of a pneumatic (including rinsing circuits for radiolysis gas analysis) and an electrical panel.

Preparation of the experiments

By the end of 1984, the whole equipment was tested in blank (outside of the reactor), and approved by the security authorities of the BR2 reactor. Before starting the experiments a blank experiment is performed, within the radiation field, to obtain data on the performance of the equipment during operation, on the generation of radiolysis products, and the evolution of the pH, E_h and dissolved O_2 in the different media (ICW and CWM in oxidizing or reducing conditions).

In a second phase selected container and waste form (glasses SM 58, SAN 60, UK 209, SON 58, SON 64, SON 68 and glass-ceramic C31) samples will be corroded separately in clay related media. The tests will be performed during 1000 h (yielding a total dose of about 10^8 rad), at 90°C, in either oxidizing (air atmosphere) or reducing (argon atmosphere) conditions. In a third phase the tests will be integrated, including the waste form, the container and the clay medium.

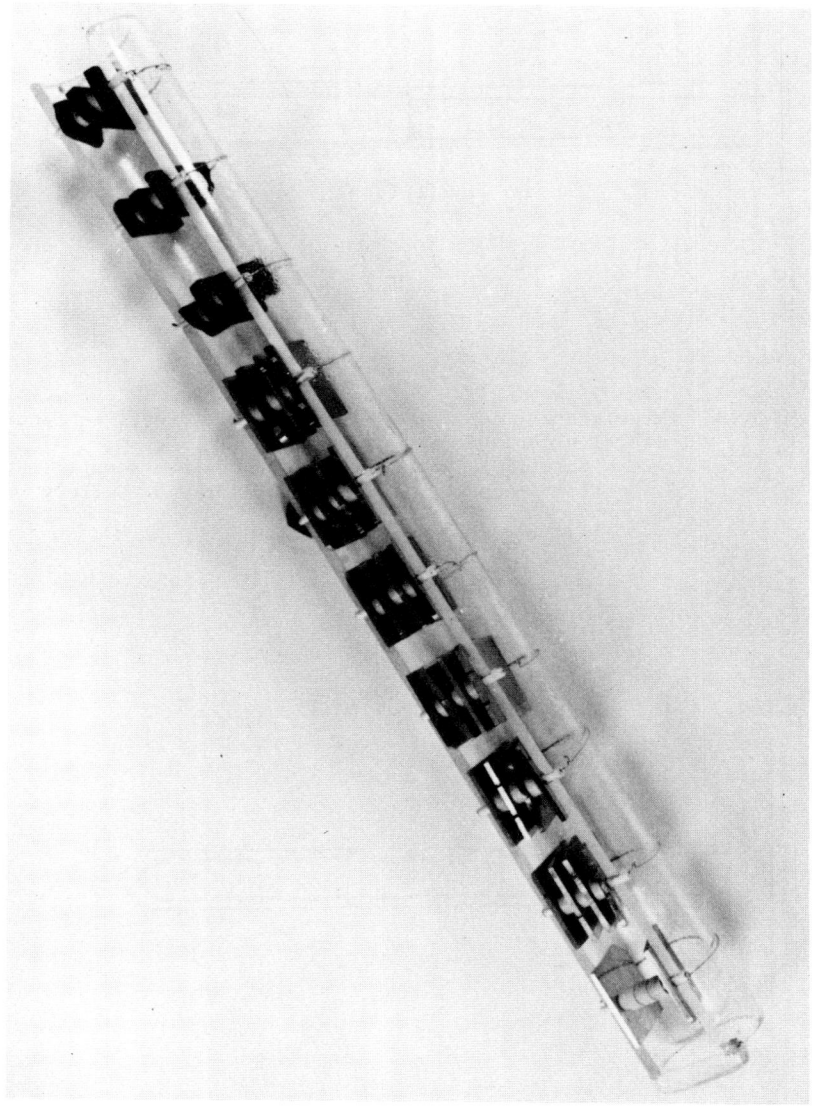

Fig. 6.9 Photograph of the barge used in the corrosion tests in contact with the clay-derived atmosphere in the Terhagen quarry

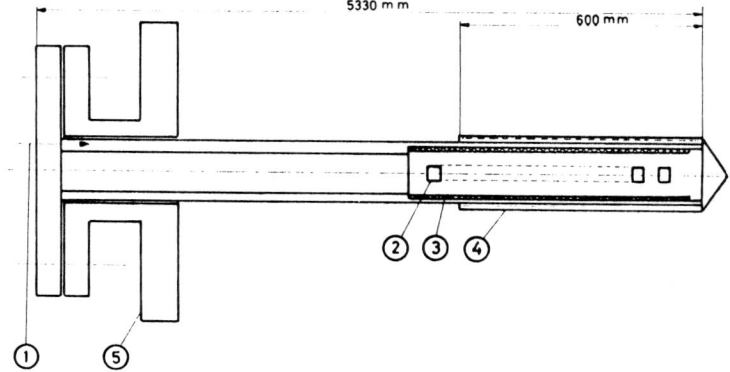

Fig. 6.10a Experimental loop for in-situ corrosion tests in contact with clay-derived atmosphere.
1. Gas inlet;   2. Container and waste from sample;
3. Heating elements;   4. Porous filter;   5. Gallery

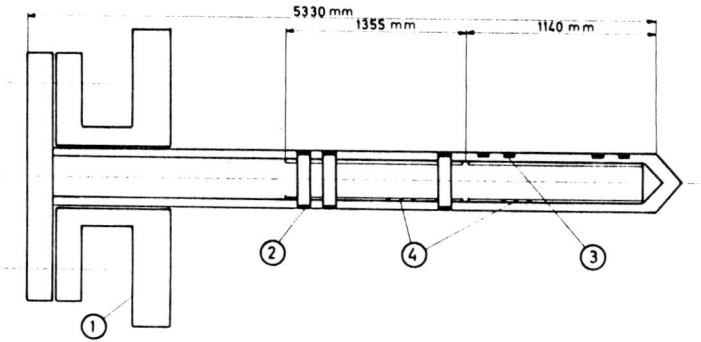

Fig. 6.10b Experimental loop for in-situ corrosion tests in direct contact with clay.   1. Gallery;   2. Container samples;
3. Waste form samples;   4. Heating elements

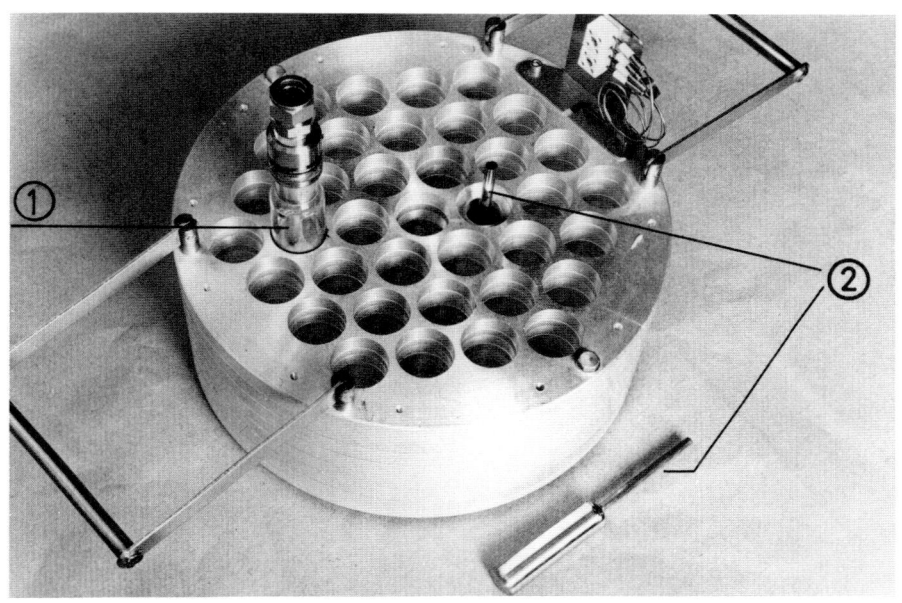

Fig. 6.11 Top view of the sample basket, showing corrosion cups
for metal (1) and waste form (2) samples

References - Chapter 6

[1] G. Malow et al., EUR 10038 EN (1985).
[2] F. Casteels et al., Jül-Conf-42, Vol. 2 (1982), Ed. R. Odoj and E. Merz, 944-977.
[3] P. Van Iseghem et al., EUR 8979, (1984) 27-42.
[4] P. Van Iseghem et al., Scientific Basis for Nuclear Waste Managemen VII, Ed. G. McVay, North-Holland (1984) 527-534.
[5] P. Manfroy et al., STI/PUB/649, Vol. 3, IAEA-CN 43/54 (1984).
[6] F. Casteels et al., Presented at the CEC/NEA Workshop on design and instrumentation of in-situ experiments in underground laboratories associated with geological disposal of radioactive waste, Brussels, May 1984.
[7] S. Izuhara et al., BLG 546 (1981).

7. RADIATION STABILITY (A.E.R.E)

7.1 Introduction

It has been pointed out by various authors (for example, Burns et al [1] that the most likely cause of any radiation effects in the vitrified waste is not the β and γ radiation from the fission products but the α-decays of the incorporated actinides. The β-particles (electrons) lose almost all their energy by ionising the atoms through which they pass and this has only a transitory effect. The α-particles on the other hand displace some atoms from their positions in the glass network whilst the recoiling actinide nuclei lose virtually all their energy in this way. Assuming a displacement energy of 25 eV, each α-particle will displace c 150 atoms and each recoiling nucleus c 1500, the latter being concentrated in a 'spike', somewhat akin to a fission spike.

To test the long-term effect of this on the glass, the most realistic, if rather laborious way, is to dope them with a short half-life α-emitter (Cm-244 or Pu-238) so that as many α-decays will occur in the glass in a few years as will occur in the real waste in many millenia.

Accordingly, as part of the first CEC sponsored programme [2], some samples of glass, with compositions suggested by the collaborating laboratories, were doped with $2\frac{1}{2}$ wt.% Pu-238. Initially the densities, leach-rates, stored energy and helium releases were investigated [2] to see if the radiation produced any effects. The stored energies were small and released over a wide temperature range such that no self sustaining temperature rise could occur. Most of the helium from the α-particles was retained in the glass and this also was not seen as a problem. Changes in leach rate and density have continued to be monitored.

After storage at room temperature for almost 6.5 years, the Pu-238 doped samples had reached a dose of about 2.8×10^{-18} α-disintegratons per gram. This is approximately equivalent to the following times in years for the real waste (assuming 0.1% of the Pu and all the Am and Cm are incorporated in the glass).

189 and 209	B1/3 and VG 98/3	SON 58 30 20
(Magnox waste)	(LWR waste)	(LWR waste)
1.5 M	70 K	10 K

The much longer equivalent times for the glasses containing Magnox waste are because of the smaller amount of fission products in the glass and the much lower fuel burn-up, leading primarily to a smalle amount of Am in the waste.

7.2 Leach Rates

The leach rates of samples of all six glasses, stored at room temperature have been measured at intervals using the Soxhlet technique to detect whether any changes had occurred. The results are plotted in Fig. 7.1 and show increases of about x4 for SON 58, x for VG 98/3 and 189 and no effect on the others. It seems to be the softer glasses that are more affected by radiation but it is not known why this should be so.

7.3 Densities

The densities of these glass and glass-ceramic samples were also measured at intervals and the values fitted to the equation

$$\Delta\rho/\rho o = A(1 - \exp(-\alpha.D))$$

where $\Delta\rho$ is the change in density, ρo is the original density, A is the saturation value of $\Delta r/\rho o$, D is the dose and α is a constant.

The values of A and α that give the best fits to the data are given in Table 7.1 and an example in Fig. 7.2: the data obey the exponential equation very closely. The borosilicte glass SON 58 30 20 and the phosphate glass contract with increasing dose whilst the other borosilicate glasses and the glass-ceramic expand.

TABLE 7.1 Fitted values of the constants for the density/dose curves

Glass	$(\rho - \rho_0)/\rho_0$ at saturation	$\alpha \times 10^{17} (g^{-1})$
UK189	$-0.0039_7 \pm 0.0001$	0.15 ± 0.01
UK 209	$-0.0075_5 \pm 0.0004$	0.051 ± 0.005
F SON 58.30.20.U2	$+0.0061_6 \pm 0.0002$	$0.061 \pm .004$
G. Celsian B1/3	-0.00484 ± 0.00006	0.118 ± 0.004
VG98/3	-0.0078 ± 0.0002	0.110 ± 0.007
Phosphate	$+0.0058 \pm 0.0006$	0.052 ± 0.011

TABLE 7.2 Values of $vr/(vr + k)$ for various values of k and vr in s^{-1}

	vr	$k=10^{-11}$	$k=10^{-10}$	$k=10^{-9}$	$k=10^{-8}$	$k=10^{-7}$
Real waste	10^{-12}	10^{-1}	10^{-2}	10^{-3}	10^{-4}	10^{-5}
Pu-238 doped glasses	3×10^{-8}	0.9997	0.997	0.97	0.75	0.23
Ion bombardment (Dran et al[6])	10^{-5} to 10^{-1}	1 1	1 1	0.9999 1	0.999 1	0.99 1

Since the radiation damage to glasses by α-decays is almost entirely in the form of heavily damaged zones round the track of the recoil nuclei, the build up of damage consists essentially of the increase in the number of such zones within the glass.

Let r = rate of damage (α-decays $cm^{-3} s^{-1}$), v = volume of damaged zone (cm^3) and F = fraction of sample volume occupied by damaged zones.

In the absence of any recovery:

$$dF/dt = vr(1 - F) \tag{1}$$

where $1 - F$ is the probability that a new damaged zone overlaps a previous one. Integrating we have:

$$F = 1 - \exp(-vrt) \tag{2}$$

If we assume that the density change $\Delta\rho$ is proportional to the volume fraction damage we have

$$\Delta\rho = \Delta\rho_{sat} \cdot (1 - \exp(-vrt)) \tag{3}$$

The dose at time t is rt/ρ decays/gram. However there is the possibility that recovery is occurring simultaneously. For simplicity this is assumed to be first order so that equation (1) becomes:

$$dF/dt = vr(1 - F) - kF \tag{4}$$

On integration:

$$F = \left(\frac{vr}{vr + k}\right)(1 - \exp(-(vr + k)t)) \tag{5}$$

Thus similar exponentially saturating behaviour is predicted as before but with the saturation value reduced by a factor $vr/(vr + k)$.

Again assuming that the recovery is first order, the irradiation induced density change $\Delta\rho$ which is assumed to be proportional to F should recover exponentially with time:

$$\Delta\rho = \Delta\rho_o \exp(-kt) \tag{6}$$

The samples were annealed isochronally after a dose of 2.8 x 10 α-decays per gram. In this technique, the samples were heated to successively higher temperatures, spaced at about 25°C intervals, for a constant time (here, 15 hours) and the densities measured (at room temperature) between each annealing: the results are given in Fig. 7.3.

In radiation damage studies of metals, for example, this technique sometimes reveals the presence of different types of defect in the structure when they anneal out at different temperatures. When this does occur, the isochronal annealing curve will have 'steps' in it. However, there is no sign of this in the present case. The annealing therefore shows either a single broad annealing stage or, perhaps more probably, the superposition of a large number of overlapping stages.

At each temperature, k was determined from equation (13) and is shown as an Arrhenius plot in Fig. 7.4. The lines drawn through the points in this figure extrapolate to values of k at 20°C as follows (in units of s^{-1}):

189	209	VG 98/3	CELSIAN	SON 58 30 20
1.2×10^{-7}	1.2×10^{-7}	5.8×10^{-8}	2.6×10^{-8}	2×10^{-11}

Using ion-bombardment techniques, Dran et al [3] found a large increase in leach rate in some glasses after a critical ion dose. This has been attributed to a dose-rate effect [1]. Equation (5) shows that if recovery is fairly rapid, i.e. the recovery constant k ≥ vr then the fraction, F, damaged will always be small since its maximum value is

$$F_{max} = vr/(vr + k)$$

Table 7.2 gives value of this fraction for various values of k and for values of vr appropriate to the real waste disposal case, to the present Pu-238 doped glass and to the ion-bombardment experiments.

For the values of k found from the isochronal annealing experiments, tabulated above, in the real situation, damage will only build up to a small extent in the French glass and hardly at all in the other substances. Partial annealing will occur in the Pu-238 doped samples and little annealing will occur in the ion bombardment experiments. Thus, on this simple theory, the ion bombardment experiments are likely to give misleading results and any changes observed in the doped glasses should be regarded with caution.

There are, however, various observations that suggest that the real situation is more complicated.

(a) At higher temperatures, the Arrhenius plots of k for some of the glasses appeared to show a second stage with a much higher activation energy.

(b) When some of the glass samples were annealed isothermally, i.e. when these were held at a constant temperature for a longer period, the densities only followed an exponential decay curve for a short period before becoming constant, i.e. there was a proportion of the radiation induced density change that was stable at each temperature [4]. The results of this experiment are shown in Fig. 7.5.

(c) After annealing, the Pu-238 doped sample of glass VG 98/3 has been held at 130°C. The density changes compared to those found during the original hold at room temperature are shown in Fig. 7.2.

The version of equation (5) that applies to density changes is:

$$\Delta\rho = \Delta\rho°SAT \cdot \frac{vr}{vr + k} (1 - \exp(-(vr + k)t))$$

where $\Delta\rho°SAT$ is the saturation value of $\Delta\rho$ at low temperatures where the recovery constant k is small compared to vr. Using values for k for 130°C and for 20°C taken from Fig. 7.4 suggests that the density changes at 130°C should be much smaller compared to those at 20°C than were actually found and also that the density changes at 130°C should approach saturation much more quickly than at 20°C: again this is not borne out by experiment as may be seen by comparing the observed values at 130°C with the calculated curve in Fig. 7.2. Because of the complicated structure of glasses it seems most likely that many types of defects will occur. Each will have its own relaxation constant, with different activation energies and will affect the density to a different extent.

The equations are difficult to deal with, and separating out the various terms may well prove impossible.

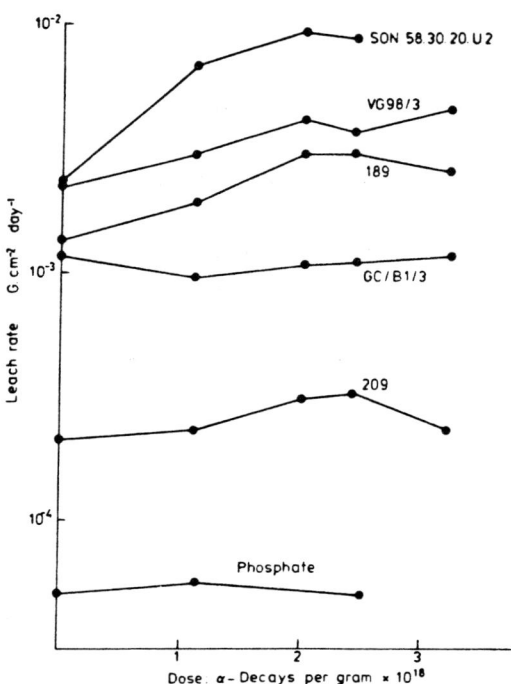

Fig. 7.1 Pu-238 spiked glasses: Soxhlet leach rate versus dose

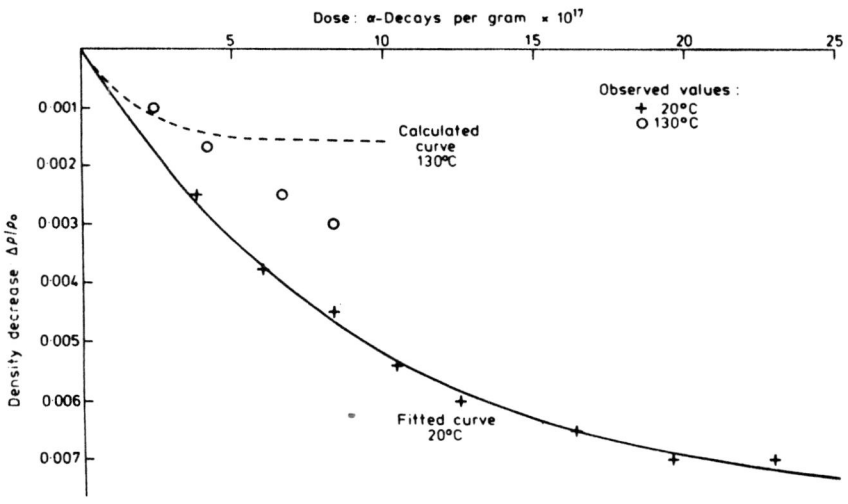

Fig. 7.2 Density changes versus dose for VG 98/3

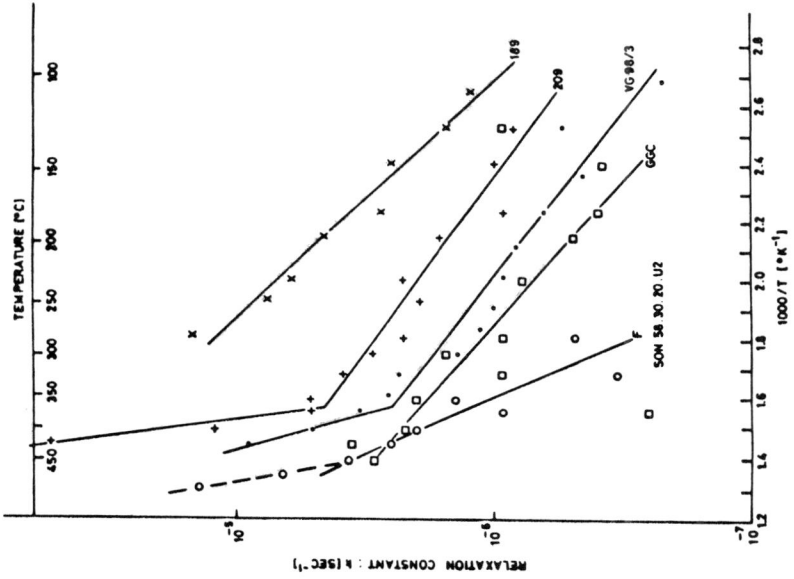

Fig. 7.4 Arrhenius plot of relaxation constant

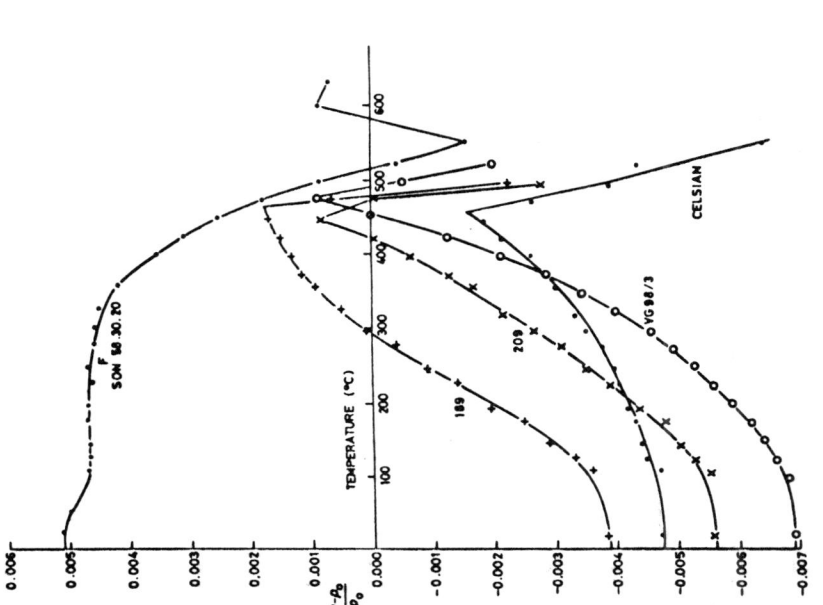

Fig. 7.3 Isochronal annealing of Pu-238 spiked glasses

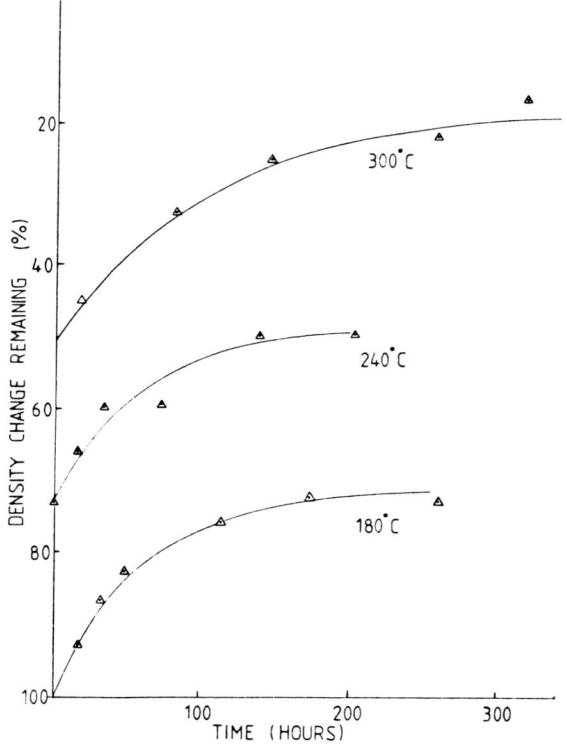

Fig. 7.5 Isothermal annealing of UK209

References - Chapter 7

[1] W.G. Burns, A.E. Hughes, J.A.C. Marples, R.S. Nelson and A.M. Stoneham J. Nuc. Mat. 107 (1982) 245.

[2] G. Malow, J.A.C. Marples and C. Sombret in Radioactive Waste Management and Disposal. R. Simon and S. Oriowski Eds. Harwood Academic Publishers Luxembourg 1980 p.341.

[3] J.C. Dran et al. Science 209 (1980), 1518 and Scientific Basis for Nuclear Waste Management, Ed. J.G. Moore, Vol. 3 Plenum Press 1981 p.449.

[4] Testing and Evaluation of Solidified High Level Waste Forms. R. de Batist Ed. European Commission Report EUR 8424 En (1983) p.103.

8. MECHANICAL STABILITY

8.1 FRACTURE STUDIES (HMI Berlin)

It is expected that most of the glass blocks will show some cracking due to the relatively short cooling time and part of the resulting thermoelastic energy will be converted to surface energy immediately.

Surface enlargement during the cooling procedure was determined experimentally and a factor of 10 to 20 was found [1].

Nevertheless, there will remain some thermoelastic stress in the glass and additional stresses will arise from the decrease of the heat production rate due to the radioactive decay. Although these stresses are probably not able to initialize spontaneous failure of the glass, the possibility of stress release by slow crack propagation has to be considered - a phenomenon well known in glasses, where crack velocities as small as 10^{-9} m/s were observed. This is a value still high enough to destroy an initially intact glass cylinder of the dimensions envisaged for radioactive waste disposal. To estimate the final surface area of the cracked glass block it is not only necessary to quantify the remaining stresses, but also to characterize the flaws in the surface of the glass as they act as initial cracks. Finally the velocity of the crack propagation has to be known which is determined by the so-called K_I - v - curve

$$v = \frac{dc}{dt} = A \cdot K_I^n \tag{1}$$

where K_I is the stress intensity factor, c the crack length and A and n are material constants. The stress intensity factor is given by the stress σ in the crack tip and the length of the crack:

$$K_I = \sigma c^{1/2} \cdot f \tag{2}$$

(f is a correction factor depending on the geometric dimensions of the specimen and of the crack itself). From formulae (1) and (2) a time can be calculated needed to lengthen a crack from a starting length c_b to a final length c_e at constant stress:

$$t = \frac{2\,(K_{I_b}^{2-n} - K_{I_e}^{2-n})}{(n-2)\,A\sigma^2 f^2} \qquad (3)$$

Unfortunately there are no direct methods of measuring the flaw distribution in a glass surface as the flaws are normally too small. However, it is possible to estimate the flaw distribution by measuring the surface fracture stress. A simple and widely used method of doing this is the Hertzian indentation test. The principle of the test is shown in Fig. 8.1.

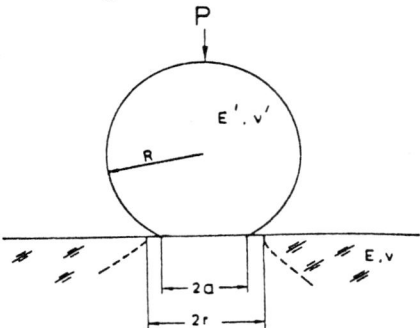

Fig. 8.1   Principle of the Hertzian indentation test

An elastic spherical indentor is pressed on the plane surface of the material to be investigated. (E-modulus and Young-modulus v should be about the same in both materials). The load P produces a stress field which is tensile in the surface outside of the contact circle (radius a_c).

In the test the load P is increased until a critical stress is reached and the material is fractured. Normally, a well defined circular crack is visible just outside the contact circle. The posision r_c of the ring crack is determined by the probability that a suitable flaw is enlarged by the stress field. Beneath the surface

- 321 -

there will develop a cone crack which follows a path approximately perpendicular to the maximum tensile stress. Due to the statistical distribution of the flaw sizes and positions the measured critical load P_c varies over a certain range. The corresponding distribution function is asymmetric. In brittle materials it is approximated by the Weibull-statistic

$$F = 1 - e^{-(\frac{\sigma - \sigma_u}{\sigma_0})^m} \qquad (4)$$

which gives the probability that a crack occurs if the applied stress is σ ($F = 0$ for $\sigma < \sigma_u$). The exponent m is called the Weibull-modulus and is a measure of the range of the flaw distribution. In brittle materials the lower limit σ_n is very small – in most cases essentially zero. Then σ_0 is the so-called characteristic stress ($F(\sigma_0) = 1 - 1/e$). The average fracture stress is normally taken as the 50%-value σ_c which is slightly smaller than the characteristic value.

Critical stresses were measured for a series of glasses containing simulated HLW (Table 8.1). A soda-lime glass was also tested for comparison. Fig. 8.2 gives an example of the failure probability as a function of the fracture stress. The data were evaluated using formula (4) with $\sigma_u = 0$. m and σ_0 were determined by a least square fit of the logarithmic data.

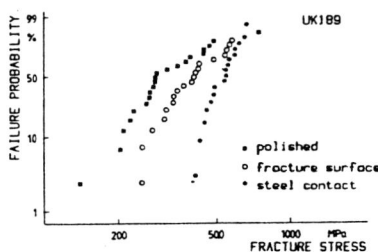

Fig. 8.2 Weibull-statistic of fracture stresses for the glass UK 189 (indenter radius R = 1.5 mm), measured on different surfaces

- 322 -

The results for glass indenters with R = 3 mm diameter are given in Table 8.1. The stresses at the contact circle are - as expected - a little higher than the stresses calculated at the position of the ring crack (for the corresponding formulae see [2]). All values measured are higher than that obtained for the soda-lime-glass indicating that the addition of the nuclear waste oxides does not decrease the surface fracture stress of common glass. The glass ceramic is the toughest material investigated. Different surface treatments were used for some of the glasses. The lowest value for the critical stress is found for polished surfaces whereas a surface which has been in contact with polished stainless steel during cooling is even tougher than a fractured surface. This may be due to the compressive stress induced by the preparation procedure. In conclusion the measurements indicate that the flaw sizes in the fracture surfaces of a glass block will not be larger than the flaws introduced by the polishing procedure where the flaws are estimated to be in the order of several microns.

TABLE 8.1 Fracture stress measured by Hertzian indentation (last column: ratio of ring crack to contact circle diameter)

Material		Fracture stress				
		Ring crack		Contact circle		
		σ_c (N/mm^2)	m	σ_c (N/mm^2)	m	r_c/a_c
Mirror glass	a	357	3.9			
VG 98/3	b	361	2.4	455	6.6	1.12
	c	364	4.9	541	15.1	1.21
	d	389	2.8	569	8.2	1.20
SON 58	b	428	3.3	541	7.9	1.12
	c	443	3.3	553	9.1	1.11
	e	531	7.2	525	15.6	1.01
UK 189	b	396	2.8	490	10.4	1.11
	c	442	4.1	579	6.0	1.14
	e	561	8.3	526	18.2	1.03
UK 209	b	397	3.0	559	6.6	1.18
	c	406	3.6	533	11.5	1.14

Surface condition: a as received float glass
                    b polished
                    c fracture surface
                    d stainless steel contact
                    e fire polished

In order to calculate the crack velocity as a function of the stress the double-torsion method was applied using a time dependent load, thereby allowing the measurement of crack velocities over a wide range in one experiment. The principle of the method is shown in Fig. 8.3. A plate with a crack of length c is loaded (force P) resulting in the torsion of two bars. The stress intensity factor (2) is directly proportional to the load and the velocity is related to the change of the displacement y which is measured as a function of the load rate. Fig. 8.4 shows two examples of K_I - v - curves measured for glasses C31-3 and B 1-3 and the corresponding glass ceramic of B 1-3. At low K_I-values the velocity increases continuously with increasing load followed by a plateau. Finally there is a further large increase of v for only small changes of K_I. The maximum value of K_I observed for very high velocities is identical to the fracture toughness K_I (Table 8.2).

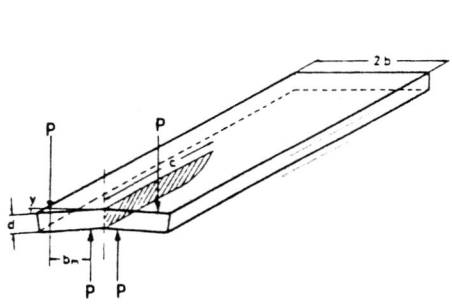

Fig. 8.3 Principle of the double-torsion test.

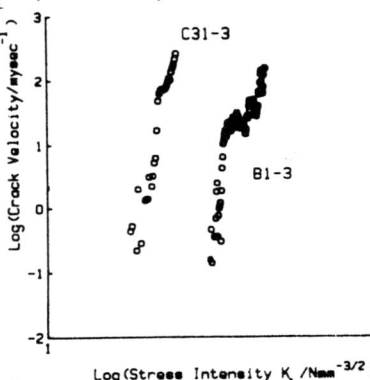

Fig. 8.4 v-K_I curves of the glass ceramic B 1-3.

For slow crack propagation only the part at low K_I-values is of interest which can be fitted by a straight line in a double-logarithmic representation leading to the empirical law of equation (1).

The parameters A and n were calculated by a least-squares fit of the logarithmic data. The results are listed in Table 8.2 together with the fracture toughness K_{I_c} (for comparison the value for $v = 10^{-3}$ m/s was chosen). The glass ceramic has the highest K_{I_c}-v curve, close to

- 324 -

the value measured for the polycrystalline mineral eucryptite.
Although the slope of the K_I-v curve is very sensitive to the
environment, all materials investigated have a much higher n-value
than the soda-lime-glass, indicating that the waste glasses are less
susceptible to slow crack growth.

According to equation (2) the critical flaw size can be calculated
from the fracture toughness and the fracture stress. f = 2 is an
upper limit for the correction factor (scratch) and f = 1.2 a lower
limit (semi-circular crack). For both values the crack length c_b was
calculated using the data in Tables 8.1 and 8.2.

As the actual stresses in the waste form are not yet known a lifetime could not be calculated using formula (3). Therefore only
maximal allowable stresses are given in Table 8.3 which will not
increase the crack surface by more than a factor of 10 (c_e/c_b = 10)
within the period 100 - 1000 years:

$$\sigma_{max} = \frac{1}{f} \left(\frac{2}{t(n-2)A}\right)^{1/n} c_b^{(2-n)/2n} \left(1 - \left(\frac{c_e}{c_b}\right)^{(2-n)/2}\right)^{1/n}$$

Because of the high values of n-typical for brittle materials the
fracture surface is increased significantly if σ_{max} is exceeded. It
has to be extrapolated to larger surfaces revealing more realistic
flaw size distributions.

TABLE 8.2 **Fracture toughness and crack growth parameters measured by the double torsion method**

Material	Fracture toughness $K_{I_c}/Nmm^{-3/2}$	v-K_I-parameters n	-log A*
Mirror glass	24	18	21
VG 98/3	17	36	45
UK 209	22	113	152
C-31-3	25	66	87
GP98/12	27	13	16
SM 513	31	29	36
B 1/3	42	76	112
B 1/3	30	67	99
Eucryptite	50	129	189

* $v/1^{-6}$ ms^{-1}

TABLE 8.3 Fracture stresses (50%-probability values) of nuclear waste glasses and ratio of ring crack and contact circle diameters; indentor radius R = 1.5 mm

Material	At ring crack		At contact circle		Ratio
	$\sigma_C(N/mm^{-2})$	m	$\sigma_C(N/mm^{-2})$	m	r_C/a_C
Soda lime-silica glass	323	3.9	465	16.6	1.20
SON 58	428	3.3	541	7.9	1.12
209	397	3.0	559	6.6	1.18
SM513	363	4.3	533	10.5	1.20
GP98/12	454	10.0	580	12.3	1.13
C31-3	447	5.5	571	13.7	1.13
B1-3	558	4.6	565	11.1	1.01

In conclusion it can be stated that nuclear waste glasses are less susceptible to slow crack propagation than soda-lime glass. However, small cracks will increase in the long run even at stresses which are much smaller than the fracture stress.

8.2 Canister/Glass Interaction (AERE)

8.2.1 Cracking of glass in cylinders

In an investigation of the amount of cracking occurring when glass cylinders are cooled, glass frit was melted in 18/9/1 stainless steel canisters, 0.25 m diameter, 0.75 m long with a wall thickness of 5 mm. The melt was 0.6 m deep and weighed about 80 kg and was allowed to stand at 950°C for at least 24 hours to clear any bubbles. The furnace power was then switched off and the glass cooled following the exponential equation:

$$\Delta T = \Delta T_o \exp(-0.0184t)$$

where ΔT is the excess temperature above ambient (°C) and t is in hours. Above T_g (446°C) the glass is soft and no stress occurs. Below this temperature, stresses (and hence fracture) only occur due to interaction with the canister or change in temperature profile.

After cooling, the canisters were cut away from the glass, the fragments were examined and their total surface area estimated by laborious measurement assisted by a computer programme. Fig. 8.5 shows the crack pattern found in two cases. In the first instance, the glass adhered extensively to the cylinder and this had apparently caused much of the cracking. The surface area had increased from 0.57 m^2 for an intact cylinder to 5.6 m^2, an increase by a factor of 10. For the second run, the cylinder was lined with a graphite mat and this decreased the adhesion except near the top of the cylinder where it had oxidised away. The surface area had decreased to 2.9 m^2, a worthwhile improvement of about a factor of two.

8.2.2 The effect of decay heat on the temperature profile

At first, it was thought that decay heat could be simulated approximately by using a central heater in the cylinder. However, this gives the wrong profile. In Fig. 8.6 curve C shows the equilibrium temperature profile for a cylinder of specific activity (Ao) of 5 kw m^{-3}, surface temperature constant at 500°C. Curve D shows the equilibrium profile for a cylinder having a central heater of 20 mm diameter, power adjusted to maintain the same centre temperature. The difference in profiles is immediately obvious. However, from an examination of the equations for temperature distribution in cylinders an easier alternative method of simulating the effects of decay heat on the temperature profile becomes apparent.

For infinite cylinders, radius a, surface cooled at rate p, with no decay heat

$$\Delta T_{r,t} = \frac{p(a^2 - r^2)}{4k} - \frac{2p}{ak} \sum^{\infty} \frac{J_o[r\alpha_n] \exp(-k\alpha_n^2 t)}{\alpha_n^3 J_1[a\alpha_n]} \qquad (5)$$

$$n = 1, 2, \ldots$$

and for the same cylinder, surface temperature constant, decay heat rate Ao.

$$\Delta T_{r,t} = \frac{Ao(a^2 - r^2)}{4k} - \frac{2p}{ak} \sum_{n}^{\infty} \frac{J_o[r\alpha_n] \exp(-k\alpha_n^2 t)}{\alpha_n^3 J_1[a\alpha_n]} \qquad (6)$$

where k is thermal diffusivity

K is thermal conductivity

α_n is the nth root of the equation $J_o[a\alpha_n] = 0$

$\Delta T_{r,t} = T_{a,t} - T_{r,t}$

$T_{r,t}$ = temperature at radius r and time t

now since $K = \kappa\rho Cp$, then by substituting $p = Ao/\rho Cp$ in equation (5), equation (6) is obtained. Thus by cooling a non-active cylinder at a rate equal to Ao/ρCp, the same temperature profile exists during the cooling period as would exist in a cylinder with constant surface temperature and specific activity Ao. Curve A on Fig. 8.6 is an example of this. When the cylinder is finally cold and at ambient temperature the stress pattern would then simulate the final condition when the radioactivity had all decayed away. It is in this condition that the maximum stresses would develop. Similar equations and argument would also apply to axial temperature distributions.

An important result of this is that stresses will arise in the fully active blocks when the activity has decayed away however slowly they are cooled shortly after manufacture.

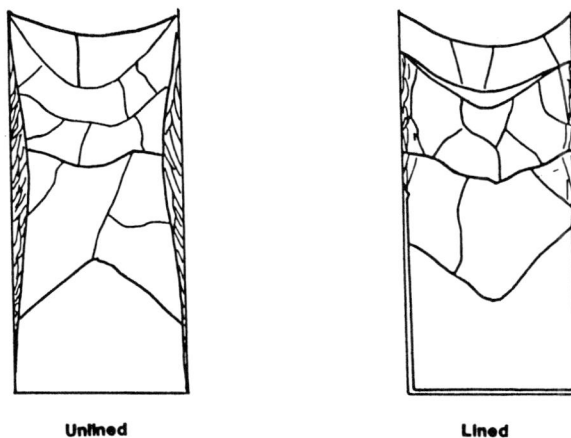

Unlined **Lined**

Fig. 8.6 Cracks in large glass cylinders

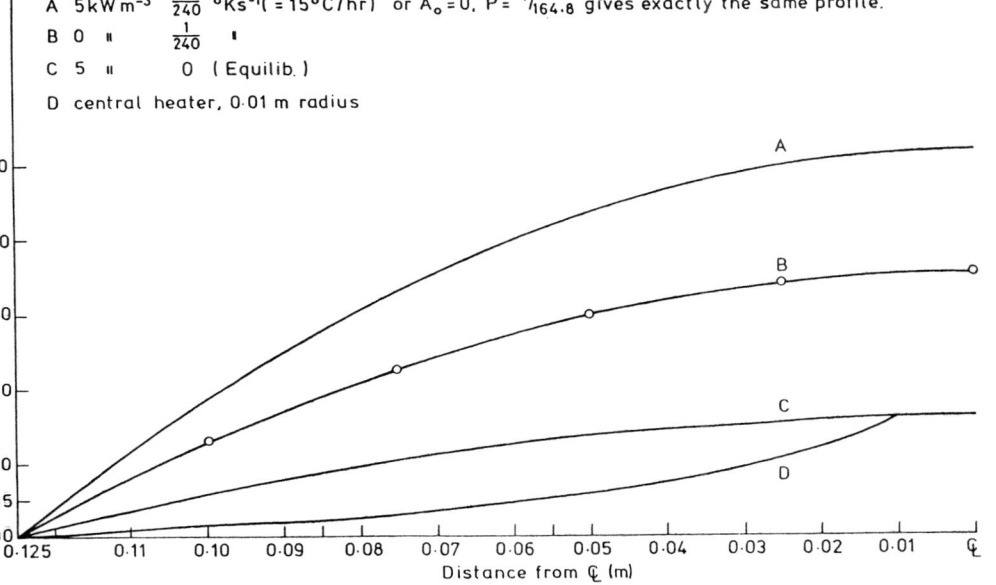

      A_o P
A 5 kW m^{-3} $\frac{1}{240}$ °Ks^{-1} ($\equiv$ 15°C/hr) or $A_o = 0$, $P = \frac{1}{164.8}$ gives exactly the same profile.
B 0 " $\frac{1}{240}$ "
C 5 "   0 (Equilib.)
D central heater, 0.01 m radius

Fig. 8.7

References - Chapter 8

[1] F. Laude, G. Vernaz, M. Saint-Gaudens. Fracture Appraisal of Large Scale Glass Blocks under Realistic Thermal Conditions, Scientific Basis for Nuclear Waste Management, V, W. Lutze Ed., North-Holland, New York (1982).

[2] Ch. Engelmann, Ed., Testing and Evaluation of Solidified High-level Waste Forms, Joint Annual Progress Report 1982, EUR 9268 EN (1984).

9. CERAMICS (LEIDEN UNIVERSITY)

The aim of the research reported here is to prepare model compounds and study their thermal stability, phase relations, crystallographic properties and behaviour in water and 10% NaCl solution at elevated temperature (300°C under pressure). Some of the model compounds are possibly present in waste glasses and Synroc phases, other model compounds can give information about the behaviour of the element Technetium in such systems.

The systems studies are:
1. hollandite phases
2. perovskite and perovskite-like phases
3. phase relations of RuO_2.

TABLE 9.1 *Composition of 1 t spent uranium fuel, originally enriched to 3.3% in ^{235}U, after 33 000 M W'd t U burn-up at a flux of 3×10^{13} n cm^{-2} s^{-1} (30 M W t U) at a cooling time of 10 y Straight figures are weight in kg, italics radioactivity in kCi. Total weight is 35 kg FP and 11 kg transuranium elements. Total radioactivity is 312 kCi from the FP and 2.35 kCi from the TU elements. Thermal power is 1.02 kW from the FP and 0.070 kW from the actinides. Most common corrosion products in HAW are shaded; P is a radiolysis product. Arrows indicate mother-daughter relations*

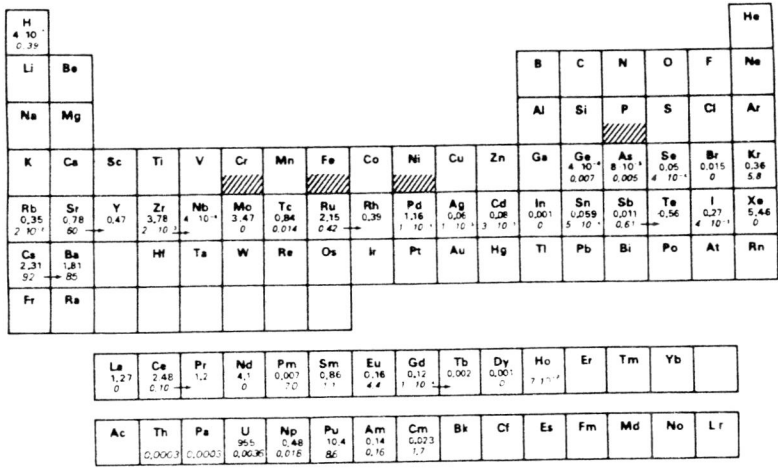

In the model systems we have chosen Nd as an average lanthanide and Ru as an average second row transition element. Table 9.1 gives the composition of the waste in kg and radioactivity in kCi from 1 ton spent uranium fuel, originally enriched to 3,3% in 235 U at a cooling time of 10 years. Also most common corrosion products in HAW are indicated. For comparison we give in Table 9.2 the activity after a cooling time of 1000 year from the long-lived fission products with activity practically independent of cooling time for thousands of years (1).

TABLE 9.2 *Long-lived fission products with activities practically independent of t_{cool} for thousands of years*
Activities at t_{cool} 1000 y, otherwise as in Table 1

Isotope	Half-life (y)	Radioactivity (Ci t L)
^{79}Se	6.5×10^4	0.39
^{87}Rb	4.7×10^{10}	2×10^{-5}
$^{93}Zr \rightarrow ^{93m}Nb$	1.5×10^6	2.6
^{99}Tc	2.1×10^5	14.3
^{107}Pd	6.5×10^6	0.11
$^{126}Sn \begin{smallmatrix} \nearrow ^{126m}Sb \\ \searrow ^{126}Sb \end{smallmatrix}$	10^5	0.54
^{129}I	1.57×10^7	0.038
^{135}Cs	2.1×10^6	0.29

The fission products of Table 9.2 are dangerous for a long period. From these Technetium is the most dangerous one, because it can be easily transported as TcO_4^- ions. The migration of heavy metals as actinides is very limited.

9.1. HOLLANDITES

Synthetic hollandite compounds of composition $Ba_xM(IV)_{4-2x}N(III)_{2x}O_8$ (M=Ti,Ge,Ru,Zr,Sn; N=Al,Sc,Cr,Ga,Ru,In) and $Sr_xM(IV)_{4-2x}N(III)_{2x}O_8$ (M=Ti,Ge,Ru; N=Al,Cr,Ga) and $(A,Ba)_xTi_yAl_zO_8$ with A=Rb,Cs,Ca,Sr have been studied by electron and X-ray diffraction and high resolution electron microscopy. They have been found to be stable only within certain ranges of x, depending on the M and N ions (see Table 9.3). The lower value of x is never less than 0.56, the upper x level is about 0.73. Higher x are correlated with larger radii of M and N. The variation in x gives rise to an incommensurate occupation by the A ions in the tunnels. Correlation among tunnel sequences varies widely, also depending on the nature of the M and N ions. The lower and upper x, the cell dimensions (see Table 9.4), the correlation between the tunnels and the stability in water at 300°C of the existing hollandites are given. It was found that Rb, Cs and Sr can only be incorporated in the hollandite phase if Ba is not present in excess. Ca cannot be incorporated in the hollandite structure.

TABLE 9.3

Composition limits of the hollandites A,M/N which could be prepared, the radii of M and N according to Shannon and Prewitt and the correlation C between the phases of the occupation waves in the tunnels (s=strong, i=intermediate, w=weak).

	x(min)	x(max)	$r(M^{4+})$	$r(N^{3+})$	C
Sr,Ti/Cr	.61(1)	.68(1)	0.605	0.615	i
Sr,Ge/Cr	.62(1)	.67(1)	0.54	0.615	i/w
Ba,Ti/Al[a]	.58(1)	.64(1)	0.605	0.53	s/i
Ba,Ti/Cr	.56(1)	.67(1)	0.605	0.615	s/i
Ba,Ti/Ga	.58(1)	.66(1)	0.605	0.62	s
Ba,Ti/Sc	.59(1)	.67(1)	0.605	0.73	i
Ba,Ge/Cr	.57(1)	.68(1)	0.54	0.615	w
Ba,Ge/Ga	.60(1)	.67(1)	0.54	0.62	i
Ba,Zr/Sc	.67(1)	b	0.72	0.73	s/i
Ba,Ru/Cr	.67(1)	.67(1)	0.62	0.615	i
Ba,Ru/Ga	.63(1)	.69(1)	0.62	0.62	i/w
Ba,Ru/In	.66(1)	.69(1)	0.62	0.79	s/i
Ba,Ru/Sc	.65(1)	.69(1)	0.62	0.73	s/i
Ba,Ru,Ru	.66(1)	.68(1)	0.62	0.68	s
Ba,Sn/Cr	.67(1)	.69(1)	0.69	0.615	w
Ba,Sn/Ga	.66(1)	.71(1)	0.69	0.62	w
Ba,Sn/In	.67(1)	.73(1)	0.69	0.79	i
Ba,Sn/Sc	.67(1)	.72(1)	0.69	0.73	s/i

a) Cheary, Hunt and Calaizis report that Ba,Ti/Al hollandite in Synroc has the composition x=0.57.
b) No experimental value available.

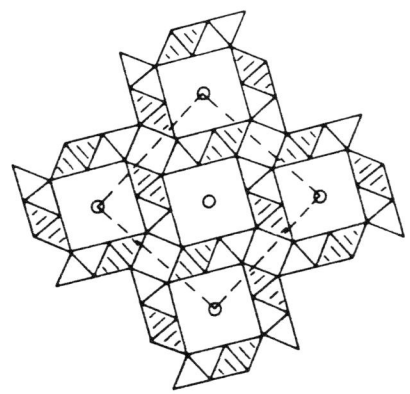

Fig.9.1. The hollandite structure.

The big ions Ba, Sr, Rb and Cs are accomodated along tunnels (see Fig.9.1). The partial filling of the tunnels gives rise to incommensurate super structures A simple vacancy modulation ware is proposed (Mijlhoff, IJdo and Zandbergen 1985) to explain the diffraction patterns (2).

Experiments in the system $(A,Ba)_x(TiAl)_4O_8$, with A is Rb, Cs or Sr, showed that Sr can be incorporated up to 20% and Rb and Cs up to 100% and that the formed hollandites are stable in water at 300°C. However, in the presence of excess of Ba no significant quantity of A was incorporated and other soluble compounds as $CsAlTiO_4$ are partly formed. This research is to be published (see Appendix I)

TABLE 9.4

Unit cell dimensions (in Å) and c/a' $(a'=(a^2+b^2\cos^2\gamma)^{1/2})$ of the prepared hollandites with x=x(min).

compound	a	b	c	γ	c/a'
Sr,Ti/Cr	10.208(11)	9.954(12)	2.954(3)	91.61(4)	0.2956(5)
Sr,Ge/Cr	9.868(2)	9.605(12)	2.892(2)	90.92(4)	0.2971(4)
Ba,Ti/Al	9.946(1)		2.923(1)		0.2939(1)
Ba,Ti/Cr	10.053(2)		2.948(1)		0.2932(1)
Ba,Ti,Cr[a]	10.133(6)	9.871(7)	2.953(1)	90.87(3)	0.2953(2)
Ba,Ti/Ga	10.048(2)		2.961(1)		0.2941(1)
Ba,Ge/Cr	9.800(3)		2.892(1)		0.2951(2)
Ba,Ge/Ga	9.795(2)		2.958(1)		0.3020(2)
Ba,Zr/Sc	10.706(3)	10.234(2)	3.170(3)	91.24(1)	0.3029(3)
Ba,Ru/Cr	9.861(5)		3.017(1)		0.3059(2)
Ba,Ru/Ga	9.851(1)		2.982(1)		0.3027(2)
Ba,Ru/In	9.844(4)		3.016(2)		0.3064(3)
Ba,Ru/Sc	9.835(4)		2.965(1)		0.3014(2)
Ba,Ru/Ru	9.821(3)		3.121(1)		0.3178(2)
Ba,Sn/Cr	10.516(2)	10.025(4)	3.104(2)	91.47(1)	0.3024(2)
Ba,Sn/Ga	10.449(4)	10.106(3)	3.135(2)	90.99(2)	0.3051(3)
Ba,Sn/In	10.837(5)	10.177(4)	3.206(1)	91.83(3)	0.3054(2)
Ba,Sn/Sc	10.694(9)	10.231(9)	3.193(4)	92.10(4)	0.3054(4)
Ba,Sn/Sc	10.694(9)	10.231(9)	3.193(4)	92,10(4)	0.3054(4)

a) with x=x(max) a monoclinic unit cell is obtained.

2. PEROVSKITE AND PEROVSKITE-LIKE PHASES

Introduction

According to Ringwood (3) the perovskite family of structures has the capacity to accept a wide range of ions into solid solution, i.e. Sr^{2+}, Ba^{2+}, Ca^{2+}, lanthanides and actinides. In this chapter we discuss first the perovskite phases.

The perovskite structure

The ideal perovskite structure $\underline{A}\underline{B}O_3$ is given in Fig.9.2. The $\underline{A}$ ions have to be sufficiently large to form a closed packed array with the oxygen ions. The $\underline{B}$ ions must fit in the octahedral holes of the array. In this structure the

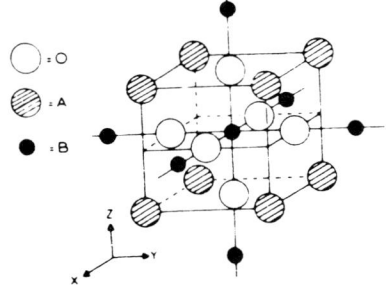

Fig.9.2: The perovskite structure, A: large black circles, O: large white circles, B: small white circles.

radius relation exists $R_A + R_O = \sqrt{2}\,(R_B + R_O)$. Examples of this structure are found in $SrTiO_3$ and $BaZrO_3$. When the above requirement is not met, the octahedra are tilted giving related structures with lower symmetry. Examples are the mineral perovskite, $CaTiO_3$, $BaCeO_3$ and the class $Ln\underline{B}O_3$, Ln = rare earth, $\underline{B}$ a small ion, i.e. Ti^{3+}, Fe^{3+}, etc. also $BaCeO_3$. In Fig.9.3 the unit cell volumes are given of a member of this compounds.

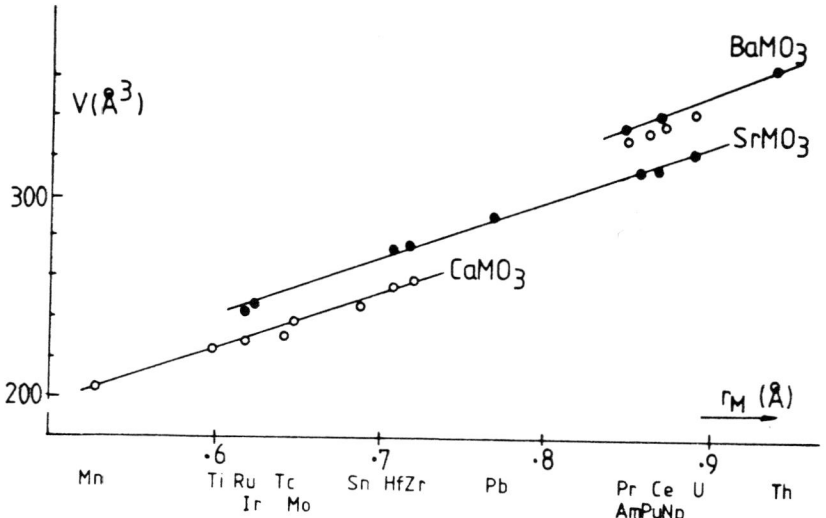

Fig.9.3: Unit cell volumes of perovskite $\underline{AB}O_3$.

Related perovskites $A_2BB'O_6$

These compounds have the same structure except the small ions B and B' are ordered as in figure 4.

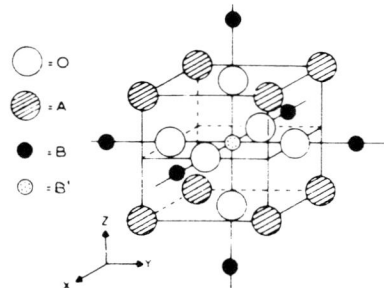

Fig.9.4: The positions of the ions B (small black circles) and B' (small white circles) in the perovskite structure.

The symmetry of these compounds is more complicated and up to now not fully known. The $\underline{B}O_6$ octahedra share corners with neighbouring $\underline{B}'O_6$ octahedra and reverse. In relation to SYNROC there exists a number of perovskites $A_2BB'O_6$ related to the mineral perovskite $CaTiO_3$, with about the same lattice constants and unit cell volume but with lower symmetry and more complicated X-ray patterns. Moreover the compounds $Ln\underline{B}(III)O_3$ have exactly the same structure. Table 9.5 gives the possible perovskites $Ca_2\underline{BB}'O_6$ and $Ln_2\underline{BB}'O_6$. There exist also some $Sr_2\underline{BB}'O_6$ and $Ba_2\underline{BB}'O_6$ compounds.

Table 9.5

Perovskites $Ca_2\underline{BB}'O_6$ and $Ln_2\underline{BB}'O_6$

$Ca_2M(II)M'(V)O_6$ $M(III)$ = Al, Cr, Fe, Rh, Ln
                             $M(V)$ = Ru, Nb, Ta, Sb, U
$Ca_2M(II)M'(VI)O_6$ $M(II)$ = Mn, Ca, (Eu, Sm)
                             $M(VI)$ = Mo, W, U
$Ln_2M(II)M'(IV)O_6$ Ln = La, Lu, Y (Ac?)
                             $M(II)$ = Mg, Mn, Co, Ni
                             $M(IV)$ = Ti, Ru (and Tc?)

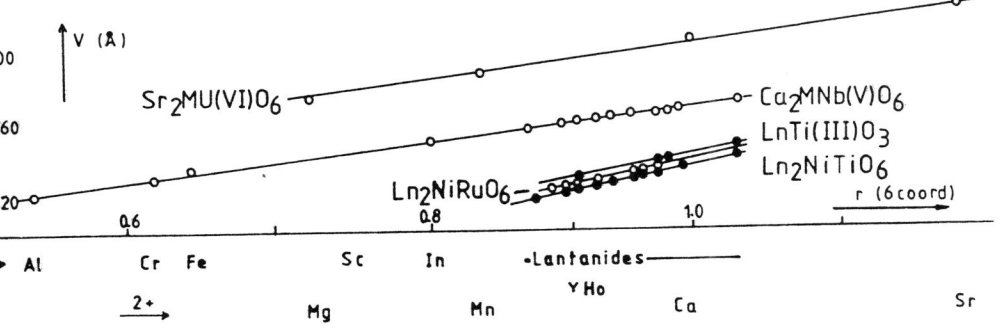

Fig. 9.5. The unit cell volumes of perovskites ABO_3 as a function of the radius of B

From the existance of this compounds it is concluded that in SYNROC t[he] perovskite phase can accomodate the small ion M(II), M(III), M(IV) M(V) a[nd] M(VI) in octahedral positions. In the large cation positions A(II): Ca, Sr, [Ba] and A(III): Lanthanides and actinides.

The question arises whether there is in SYNROC one perovskite phase or ma[ny] phases with limited intergrowth. From Fig.9.5 and Table 9.6, it is seem th[at] there are considerable differences in unit cell volumes for the disti[nct] phases. For geometrical reasons it is impossible that continuous sol[id] solution can occur. It can be concluded that it is not expected than only o[ne] perovskite phase is present in SYNROC.

Solubility tests

Most of the perovskite phases mentioned above are very insoluble compound[s]. The compounds Ca_2CaUO_6 (=Ca_3UO_6) and Sr_2MUO_6 (M=Mg, Mn, Ca) are very solub[le] in water.

Neutron powder diffraction data of some $A_2BB'O_6$ compound

As an illustration of the ideas represented a number of compounds were studi[ed] with use of neutron powder diffraction (E.C.N., Petten, The Netherlands). T[he] space group of the compounds is lowered from Pbnm ($CaTiO_3$) to $P2_1/n$ wi[th] β about equal to 90°. The lattice constants are given in Table 9.6. This stu[dy] is to be published (see Appendix II)

Table 9.6

Lattice parameters of the investigated compounds $A_2\underline{BB}'O_6$ and for comparis[on] some compounds $Ca\underline{B}O_3$

	a(A°)	b(A°)	c(A°)	(β°)	V(A°^{3})
Ca_3UO_6	5.7292(2)	5.9562(2)	8.2991(3)	90.56(1)	283.19(2)
Sr_2MnUO_6	5.8422(5)	5.8886(6)	8.2691(7)	90.05(2)	284.48(2)
Sr_2CaWO_6	5.7697(2)	5.8516(2)	8.1967(4)	90.15(1)	276.74(2)
Nd_2MgTiO_6	5.4661(2)	5.5705(2)	7.7768(3)	90.01(1)	237.64(1)
$CaTiO_3$	5.3829(3)	5.4458(3)	7.6453(4)	90	224.12(4)
$CaZrO_3$	5.5912(1)	5.7616(1)	8.0171(2)	90	258.26(2)

3. THE PHASE RELATIONS OF RuO_2

Introduction

High level radioactive fission waste consists for a substantial part of second row transition elements, together with Sr, Cs, Ba and lanthanides. The oxides of these elements form such a complicated multicomponent system that for reason of simplicity we select some simple systems. The system Nd-Mo-O (1000-1300°C) is studied in detail (4) but the systems with Ru were fully unknown. We investigated the systems $\underline{A}O-RuO_2-\underline{B}_2O_3$, $\underline{A}$ being Sr or Ba, $\underline{B}$ being Al, Fe or Nd. The Fe is added because it is found in the waste from corrosion products and Al for its use by the immobilisation of highly radioactive waste in ceramic material. Nd stand for an average Lanthanide. The system $SrO-TiO_2-Al_2O_3$ was studied in relation to SYNROC.

Experimental

There were several experimental difficulties met in these phase studies. To prevent the loss of Ru as volatile RuO_3 or RuO_4 in air, all mixtures were heated, after thoroughly mixing in an agate mortar, in sealed Pt capsules at 1300°C for 3 or 4 days. The heating was repeated after grinding till equilibrium was obtained. It was assumed that a reaction mixture was in equilibrium if three or less compounds could be detected.
In the system with Fe_2O_3 and the systems with Nd_2O_3 the valence of Ru is not fixed, so in principle it is possible to get four compounds, but we never found this in our investigations. The research of the systems was hindered by the fact that a large number of Pt capsules ruptured during the heating. This might be caused by a reaction of Pt with Ru ions followed by recrystallisation of Pt and loosing its elasticity.
The reaction products were examined with a Philips PW 1050 X-ray diffractometer using graphite monochromated CuKα radiation and the Siemens Elmiscope 102 electron microscope, fitted with a 40° double tilt and a lift cartridge, operating at 100 kV.
The stability in water at 300°C and 2 kb of the compounds occurring in the system were determined by means of an equipment for hydrothermal crystal growth. For this purpose about 200 mg sample and 0.3 ml water were sealed in a golden tube and heated for a week.
The equilibrium diagrams found are presented in figures 9.5 to 9.11.

The system $BaO-RuO_2-Fe_2O_3$

The system $BaO-RuO_2-Fe_2O_3$ has been investigated at 1300°C in sealed Pt capsules. The mol % BaO was less than about 50%. It was found that the valence of Ru was not constant and has to be taken as an extra variable. Because of this the system has to be considered as a description of the equilibria and not as a ternary phase system. A dashed line in the composition diagram means that there is at least one compound at the end of this line with a deviating valence for Ru.

The unit cell parameters of all compounds investigated part of the $BaO-RuO_2-Fe_2O_3$ system are given in Table 9.7

Table 9.7

List of all Compounds determined in the Part of the $BaO-RuO_2-Fe_2O_3$ System with Unit Cell Parameters and in Parentheses the Symbol used in Fig.9.1

Compound	a(Å)	b(Å)	c(Å)	S.G.	z
RuO_2(R)	4.4902	4.4902	3.1059	$P4_2/mnm$	2
Fe_2O_3(Fe)	5.038	5.038	13.772	$R3c$	6
$BaFe_{12}O_{19}$(M)	5.892	5.892	23.198	$P6_3/mmc$	2
$BaFe_2O_4$(B)	19.056	5.387	8.458	$Bb2_1m$	8
$BaRuO_3$(P)	5.75	5.75	21.60	$R3m$	9
$Ba_3FeRu_2O_9$	5.726(1)	5.726(1)	14.06(1)	$P6_3/mmc$	2
$Ba_3FeRu_2O_9$(F)	5.729(1)	5.729(1)	14.074(3)	$P6_3/mmc$	2
$BaRu_2Fe_4O_{11}$	5.853(1)	5.853(1)	13.587(2)	$P6_3/mmc$	2
$BaFe_2Ru_4O_{11}$(G)	5.867(1)	5.867(1)	13.489(5)	$P6_3/mmc$	2
$Ba_{0.67}Ru^{IV}_{2.67}Ru^{III}_{1.33}O_8$(H)	9.837(1)	9.837(1)	9.374(2)	$P\bar{1}/m$	6
$BaFe_{0.1}Ru_{0.9}O_3$	5.734(2)	5.734(2)	9.510(5)	$P6_3/mmc$	4
$BaFe_{0.08}Ru_{0.92}O_3$(I)	5.734(1)	5.734(1)	9.510(1)	$P6_3/mmc$	4

The value of the c-axis for $Ba_{0.67}Ru^{IV}_{2.67}Ru^{III}_{1.33}O_8$ is the value for the cell with the tripled c-axis.

Leaching stability

X-ray powder diffraction analysis of samples tested on their stability at 300°C in a 10% NaCl-solution showed that RuO_2, $Ba_3FeRu_2O_9$, $BaFe_2Ru_4O_{11}$, $Ba_{0.67}Ru^{IV}_{2.67}Ru^{III}_{1.33}O_8$, $BaFe_{0.08}Ru_{0.92}O_3$, $BaFe_{12}O_{19}$ and $BaRuO_3$ are stable under these conditions. Fe_2O_3 and $BaFe_2O_4$ decompose completely.

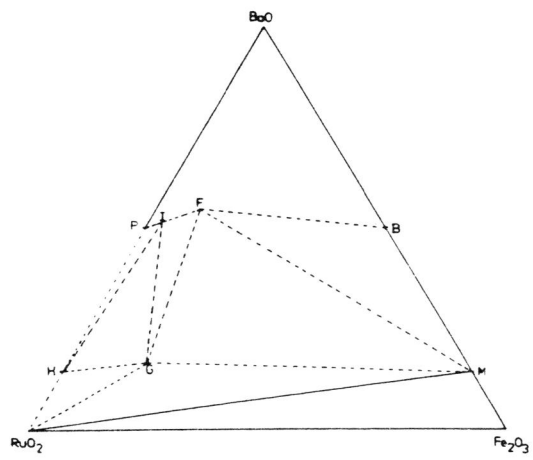

Fig.9.6 The phase diagram of $BaO-RuO_2-Fe_2O_3$

The system $SrO-RuO_2-Nd_2O_3$

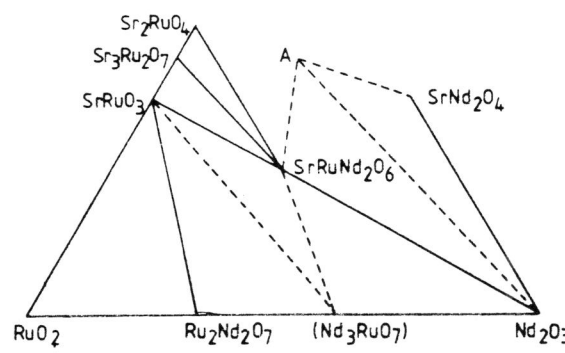

Fig.9.7 The phase diagram of the system $SrO-RuO_2-Nd_2O_3$ at $1300°C$.

The system $SrO-RuO_2-Nd_2O_3$ has been investigated at 1300°C in sealed P capsules. The mol % SrO was less than about 50%. It was found that the valenc of Ru was not constant and has to be taken as an extra variable. Because o this the system has to be considered as a description of the equilibria an not as a ternary phase system. A dashed line in the composition diagram mean that there is at least one compound at the end of this line with a deviatin valence for Ru. We were not able to determine the structures of the quaternar compounds. Further research will be done in the future.

The unit cell parameters of all compounds investigated part of the $SrO-RuO_2$ Nd_2O_3 system are given in Table 9.8

Table 9.8 List of all compounds in the equilibrium system $SrO-RuO_2-Nd_2O_3$

	a(A)	b(A)	c(A)	S.G.
RuO_2	4.4902		3.1059	P42/mmc
Nd_2O_3	3.831		5.999	P321
$Nd_2Ru_2O_7$	10.342			Fd3m
Nd_3RuO_7	10.902(3)	7.379(2)	7.493(2)	Cmcm
$SrRuO_3$	5.56	7.87	5.55	Pnma
$Sr_3Ru_2O_7$	3.88		20.68	I4/mmm
Sr_2RuO_4	3.870		12.74	I4/mmm
$SrNd_2RuO_6$	11.48		28.05	R3 ?
$Sr_{15.33}Ru_{5.67}Nd_{8.67}O_{40}$ A	5.707		18.15	R3 ?
$SrNd_2O_4$	9.217	3.018	10.702	Pnma

Leaching stability.

X-ray powder diffraction analyses of samples test or on their stability a 300°C in water and 10% NaCl solution showed that RuO_2, $SrRuO_3$ and $Nd_2Ru_2O_7$ ar stable under there conditions. All other compounds decompose completely.

The system $BaO\text{-}RuO_2\text{-}Nd_2O_3$

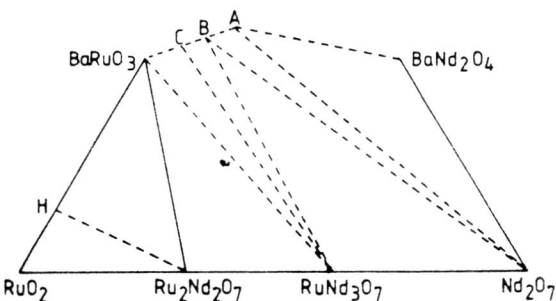

Fig.9.8   The phase diagram of the system $BaO\text{-}RuO_2\text{-}Nd_2O_3$

The system $BaO\text{-}RuO_2\text{-}Nd_2O_3$ has been investigated at 1300°C in sealed Pt capsules. The mol % BaO was less than about 50%. It was found that the valency of Ru was not constant and has to be taken as an extra variable. Because of this the system has to be considered as a description of the equilibria and not as a ternary phase system. A dashed line in the composition diagram means that there is at least one compound at the end of this line with a deviating valence for Ru.

The unit cell parameters of all compounds investigated part of the $BaO\text{-}RuO_2\text{-}Nd_2O_3$ system are given in table 9.9

Table 9.9

List of all compounds in the equilibrium system $BaO\text{-}RuO_2\text{-}Nd_2O_3$

	a(Å)	b(Å)	c(Å)	S.G.
RuO_2	4.4902		3.1059	$P4_2/mmc$
Nd_2O_3	3.831		5.999	P321
$Nd_2Ru_2O_7$	10.342			$Fd\bar{3}m$
$Ba_{0.67}Ru_4O_8$	10.902	7.379(2)	7.493(2)	Cmcm
$BaRuO_3$	5.75		21.60	$R\bar{3}m$
$BaNd_2O_4$	10.592(5)	3.606(1)	12.458(3)	Pnma
$Ba_2NdRu(V)O_6$ (A)	4.2347(14)			?
$Ba_3NdRu_2O_9$ (B)	5.9279(10)		14.7399(28)	$P6_3/mmc$
$Ba_4NdRu_3O_{12}$ (C)	5.90(1)		29.20(1)	$R\bar{3}m$

-343-

Leaching stability.

X-ray powder diffraction analyses of samples test or on their stability at 300°C in water and 10% NaCl solution showed that RuO_2, $Ba_{0.67}Ru_4O_8$, $BaRuO_3$ and $Nd_2Ru_2O_7$ are stable under there conditions. All other compounds decompose completely. The compounds (A), (B) and (C) are perovskite like compounds, but only (A) is derived from the cubic stacking of layers. This indicate again that the incorporation of Ru(V) in the perovskite phase gives a increased solubility for Ba and the Lanthanides. We found the same behaviour for $Ba_3SrRu_2O_9$.

Description of the quarternary compounds.

Nd_3RuO_7 see appendix II

$Ba_{0.67}Ru_4O_8$ see report january-june 1985

Ba_2NdRuO_6: The spacegroup of this perovskite compound could not uniquely be determined. The X-ray diffraction we get a pseudo cubic pattern. From electron diffraction pattern suggests a (nearly) rhombohedral lattice with a_{hex} = 5.9888 and c = 14.670 A, indicating a 6 layer system. This can be explained by the fact that because there is only a small difference between the coherent scattering length of Nd and Ru the diffraction patterns look like a simple perovskite.

$Ba_3NdRu(IV)Ru(V)O_9$: The structure of this compound is the same has hexagonal $BaTiO_3$. Electron diffraction pattern are in agreement with spacegroup $P6_3/mmc$. A list of other compound $Ba_3M(III)Ru_2O_7$ is given in ref (5)

$Ba_4NdRu_3O_{12}$: The electron diffraction patterns of this compound indicate a rhombohedral lattice. The unit-cell parameters suggest the compound has the same structure as $Ba_4ZrRu_3O_{12}$ with 12 layers (6).

The system $SrO-RuO_2-Al_2O_3$

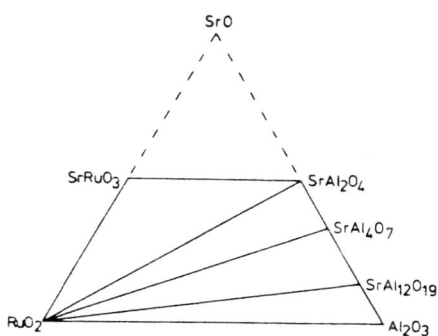

Fig.9.9   The phase diagram of the system $SrO-RuO_2-Al_2O_3$ at 1300°C.

The system $SrO-RuO_2-Al_2O_3$ has been investigated at 1300°C in sealed Pt capsules. The mol % SrO was less than about 50%. It was found that the valency of Ru was not constant and has to be taken as an extra variable. Because of this the system has to be considered as a description of the equilibria and not as a ternary phase system. A dashed line in the composition diagram means that there is at least one compound at the end of this line with a deviating valence for Ru. We were not able to determine the structures of the quaternary compounds. Further research will be done in the future.

The unit cell parameters of all compounds investigated part of the $SrO-RuO_2-Al_2O_3$ system are given in Table 9.10

Table 9.10 List of all compounds in the equilibrium system $SrO-RuO_2-Al_2O_3$

	a(Å)	b(Å)	c(Å)	β	S.G.
RuO_2	4.4902		3.1059		$P4_2/mmc$
Al_2O_3	3.831		5.999		$P32^\cdot$
$SrRuO_3$	5.56	7.87	5.55		Pnma
$Sr_3Ru_2O_7$	3.88		20.68		$I4/mmm$
Sr_2RuO_4	3.870		12.74		$I4/mmm$
$SrAl_{12}O_{19}$	5.585		22.07		$P6_3/mmc$
$SrAl_4O_7$	13.039	9.011	5.536	106.12	$C2/c$
$SrAl_2O_4$	5.13		8.44		$P6_322$

Leaching stability.

X-ray powder diffraction analyses of samples test or on their stability at 300°C in water and 10% NaCl solution showed that RuO_2, $SrRuO_3$ and $SrAl_{12}O_{19}$ are stable under there conditions. All other compounds decompose completely.

The system $BaO-RuO_2-Al_2O_3$

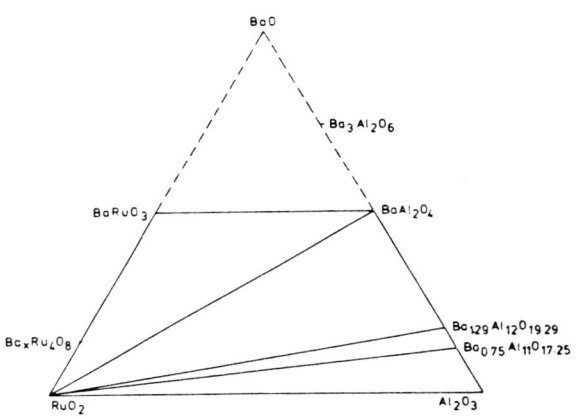

Fig.9.10. The phase diagram of the system $BaO-RuO_2-Al_2O_3$

The system $BaO-RuO_2-Al_2O_3$ has been investigated at 1300°C in sealed Pt capsules. The mol % BaO was less than about 50%. It was found that the valency of Ru was not constant and has to be taken as an extra variable. Because of this the system has to be considered as a description of the equilibria and not as a ternary phase system. A dashed line in the composition diagram means that there is at least one compound at the end of this line with a deviating valence for Ru.

The unit cell parameters of all compounds investigated part of the $BaO-RuO_2-Al_2O_3$ system are given in Table 9.11.

Table 9.11 List of all compounds in the equilibrium system $BaO-RuO_2-Al_2O_3$

	a(A)	b(A)	c(A)	S.G.
RuO_2	4.4902		3.1059	P42/mmc
Al_2O_3	3.831	5.999		P321
$BaRuO_3$	5.75		21.60	R3m
$Ba_{0.67}Ru_4O_7$	10.902	7.379(2)	7.493(2)	Cmcm
$BaAl_2O_4$	5.224		8.795	$P6_322$
$1.31 BaO.6Al_2O_3$	5.600		22.910	?
$0.82 BaO.6Al_2O_3$	5.582		22.715	$P6_3/mmc$

Behaviour in water at 300°C and 2kb.
X-ray powder diffraction analysis showed $BaRuO_3$, $Ba_{0.67}Ru^{4+}_{2.67}Ru^{3+}_{1.33}O_8$, RuO_2, $Ba_{1.29}Al_{12}O_{19.29}$ and $Ba_{0.75}Al_{11}O_{17.25}$ to be stable in water at 300°C and 2kb. All other compounds of the system react under these conditions and decompose completely. A full account of this work is to be published (see Appendix III)

The system $SrO-TiO_2-Ga_2O_3$

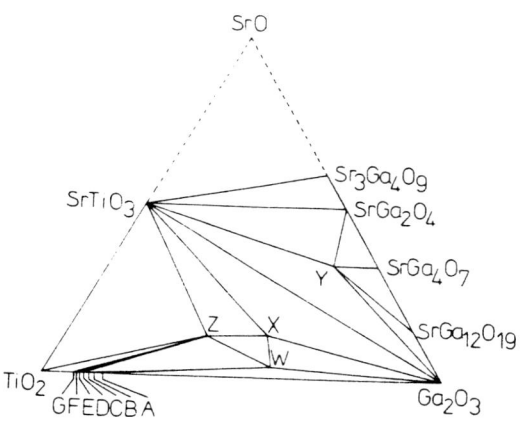

Fig.9.11   The phase diagram of the system
            $SrO-TiO_2-Ga_2O_3$

The system $SrO-TiO_2-Ga_2O_3$ has been investigated at 1300°C because of the great resemblence between Ga and Fe. The mol % SrO was less than about 50%. The unit cell parameters of all compounds investigated part of the $SrO-TiO_2-Ga_2O_3$ system are given in Table 9.12

Table 9.12

Unit Cell Parameters, Cell Volume and Space Group of the Four Quaternary Compounds

Compounds	a (Å)	b (Å)	c (Å)	β (°)	V (Å^3)	S.G.
W	13.170(1)	13.170(1)	27.913(8)		4193.01(2)	R3 ?
X	23.261(1)	11.345(3)	10.065(4)	98.715(2)	2624.95(2)	C2/c
Y	14.663(7)	11.640(6)	5.081(3)	90.560(2)	867.17(1)	I2/m
Z	11.339(2)	9.938(2)	22.810(4)		2570.20(1)	Pbcm

Behavior in a 10% NaCl solution at 300°C and 2kb

All compounds occuring in the system studied were tested on their stability, (except for the compounds in the homologous series $Ga_4Ti_{m-4}O_{2m-2}$). $Ga_4Ti_{11}O_{28}$ (A) was taken as representative for the homologous series, because when this compound decomposes into the compounds TiO_2 and Ga_2O_3, the whole series, therefore must be unstable. After leaching, the X-ray diffraction patterns of $SrTi_3Ga_8O_{19}$ and $Sr_2Ti_9Ga_{12}O_{38}$ show some extra reflections of an unknown contamination, with an estimated amount of about 5%. Compound A decomposed forming the compounds $Ga_4Ti_{13}O_{32}$ (B) and Ga_2O_3. W, X, Z, B and Ga_2O_3 were stable in a 10% NaCl solution at 300°C and 2kb. All other compounds of the system decomposed under these conditions. A full account is to be published (see Appendix IV).

The system $SrO-TiO_2-Al_2O_3$

This system has already been discussed in our previous reports and in literature (7).

Discussion

Leaching experiments. From all the leaching experiments we found that water and 10% NaCl solution have the same behaviour against the reaction products. The Ru compounds of the studied systems generally do not react with water or 10% NaCl solution. It follows from the equilibrium diagrams that already small amounts of Al_2O_3 gives in the equilibrium state the soluble compounds $SrAl_2O_4$, $BaAl_2O_4$ or $SrAl_4O_7$. In other words, it is possible that addition of Al_2O_3 increases the solubility of Sr and Ba in glassy or solid waste forms. To a lesser degree the same is possible for Fe_2O_3 and Nd_2O_3 giving the soluble compounds $SrFe_2O_4$ and $SrNd_2O_4$.

CONCLUSIONS

The important fission products in the wastes are the oxides of Rb, Cs, Sr, Ba, and lanthanides and the second row transition metal oxides. Noble metals as Rh and Pd form a less important separate group. Fixation of these elements in host lattices to avoid the migration of radioactive ions led to solid state proposals as SYNROC or alumina based materials.

Ringwood claims that the important phases in SYNROC are hollandite, perovskite and zirconolite. Alumina based materials consist always of a considerable amount of magnetoplumbite phases and spinels.

In the 1981-1984 program we studied hollandites and perovskites. Special attention is given to Ru compounds. The chemistry of Zr, Mo, Tc and Ru in the solid state for the valencies 4+ and 5+ respectively is very similar because the radii of the ions are alike.

Of these metals only Zr and Tc have radioactivities practically independent for thousands of years. No experiments with Tc were done because this material was not available. The chemistry of Zr is relatively simple because it has a fixed valence, 4+. The behaviour of Ru is complicated because Ru forms volatile oxides RuO_3 and RuO_4. This is quite similar to the Tc chemistry were the volatile Tc_2O_7 is formed. Therefore we studied the behaviour of Ru starting from the solid oxide RuO_2. Molybdenum compounds were not studied yet.

Details were already published in the contract reports 1981-1984 and in the open literature.

The conclusions of the research are:

1. Hollandites $A_xM(IV)_{4-2x}M(III)_{2x}O_8$ with A is Ba or Sr are stable in water and 10 % NaCl solution at elevated temperature, 300°C and 2 kbar. Very little Sr hollandites and only one Zr containing Ba hollandite are formed.

2. Cs and Rb can be incorporated in the tunnels of hollandites. If excess Ba is present, Cs and Rb are found in the soluble compounds and not in the hollandite phase.
   The leaching stability of Cs,Ba and Rb,Ba hollandites is not reproducible and depends on the microcrystalline structure.

3. In the solid state a great number of perovskites and perovskite-like phases are formed. Extended solid solutions, giving only one perovkite phase, are not expected.
   Most perovskites are stable in water and NaCl solution. In the case of special crystal structures the compounds can be very soluble: Ca_3UO_6, Sr_2CaUO_6, and $Ba_3SrRu_2O_9$. These compounds were studied in detail.

4. The phase studies with Ru indicate that most Ru(IV) and Ru(V) compounds are stable against water and 10 % NaCl solution. Simple compounds $ARuO_3$ (A=Ca,Sr,Ba) are not in equilibrium with Al_2O_3, Fe_2O_3 or Nd_2O_3. The equilibrium diagrams show that soluble compounds AB_2O_4 with B=Al,Fe,Nd are formed. This leads to an increased leachability of Sr and Ba in the waste.

5. Zr, Mo, Tc and Ru give, beside perovskite phases, also fluorite like. The valence of Mo and Ru is in such phases not fixed. We studied some $Ln_3M(V)O_7$ phases in detail. There is no reason to expect Tc to behave differently. The leaching stability of such Tc compounds has to be studied in the future because the migration of Tc is high as TcO_4^-. Fixation as insoluble Tc(IV) oxide compounds like the Ru(IV) oxides is expected to be a good solution.

REFERENCES

1. Chopin and Rydberg, Nuclear Chemistry, Pergamon press, 1980.
2. F.C. Mijlhoff, D.J.W. IJdo and H.W. Zandbergen, Acta Cryst. B41, 85 (1985)
3 A.E. Ringwood, Safe disposal of High-level Radioactive Wastes, A new Strategy, N.N.U. Press, Canberra, 1978.
4. H. Czeskleba-Kerner, B. Cros and G. Tourne, J. Sol. State Chem. 37, 94 (1981)
5. H.W. Zandbergen, D.J.W. IJdo, 1983, Acta Cryst C39, 829-832.
6. C.H. de Vreugd, H.W. Zandbergen and D.J.W. IJdo, 1984 Acta Cryst C40, 1984
7. H.W. Zandbergen, and D.J.W. IJdo, Mat. Res. Bull. 18, 371 (1983)

APPENDIX I

COMPOSITION, CONSTITUTION AND STABILITY OF THE SYNTHETIC HOLLANDITES
$A_xM_{4-2x}N_{2x}O_8$, M=Ti,Ge,Ru,Zr,Sn and N=Al,Sc,Cr,Ga,Ru,In
AND THE SYSTEM $(A,Ba)_xTi_yAl_zO_8$ WITH A=Rb,Cs,Sr

H.W. Zandbergen, P.L.A. Everstijn, F.C. Mijlhoff, G.H. Renes and D.J.W. IJdo
Gorlaeus Laboratories, State University, Leiden,
P.O. Box 9502, 2300 RA Leiden, The Netherlands

ABSTRACT

Synthetic hollandite compounds of composition $Ba_xM(IV)_{4-2x}N(III)_{2x}O_8$ (M=Ti,Ge,Ru,Zr,Sn; N=Al,Sc,Cr,Ga,Ru,In) and $Sr_xM(IV)_{4-2x}N(III)_{2x}O_8$ (M=Ti,Ge,Ru; N=Al,Cr,Ga) and $(A,Ba)_xTi_yAl_zO_8$ with A=Rb,Cs,Ca,Sr have been studied by electron and X-ray diffraction and high resolution electron microscopy. They have been found to be stable only within certain ranges of x, depending on the M and N ions. The lower value of x is never less than 0.56, the upper x level is about 0.73. Higher x are correlated with larger radii of M and N. The variation in x gives rise to an incommensurate occupation by the A ions in the tunnels. Correlation among tunnel sequences varies widely, also depending on the nature of the M and N ions. The lower and upper x, the cell dimensions, the correlation between the tunnels and the stability in water at 300°C of the existing hollandites are given. It was found that Rb, Cs and Sr can only be incorporated in the hollandite phase if Ba is not present in excess. Ca cannot be incorporated in the hollandite structure.

Introduction

A number of authors (1-4) have suggested that high level radioactive waste can be immobilized in highly stable synthetic minerals. One of the minerals suggested for this purpose is hollandite. We tried to prepare a large number of hollandites to investigate the structural properties and the stability in water at elevated temperatures.
Hollandites may be characterized as oxides of composition $A_xM(IV)_{4-y}N_yO_8$ (5). The M and N ions are octahedrally coordinated. The structure consists of corner-sharing double chains of edge-sharing octahedra, thus forming square tunnels, which are occupied by large A ions.
The difference between the charges of M and N determines the amount x of A, necessary to obtain electroneutrality. In this paper we will use the shorthand notation A,M/N hollandites.

APPENDIX II

THE ORTHORHOMBIC FLUORITE RELATED COMPOUNDS Ln_3RuO_7
Ln=Nd, Sm AND Eu.

F.P.F. van Berkel and D.J.W. IJdo
Gorlaeus Laboratories, State University Leiden,
P.O. Box 9502, 2300 RA Leiden, The Netherlands.

Abstract. Fluorite-related Ru(V) compounds with composition Ln_3RuO_7 have been found. These compounds with space group Cmcm adopt a superstructure of the cubic fluorite structure with $a_{orth}=2a_c$, $b_{orth}=c_{orth}=a_c\sqrt{2}$. These compounds have the same structure as La_3NbO_7.

Materials index: Lanthanides, ruthenium, fluorite-like compounds.

Introduction

In a program to investigate ceramic materials for their use as potential host-lattice for the disposal of nuclear wastes (1-4) we studied the equilibrium at 1300°C in the systems $BaO-RuO_2-Nd_2O_3$ and $SrO-RuO_2-Nd_2O_3$ with low alkaline-earth content(5,6).
In the system $Nd_2O_3-RuO_2$, $Nd_2Ru_2O_7$ with the Pyrochlore-structure is known(7). Dixon et al.(8) reported also the orthorhombic phase Nd_2RuO_5. We were unable to prepare this compound but found a new one with higher Nd-content. The lattice constants given by Dixon et al.(8) suggest a relationship to La_3NbO_7(9,10) and consequently a phase Nd_3RuO_7 is possible It was possible to prepare other compounds Ln_3RuO_7.

APPENDIX III

THE SYSTEM $BaO-RuO_2-Al_2O_3$ WITH LESS THAN 50 MOL% BaO AT 1300°C

F.P.F. van Berkel, H.W. Zandbergen and D.J.W. IJdo
Gorlaeus Laboratories, State University Leiden,
P.O. Box 9502, 2300 RA Leiden, The Netherlands.

ABSTRACT:

The Ba-poor part of the system $BaO-RuO_2-Al_2O_3$ at 1300°C in sealed Pt-capsules has been investigated. No quaternary compounds were observed in the system, but a hollandite phase $Ba_xRu_4O_8$ (0.66<x<0.68) was formed indicating the partial reduction of Ru^{4+} to Ru^{3+}. X-ray analysis showed $BaRuO_3$, $Ba_xRu_4O_8$, RuO_2, $Ba_{0.75}Al_{11}O_{17.25}$ and $Ba_{1.29}Al_{12}O_{19.29}$ to be stable in water at 300°C and 2 kbar. $BaRuO_3$ is found to form soluble compounds in the presence of Al_2O_3.

Introduction

Research on ceramic materials for their use as potential host-lattice for the disposal of nuclear wastes (1-4) and research on new quaternary Ru-compounds led us to investigate the systems $AO-RuO_2-B_2O_3$, A being Ba or Sr and B being Al, Fe or Nd.
The system with A being Sr and B being Al has been described already (5). In this work the system $BaO-RuO_2-Al_2O_3$ is reported.
In the pseudo binary system $BaO-Al_2O_3$ the compounds $Ba_3Al_2O_6$, $BaAl_2O_4$ and bariumhexaaluminate are known (6). In a recent phase study (7) Kimura, Bannai and Shindo report two phases $0.82BaO.6Al_2O_3$ and $1.29BaO.6Al_2O_3$. $BaAl_{12}O_{19}$ does not exist. Van Berkel, Zandbergen and IJdo (8) determined the structure of the first compound, formulated as $Ba_{0.75}Al_{11}O_{17.25}$, strongly related to $NaAl_{11}O_{17}$. The second compound has probably a superstructure of $NaAl_{11}O_{17}$ (9).
The binary system $RuO_2-Al_2O_3$ contains no new compounds as already observed by Zandbergen et al. (5).
In the system $BaO-RuO_2$ the phase $BaRuO_3$ was identified by Donohue et al. (10).

Experimental

Because of the ease of preparation and to prevent the loss of Ru by the formation of RuO_4, the starting materials were $BaAl_2O_4$, RuO_2, $BaRuO_3$ and Al_2O_3. These starting materials were prepared from the compounds $BaCO_3$, Ru and Al_2O_3 (all AR quality); RuO_2 by oxidation of Ru at 800°C in air for two days; $BaRuO_3$ by oxidation of a stoechiometric mixture of $BaCO_3$ and Ru at 800°C and further heating at 1000°C for four days; $BaAl_2O_4$ was prepared from $BaCO_3$ and Al_2O_3 at 1300°C for four days.

APPENDIX IV

THE Sr-POOR PART OF THE FASÉSYSTEM $SrO-TiO_2-Ga_2O_3$ AT 1300°C, A CRYSTALLOGRAPHIC AND LEACHING STUDY

C.H. de Vreugd, H.W. Zandbergen and D.J.W. IJdo
Section of Solid State Chemistry,
Gorlaeus Laboratories, State University Leiden,
P.O. Box 9502, 2300 RA Leiden, The Netherlands

ABSTRACT

The Sr-poor part of the system $SrO-TiO_2-Ga_2O_3$ has been investigated after heating at 1300 °C in air. In this system four quaternary compounds were determined:
$SrTi_3Ga_8O_{19}$ (a=23.261(1), b=11.345(3), c=10.065(4) A, β=98.715(2)°, C2/c, Z=8), probably having a pseudo-magnetoplumbite structure,
$Sr_3TiGa_{10}O_{20}$ (a=14.663(7), b=11.640(6), c=5.081(3) A, β=90.560(2)°, I2/m, Z=2), isostructural with the compound $Pb_3GeGa_{10}O_{20}$,
$Sr_2Ti_9Ga_{12}O_{38}$ (a=11.339(2), b=9.938(2), c=22.810(4) A, Pbcm, Z=4) also probably having a pseudo-magnetoplumbite structure and $SrTi_{14}Ga_{36}O_{83}$ (a=13.170(1), c=27.913(8) A, Z=3) which has a rhombohedral structure. The triangulation found in this system is discussed and compared with the $SrO-TiO_2-Al_2O_3$ system. The stability in a 10% NaCl solution at 300°C and 2 kb was tested for all compounds in the phase diagram.

Introduction

Research on a number of ceramic materials has been done for their use as potential host-lattices for the disposal of nuclear wastes (1).
One of the potential materials is the hollandite phase, because of its high stability. Ringwood (2) reported a hollandite-phase in the system $SrO-TiO_2-Al_2O_3$. We observed that the hollandite $Sr_xTi_{4-2x}Al_{2x}O_8$ could not be synthesized, and therefore the system $SrO-TiO_2-Al_2O_3$ was investigated (3). Because of the great resemblances of the Al^{3+} and the Ga^{3+} ions and the existence of two isostructural, quaternary Ga and Al compounds, the system $SrO-TiO_2-Ga_2O_3$ was investigated and the results are presented here.
In the pseudo-binary system $SrO-TiO_2$ Cocco et al. (4) reported the following compounds: $SrTiO_3$, $Sr_3Ti_2O_4$ and Sr_2TiO_4. Batto et al. (5) reported the existence of the compounds $SrGa_{12}O_{19}$, $SrGa_4O_7$, $SrGa_2O_4$ and $Sr_3Ga_4O_9$, in the pseudo-binary system $SrO-Ga_2O_3$.
In the pseudo-binary system $TiO_2-Ga_2O_3$ (6), Bursill and Stone (7) and Kamiya and Tilley (8) reported the existence of a homologous series $Ga_4Ti_{m-4}O_{2m-2}$ (13<m<33, m=odd).